MATERIALS SCIENCE AND TECHNOLOGIES

BRICK AND MORTAR RESEARCH

Materials Science and Technologies

Additional books in this series can be found on Nova's website under the Series tab.

Additional e-books in this series can be found on Nova's website under the e-book tab.

Engineering Tools, Techniques and Tables

Additional books in this series can be found on Nova's website under the Series tab.

Additional e-books in this series can be found on Nova's website under the e-book tab.

MATERIALS SCIENCE AND TECHNOLOGIES

BRICK AND MORTAR RESEARCH

SANTIAGO MANUEL RIVERA
AND
ANTONIO L. PENA DIAZ
EDITORS

nova publishers
New York

For permission to use material from this book please contact us:
Telephone 631-231-7269; Fax 631-231-8175
Web Site: http://www.novapublishers.com

Library of Congress Cataloging-in-Publication Data

Brick and mortar research / editors, Santiago Manuel Rivera and Antonio L. Pena Diaz.
p. cm.
Includes bibliographical references and index.
ISBN 978-1-61942-927-7 (hardcover)
1. Bricks--Testing. 2. Cement--Testing. 3. Masonry--Testing. 4. Brickworks. I. Rivera, Santiago Manuel. II. Pena Diaz, Antonio L.
TA432.B726 2011
624.1'836--dc23

2012001047

Published by Nova Science Publishers, Inc. † New York

CONTENTS

PREFACE

This book is a study of brick and mortar research technologies. Topics include the corrosion and environmental aspects of cements and reinforced concrete; in-plane behavior of CRFP retrofitted masonry; high temperature effects on masonry materials; raman spectroscopic characterization of brick and mortars; preparation of colored facing brick from low melting clay under a water vapor atmosphere; negative effects of the use of white portland cement as an additive to aerial lime mortars; the anti-corrosive behavior of four different types of organic coatings; the modeling of brick-mortar interface; and ultrasonic characterization of mortar using micromechanical and multiple scattering models.

Chapter 1 - Alkaline cements are the object of growing interest, as new cementitious materials, given their potential to enable the industry to operate within CO_2 emissions ceilings. The alkali activation of fly ash is achieved by mixing the ash with highly alkaline solutions (pH>13) and subsequently curing the resulting paste at a certain temperature to produce a solid material. The most prominent key factors affecting activation reactions include the nature of the fly ash, the nature and concentration of the activator and the curing conditions (temperature, time and relative humidity).

Cement is considered to show high capacity to fix hazardous components, either heavy metals or radioactive wastes. This ability depends on both the chemical and physical composition of cement matrix, but in both cases the leaching of soluble materials from the cementitious compounds will result in the consequent decrease of confinement properties. In addition, extensive leaching will finally result in loss of mechanical properties, although experimental evidence is seldom found in literature, except that of crack formation in the degraded zone. The solid hydrates of cement paste are more persistent at pH above 12–13, but at lower pHs the hydrated phases are no longer remain stable and dissolve. The ordinary used criteria to evaluate the performance of cementitious materials for stabilization/solidification of hazardous wastes are the results of leaching tests at laboratory scale, which should allow the identification of the main chemical and mass transfer mechanisms and the competition between different dynamic processes and their relative importance over a given time scale.

Concrete corrosion has serious, societal and economic impact each year, 3-4% gross national product (GNP) for direct and indirect cost in the developed countries in repair, replacement, and environmental impact. The major cause of deterioration is corrosion of steel reinforcing bar in concrete that occurs when concrete pH is reduced by chloride ions penetration and carbonation attack, destroying the passive layer formed on the reinforcing bar.

The very high temperatures (~1500 ºC) required to manufacture ordinary Portland cement (OPC), which make it responsible for 40% of all energy consumed (~4000 kJ per kg of cement), account for the extremely high costs of this process. The cement industry is regarded to be responsible for 6-7% of all greenhouse gases emitted world-wide (0.85-1 ton of CO_2 per ton of cement).

Use of new corrosion prevention strategies is reported as a palliative method to increase the reinforced concrete service life, thus contributing to diminish the environmental impact and economic cost.

Stainless steel (SS) reinforcements were first used many decades ago and have proved their ability to prevent corrosion for a very long time, even in very aggressive environments. Hence the use of SS reinforcements is one of the most reliable methods for ensuring RCS durability.

Chapter 2 - Earth is one of the most commonly used building materials on the planet today. Ancient cultures used earth for building houses, fortresses, palaces and religious buildings and it is estimated that at the present time one third of the world's population lives in houses constructed from Earth. Although the majority of its utilisation occurs in developing countries houses are still built from earthen materials in developed countries. Indeed, there are well known examples in countries such as the "The Tucson residence" in the United States of America, for instance and "The Residence Korbeek" in Belgium. Forms of Earth construction have also been used within religious buildings such as the Chapel of reconciliation in Berlin and within industrial architecture, as illustrated in "The Bodegas La Raia" in Piemonte, Italy.

Chapter 3 - Masonry buildings are designed to serve a certain lifetime. However, there are several masonry buildings that have been damaged in a shorter time than it was expected, because of different external actions like earthquakes, impact loads, changes in their use or aggressive agents. Moreover, there are many historical buildings that should be preserved as cultural heritage.

Research activities carried out a few years ago concerning the use of fiber reinforced polymers as external reinforcement of masonry walls have shown that this system considerably improves structural stability with a minor impact over foundations. The use of polymeric fiber composites has also proved to be an efficient repairing technique for historical buildings. However, different aspects of this retrofitting system should still be analyzed. This analysis involves placement techniques, anchorage length, amount and layout of the reinforcement, failure modes of the reinforced element and the behaviour of damaged and even collapsed and then repaired masonry elements.

This chapter is concerned with the in-plane mechanical behaviour of solid clay masonry panels and the same panels but reinforced or repaired with carbon fiber reinforced polymer laminates externally bonded to the wall surface. The experimental program involves; compression normal to bed joints and diagonal compression tests on unreinforced, reinforced and damaged and then repaired panels. Numerical simulation is performed using an existing coupled damaged-plasticity model that is calibrated with the experimental results obtained by the authors. Such model allows simulating the behaviour of masonry elements using the mechanical properties of constitutive materials and their layout.

First a summary of the main characteristics of the mechanical behaviour of unreinforced and retrofitted masonry, subjected to in plane loads is presented. Then the experimental tests and their results are discussed and contrasted with existing design models. After that, the numerical model is summarized and the results of several simulations carried out to verify the efficiency of the reinforcement with carbon fiber reinforced polymers as a retrofitting system for masonry elements are presented. Finally the behaviour of full scale masonry walls externally reinforced with bonded carbon fiber reinforced polymer (CFRP) sheets and subjected to in plane shear load are numerically reproduced using the same numerical model.

Chapter 4 - Research on masonry structures is very complex and manifold. A quite novel branch of research for masonry materials is here addressed, i.e. the residual behavior of brick, cement mortar and masonry after exposure to high temperatures. Masonry buildings – especially old and historic ones – are often very vulnerable to the attack of fire, and the need for fire protection may be in conflict with preservation of architectural heritage.

The whole matter of high temperature exposure is rich in physical, mechanical and chemical issues, which are mutually connected. Moreover, masonry material is composite and involves a variety of combinations of materials, geometry and textures which are identifiable in masonry buildings through different ages and countries; this leads to possible expensiveness in testing and difficulties in theoretical and experimental modeling. The point of view of the mechanical characterization of materials after high temperature exposure is here taken into consideration; about this peculiar subject, very few theoretical as well as experimental studies are currently available.

First, the state-of-the-art of such research is briefly outlined. Real events as well as standard fire tests often demonstrate that masonry walls and structures can excellently withstand the high temperatures that can be reached during a fire event; on the other hand, the residual mechanical performance of a structure after exposure may need to be evaluated, if high levels of fire safety are required.

Then, the results of recent investigations are here reported, which have given a first contribution to the experimental knowledge of the residual behavior of masonry and separate components (solid clay bricks and cement mortar) after high temperature exposure. The theoretical elaborations of such outcomes have been useful to set a first basis for the establishment and calibration of analytical tools for the prediction of the residual mechanical performance of masonry, brick and mortar.

Chapter 5 - Traditional building materials provide valuable information on the past of ancient and historical buildings. Studies on the characterisation of such materials provide information on the origin of raw materials, manufacture and building technologies, or decay of materials. Also, they provide information on different building phases and periods (chronology). Among traditional materials, bricks and mortars are particularly interesting.

One of the most useful information provided by bricks and mortars is the possibility of date them (relative and absolute dating). This is very useful as information from written historical documents is unusual. Different procedures have been developed for the study of building chronologies. Relative dating methods are the study of building stratigraphy, chronotypology, mensiochronology and chemical analysis of building materials. Other methods provide absolute ages. Luminescence dating is the most used method for dating bricks but it has been applied on mortar dating in the last decade. Archaeomagnetism has been tested to date bricks in some specific cases. Other methods, such as radiocarbon, have been tested on lime mortars with relative success. Different approaches have provided useful information on the history of ancient buildings due to brick dating. Dating mortars include some methodological problems partially overcome in the last years. This paper reviews the advances occurred on the use of dating methods on dating bricks and mortars, and specialy focuses on luminescence methods.

Chapter 6 - Raman spectroscopy is becoming a popular technique to perform molecular analysis on building materials in a non destructive way, which permits the preservation of samples. Nowadays, the technical development of Raman achieved from high sensitive microscopic instruments till handheld spectrometers. This fact sets Raman spectroscopy as the technique of choice for the characterization and diagnosis works in Cultural Heritage.

In our researches different kind of building materials, including bricks and mortars belonging to several sampling locations, have been analysed. All of them were affected by different environmental stressors, such as atmospheric acid gases and particles, infiltration waters and biodecaying. Generally, calcite ($CaCO_3$) and diverse iron oxides plus quartz (sands) were identified as original compounds in mortars samples. Silicates and oxides like rutile (TiO_2) and hematite (Fe_2O_3) were the common original compounds in bricks.

Degradation products of those original compounds were identified in environments with high atmospheric pollution; gypsum ($CaSO_4{\cdot}2H_2O$) is the principal compound formed by the attack of the SOx acid gases over calcite in the mortars. Calcite, is an original compound in mortars but in the case of the bricks it comes from a decaying (hydration and subsequent carbonation of the original calcium oxide, CaO); other original compounds can react analogously with atmospheric acid gases (CO_2, SO_2, NOx) to form the respective carbonates, sulphates and/or nitrates. The formation of nitrates is also due to the impact coming from the organic matter degradation that produces ammonium nitrate, which is the precursor of the formation of other nitrate salts such as niter (KNO_3) and nitrocalcite ($Ca(NO_3)_2{\cdot}4H_2O$). The study of the biodeterioration markers, permits the identification of several pigments and other organic compounds (calcium oxalates) associated to the metabolism of the colonizing microorganisms.

In conclusion, the Raman identification of original and deterioration compounds allow us to define the conservation state and the impacts suffered by building materials like bricks and mortars. The in situ capability of the Raman technique increases its potentiality.

Chapter 7 - Preparation of coloured facing brick by the traditional method requires high quality clay minerals, high temperature firing and occasionally colouring pigments. In this chapter the authors report their research on the preparation of coloured facing brick from locally available low melting clay (Tolgoit deposit, Mongolia) under a pressured water vapour atmosphere. It was found that colour change and mechanical properties of the brick samples strongly depends on chemical and mineralogical composition of the clay. The water vapour pressure and the chemical composition of the vapour also affects the properties of the brick after firing. The clay specimens were fired at 600, 800, 900 and 1000°C for various times. A water vapour atmosphere improved the structure formation of the ceramics and increased the crushing strength of the fired clay by up to three times. Ilmenite formed from the $FeO+TiO_2$ oxides of the original clay during firing under the water vapour atmosphere colours the brick a blue to grey colour, while $CaO+MgO+SiO_2$ oxides may form a green-coloured compound. Based on the experimental results, coloured facing bricks of acceptable quality were produced on a semi-industrial pilot plant scale.

Chapter 8 - In recent years the use and study of lime mortars are being improved because lime shows a high compatibility with the ancient materials of the Architectural Heritage. On the other hand, the scarce knowledge of the fabrication techniques, the application methods used in the past and the long time of setting have driven scientists and restorers to add other binders (i.e. cement, pozzolans) to the lime, which claim to obviate the inconveniences of the use of this material alone. The use of Portland cement as additive in lime-based mortars seems to be an easy and diffused solution for the industrial production of repair mortars. In this chapter, the authors have compared the chemical-mineralogical and physical-mechanical properties of mortars prepared with calcitic lime and a calcareous aggregate, with mortars in which a white Portland cement (WPC) has been used in substitution of a 25 wt.% of the lime. This chapter shows that mortars prepared with Portland cement do not provide satisfactory results after two months since their elaboration. From mineralogic data and porosimetric analyses no improvements were observed by the addition of cement. In fact, even if this percentage of cement produced lower shrinkage than lime mortars, it did not improve the hydric and mechanical properties. Most importantly, the WPC caused the formation of ettringite that provoked fissures in mortars samples. This chapter demonstrates that the use of cement, even if only as additive in lime mortars, would impair the longevity of the masonry that encloses it.

Chapter 9 - The anti-corrosive behavior of four different types of organic coatings was examined in the present study. High performance application coatings including epoxy, polyurethane, chlorinated rubber and solvent based coatings were exposed to highly corrosive environment and their behavior against induced chloride ion corrosion was evaluated with electrochemical measurements such as Linear Polarization Technique (LPR), Half-cell Potential and Electrochemical Impedance Spectroscopy (EIS). The durability of the organic coatings was evaluated by assessing basic physicochemical properties including water vapor and liquid water absorption and chloride ions permeability. From the results obtained it is shown that cement mortar specimens that were covered with the polyurethane coating perform better, in both electrochemical and physicochemical measurements, compared to the other three high performance application coatings. Nevertheless, all the coating tested in the present study, are able to protect steel rebars from corrosion more effectively that conventional coatings.

Chapter 10 - Traditionally, for thousand of years, masonry techniques were commonly used for the building of homes, monuments, walls, and retaining walls. Masonry is one of the oldest construction materials. Since masonry is a composite structure, however, failure of these structures will depend on the properties of the materials (mortar, bricks, etc.), as well as on the characteristics of the bonding between the various components. Two main methods of modelling masonry structures have been developed in the literature. The first method involves macroscopic models and homogenization techniques: the wall is assumed to be a single structural element characterized by a non-linear response when it is exposed to external forces. The second method has been developed for predicting the evolution of cracks and damage at the interface brick-mortar or brick-brick. This chapter deals with two families for the modelling of brick-mortar interface. The first part of this chapter deals with the experimental characterization of the materials (bricks and mortar) and the brick-mortar interface. Hereafter, we describe experimental studies on the shear behavior of masonry on the local scale, in the case of two different assemblies composed of two and three full/hollow bricks, with or without confinement. Two other structures are presented: a small wallmade by seven bricks and the classical problem of a small wall under diagonal compression. The second part of this chapter deals with a phenomenological model of interface. The mechanical modelling approach (and in particular the RCCM adhesive model adopted) is first presented. The model is based on the concept of adhesion variable due to Frémond. A variation equation of this variable is introduced and the coefficients are identified experimentally. The numerical procedure used is then described. Lastly, some numerical examples are given and compared with the experimental data. The third part presents a multi-scale model of interface. The method is based on homogenization theories and asymptotic analysis. In the present work, we assume the existence of a third material: the brick-mortar interface, which is considered as a mixture of brick and mortar with a crack density. To obtain the effective properties of the damaged intermediate material, three aim steps are performed. First, we calculate the exact effective properties of the crack-free material using homogenization techniques for laminate composites, and thus define a first homogeneous equivalent medium. In the second step, we assign a crack density to the material. To model the macroscopic behavior of the cracked material, we use the Kachanov model and then define a new homogeneous equivalent medium. Finally, in order to be sandwiched between the brick and mortar, this material is given a small thickness, and its mechanical behavior is derived using asymptotic techniques to shift from the micro to the macro level. A variation law of the crack length is introduced. The numerical procedure used is then described. Lastly, some numerical examples are given and compared with the experimental data.

Chapter 11 - It is well-known that mortar is one of the most basic elements of masonry construction. This fact has motivated to improve the understanding of relations between the strength, quality and durability of mortars with their mix proportions and the different mechanical properties of their material constituents. Several research works have been conducted to investigate the mortar microstructure by using both destructive and non-destructive testing, highlighting the ultrasonic non destructive testing. These ultrasonic techniques have proven to be effective in evaluating the mechanical and physical properties of a large number of strongly heterogeneous materials. Different modeling approaches have been used to establish relations among the microstructural properties of mortars and ultrasonic wave propagation. Micromechanical models and multiple scattering theories are addressed in this chapter. These homogenization techniques, which provide different information from materials, are used to describe the influence of microstructural properties on the ultrasonic parameters. By using micromechanicalmodels, the influence of geometry, volume fraction and elastic properties of constituent phases, namely matrix, aggregates or pores on the ultrasonic velocity is evaluated. On the other hand, the multiple scattering models provide the influence of the size and volume fraction of aggregates on mortar, as well as the interaction, contribution, and influence of entrapped air voids together with the aggregates on frequency-dependent parameters such as the phase velocity and the attenuation coefficient. Both theoretical and experimental investigations are performed throughout the present chapter.

In: Brick and Mortar Research
Editors: S. Manuel Rivera and A. L. Pena Diaz

ISBN: 978-1-61942-927-7

Chapter 1

CORROSION AND ENVIRONMENTAL ASPECTS OF CEMENTS AND REINFORCED CONCRETE

David M. Bastidas*, Irene Llorente and María Criado
Dept. of Surface Engineering, Corrosion and Durability
National Centre for Metallurgical Research (CENIM–CSIC)
Madrid, Spain

ABSTRACT

Alkaline cements are the object of growing interest, as new cementitious materials, given their potential to enable the industry to operate within CO_2 emissions ceilings. The alkali activation of fly ash is achieved by mixing the ash with highly alkaline solutions (pH>13) and subsequently curing the resulting paste at a certain temperature to produce a solid material. The most prominent key factors affecting activation reactions include the nature of the fly ash, the nature and concentration of the activator and the curing conditions (temperature, time and relative humidity).

Cement is considered to show high capacity to fix hazardous components, either heavy metals or radioactive wastes. This ability depends on both the chemical and physical composition of cement matrix, but in both cases the leaching of soluble materials from the cementitious compounds will result in the consequent decrease of confinement properties. In addition, extensive leaching will finally result in loss of mechanical properties, although experimental evidence is seldom found in literature, except that of crack formation in the degraded zone. The solid hydrates of cement paste are more persistent at pH above 12–13, but at lower pHs the hydrated phases are no longer remain stable and dissolve. The ordinary used criteria to evaluate the performance of cementitious materials for stabilization/solidification of hazardous wastes are the results of leaching tests at laboratory scale, which should allow the identification of the main chemical and mass transfer mechanisms and the competition between different dynamic processes and their relative importance over a given time scale.

Concrete corrosion has serious, societal and economic impact each year, 3-4% gross national product (GNP) for direct and indirect cost in the developed countries in repair,

* Corresponding author, Tel.: +34 91 553 8900; Fax: +34 91 534 7425; E-mail address: david.bastidas@cenim.csic.es (David M. Bastidas).

replacement, and environmental impact. The major cause of deterioration is corrosion of steel reinforcing bar in concrete that occurs when concrete pH is reduced by chloride ions penetration and carbonation attack, destroying the passive layer formed on the reinforcing bar.

The very high temperatures (~1500 ºC) required to manufacture ordinary Portland cement (OPC), which make it responsible for 40% of all energy consumed (~4000 kJ per kg of cement), account for the extremely high costs of this process. The cement industry is regarded to be responsible for 6-7% of all greenhouse gases emitted world-wide (0.85-1 ton of CO_2 per ton of cement).

Use of new corrosion prevention strategies is reported as a palliative method to increase the reinforced concrete service life, thus contributing to diminish the environmental impact and economic cost.

Stainless steel (SS) reinforcements were first used many decades ago and have proved their ability to prevent corrosion for a very long time, even in very aggressive environments. Hence the use of SS reinforcements is one of the most reliable methods for ensuring RCS durability.

INTRODUCTION

During the last decades, an enormous effort is being done to minimize energy consumption and pollution. In an attempt to overcome these problems, the construction sector is very interested in the development of new cement binder materials as an alternative to conventional Portland cement. The environmental impact attributed to the manufacturing of Portland cement is largely due to the energy-intensive processes involved. This heralds a new worldwide environmentally-friendly scenario for cement industry markets where the Kyoto Agreement, in force since 2005, the need to cut CO_2 emissions by 60% by 2050, with demonstrated progress by 2020, electricity prices, fossil fuel and cement production are significant drives for such directives [1].

The main constituents of exhaust gases from a cement kiln are nitrogen, carbon dioxide, water, and oxygen. The exhaust gases also contain small quantities of dust, chlorides, fluorides, sulfur dioxide, NOx, carbon monoxide, and still smaller quantities of organic compounds and heavy metals.

The largest portion of greenhouse gas emissions from production of cement worldwide (about 50%) originates from the process reaction that converts limestone ($CaCO_3$) to calcium oxide (CaO), the primary precursor to cement. About 40% of the industry's emissions come from fossil fuel combustion at cement manufacturing operations with remaining emissions coming from transport of raw materials and combustion of fossil fuel required to produce the electricity consumed by cement manufacturing processes. The most serious problem with our industry is that it is a major CO_2 emitter causing global warming. All of this amounts to about 7% of the total CO_2 generated worldwide. Enhanced efficiency is not likely to change this but replacement of some of the cement by a supplementary cementing material not associated with CO_2 emission can substantially reduce these emissions [2].

The emissions of nitrous oxide coming from burning benzene, coal and other fossil fuel and SOx, produced by oxidation of volatile sulphurs present in raw materials, represent an important source of pollutants. Greenhouse gases emissions are of concern as they can detrimentally affect air quality and human health, some examples of which are the production of acid rain, reduced atmospheric visibility (smog) and aggravation of respiratory systems.

Hazardous material containment structures require secondary containment and, sometimes, leak detection systems. Because of the economic and environmental impact of even small amounts of leakage of hazardous materials, both primary and secondary containment systems must be virtually leak free. Therefore, when primary or secondary containment structures involve concrete, special design and construction techniques are required.

The capacity to fix hazardous components, heavy metals or radioactive wastes depends on both the chemical and physical composition of cementitious matrix, but in both cases the leaching of soluble materials from the cemenetitious compounds will result in the consequent decrease of confinement properties. Therefore, it will be necessary to evaluate the performance of cementitious materials for stabilization/solidification of hazardous wastes by means of leaching test which is tackled below.

CONTAINMENT OF HAZARDOUS WASTES IN CEMENT

Cement is considered to show high capacity to fix hazardous components, either heavy metals or radioactive wastes [3-6]. This ability depends on both the chemical and physical composition of cement matrix, but in both cases the leaching of soluble materials from the cementitious compounds will result in the consequent decrease of confinement properties. Leaching is the process by which soluble constituents are dissolved from a solid material into a fluid by percolation or diffusion.

Thus, when solid comes into contact with the liquid (including percolating rainwater, surface water, groundwater, etc), constituents in the solid phase will dissolve into the liquid forming a leachate. The potential hazard for environment and health may arise in different stages of the product lifecycle, i.e. manufacturing, distribution, construction, service-life and end-of-life.

During the service life stage, which usually spreads over many decades, the release of substances due to contact with water is one of the main potential hazard source, as a consequence of the leaching phenomenon [7]. For example, in the case of cement based products, the use of alternative fuels for cement production, the admixture addition or the natural composition of some raw material may be the cause of several undesirable environmental and health effects (ranging from discomfort and sensorial annoyance to severe health injuries). In addition, this phenomenon is critical in regards with the durability of nuclear waste containments for which the required service life ranges from a few hundred years to several thousand years.

The extent to which the constituents dissolve into the contact liquid will depend upon several factors concerning the leaching system, the leachant or the solid but an extensive leaching will finally result in an important increase in porosity and an important decrease of mechanical properties [8], and it is possible to observe crack formation in the degraded zone. These both processes (increase in porosity and crack formation) produce an important loss of the confinement capacity of the cement based material.

Cement based materials have a network of pores produced by the excess of water needed to make the mix workable. These pores are filled with a liquid phase which has a very alkaline pH and is rich in different ions, in equilibrium with different solid phases: some of them, such as the C–S–H gel, are amorphous, some of them are crystalline (portlandite) and some minor phases contain iron and aluminium.

The major part of these phases present variable stoichiometry and change with time. When the cement paste is put in contact with water, the solid phases which are stable at the alkaline pH of the concrete, of approximately 13, are no longer in thermodynamic equilibrium and get dissolved.

The pore solution of a typical Portland cement paste is highly alkaline, so that the leaching process starts by removing alkalis (Na^+ and K^+), followed by dissolution of portlandite and subsequently by the leaching of calcium from silicates, e.g., C–S–H [9-11]. Aluminate phases are also affected, dissolution/precipitation processes of AFm (calcium monosulfoaluminate), ettringite and calcite are observed [9,12], and aluminium might be incorporated in the C–S–H gel. Silica gel formation has also been detected in the outer layer of cement paste exposed to leaching in deionised water.

Generally calcium is used as an indicator of chemical deterioration because it is the main element of hydrates and plays an essential role in chemical reactions [13-15], and most results already published have been related to the decalcification of the solid skeleton.

Figure 1 shows the effect upon the properties of the cement paste due to the leaching of the portlandite and C–S–H, the main hydrates of the cement paste. However, some authors have stood out the important role that other elements and compounds such as Fe, Al, sulphates and Mg play [16,17]. It has been postulated [18] that it could be one or several species acting as remediators of the damaged C–S–H gel stabilising the structure generated by the loose of calcium.

As mentioned above there are numerous factors that can affect the leaching process and the first step in understanding mechanisms, needed for prediction of long-term leaching behaviour, is to gain knowledge of the effects of various factors on leachability. These factors are categorized as related to the leaching system, the leachant or the solid being leached [19,20].

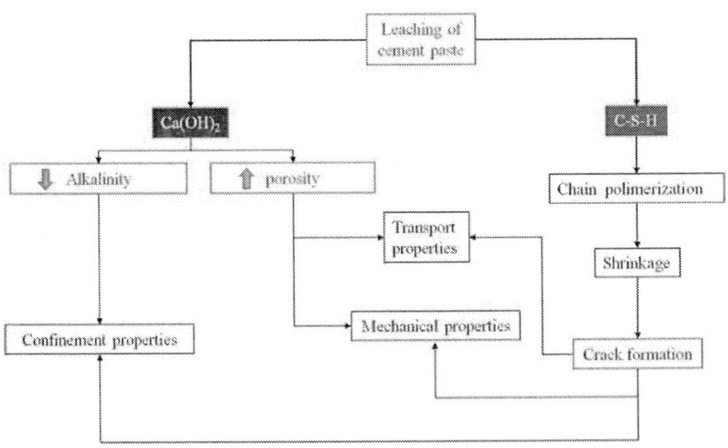

Figure 1. Effect of the leaching process upon the properties of the cement paste.

System Factors

Time

Leachability is a function of time and it is much related to the leaching mechanism. So, in the literature it is very frequent to find that the cumulative amount of calcium has square-root time dependence with time and usually it is associated with a diffusion mechanism.

Temperature

Leachability usually increases with increasing temperature, but it is not true for one of the most important hydrates of the cement paste, portlandite, which solubility decreases with increasing temperature.

Ratio of Volume of Leachant to Volume of Solid (V/Vs)

When the ratio V/Vs increases, generally the leachability increases, but it is necessary to take into account if the solid is powdered or monolithic form because the leachability of a powdered material could increase because it has more exposed area, even with identical volumes of leachant.

Leachant Factors

Leachant Composition

Leachant composition has a great influence on leachability. Deionised water is one of the most frequently leachants employed in laboratory tests, and is more aggressive than highly salted leachant. However trends in leachability with different leachants could be difficult to predict especially if the leachant contains complexing agents such as carbonate, humic acid or tannic acid, often present in groundwaters.

pH

Leachability may be a strong function of pH for some materials. Cement-based materials show larger leachabilities at low pH. It is very frequent the use of deionised water acidified to pH= 4 as leachant agent in normalized leaching tests.

Flow

Leachability of a solid in a flowing leachant increases with flow rate due to exposition of the solid to fresh leachant, so that there are not concentration effects. Agitation of a leaching system is a form of flow effect that may give larger leachabilities than static systems.

Solid Factors

Composition of the Solid. Porosity

The amount of cement and the presence of additions such as fly-ash or silica fume should have an influence on the leaching behaviour, since they lead to a different porosity. In addition to this, the compaction, the curing, as well as the achieved degree of hydration are also of importance. A higher porosity implies a surface area exposed to the leachant much greater than the geometric surface area. In this context the study of the leaching behaviour of high and ultra high resistance concretes with very low porosities has a notable interest [17,21].

Surface Condition

Leachability is generally calculated from geometric or superficial surface areas. Thus, a higher surface roughness could have greater apparent leachability from this effect alone. The presence or absence of a protective film at the surface can have a profound influence on leachability. A compact calcite layer on concrete surface can slow down the leaching process [21].

The ordinary used criteria to evaluate the performance of cementitious materials for stabilization/solidification of hazardous wastes are the result of leaching tests at laboratory scale, which should allow the identification of the main chemical and mass transfer mechanisms and the competition between different dynamic processes and their relative importance over a given time scale [22].

There are several classifications of leaching tests. CEN TC 292 establishes three ranges of forms for standardised leaching tests:

Characterisation: These tests consists on a availability (granular or pulverized specimen) procedure and a sequential/periodic tank (monolithic specimen) procedure which together provide the means for discriminating between the several transport processes such as dissolution, wash-off, diffusion, and for predicting the rate of leaching and long term behaviour of a material. In addition, physical characteristics such as tortuosity, which is a measure of the prolonged path along which leached components have to travel, can be calculated.

Compliance leaching tests: Consists of single extractions of short duration, generally without agitation, and which permit a direct comparison with regulatory limits for individual analytical components. Such tests use the prior output from characterisation tests to establish and optimise their parameters.

Verification: These tests are essentially second order compliance tests, modified for operation in the field and used to identify/assess changes in established performance or batches of a material.

However, the most frequent classification of leaching tests in the literature divides them into two main groups:

Extraction tests: These tests generally involve mixing a sample of the cement-based material with a specific amount of leaching solution without renewal of the leachant. The mixing is performed over a relatively period of time (hours to days) with the aim of reaching equilibrium conditions. The mixing is followed by filtration and analysis of the filtered liquid phase. Figure 2 provides a simplified schematic of an extraction test.

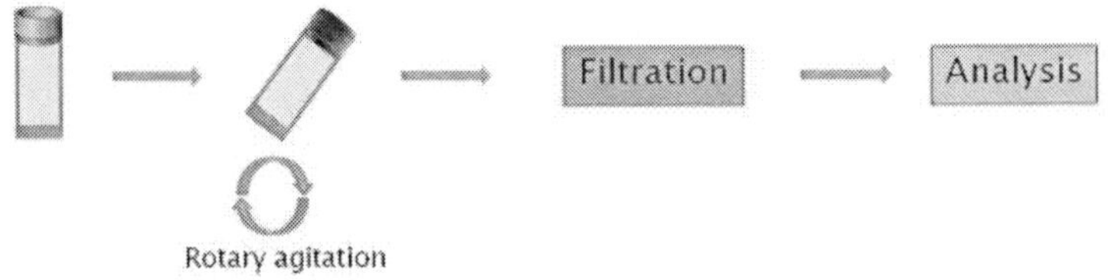

Figure 2. Simplified schematic of an extraction leaching test.

The aim of these tests is the determination of the maximum concentrations of the elements of interest in the leachant at the end of the period of time. Reaching equilibrium in extraction leaching testing is critical to predicting leaching behaviour over long periods of time. If tests are conducted at non-equilibrium conditions, leaching behaviour does not reach capacity, and predictions for long term leaching behaviour will be based on leachate concentrations that are too low or too high. Extraction tests could be classified into 3 groups:

- Extraction tests with agitation: e.g.: TCLP, SPLP, EPTOX
- Extraction tests without agitation: MCC-1 and 2
- Tests with sequential extraction: NEN 7341, ASTM D 4793, ASTM D 5744.

An example of an extraction test is the TCLP: "Toxicity Characteristic Leaching Procedure". This test involves the extraction of the elements of interest from a 100g size-reduced sample of cement-based material. The maximum size of particle is 9.5 mm and the leachant generally used is acetic acid (pH=2.88±0.05) and a 20:1 liquid to solid (L/S) ratio (mass/mass) is employed. The mixture is rotated for 18±2 hours at 30 rpm using a rotary agitation apparatus. After rotation, the final pH is measured and the mixture is filtered using a glass fibre filter. Then, the filtrate is analyzed for a number of constituents and its toxicity is determined.

Dynamic tests: These tests typically address some aspect of leaching in which time is an important variable. In these tests, a specific amount of leaching solution and test material are mixed and the leachant is periodically or continuously renewed. The mixing is performed over a relatively long time period (days to months) compared with extraction tests. The objective of these tests is to characterize the cement degradation mechanism and to develop a model of the attack in order to predict the long-term behaviour of the material. Dynamic tests could be classified into two groups:

- Multiple extraction tests: NEN 7345 (now 7375), ANSI/ANS 16-1.
- Continuous renewal tests or column leaching tests: NEN 7343, ASTM D4874, PCLT, prEN 14405.

Multiple extraction tests are usually applied to monolithic samples. The sample is placed in the test vessel and leaching fluid is added for a specified period of time. The leachate is then separated from the solids and replaced with a fresh leaching fluid until the desired number of leaching periods has been completed. Figure 3 shows the simplified schematic diagram of a multiple extraction test.

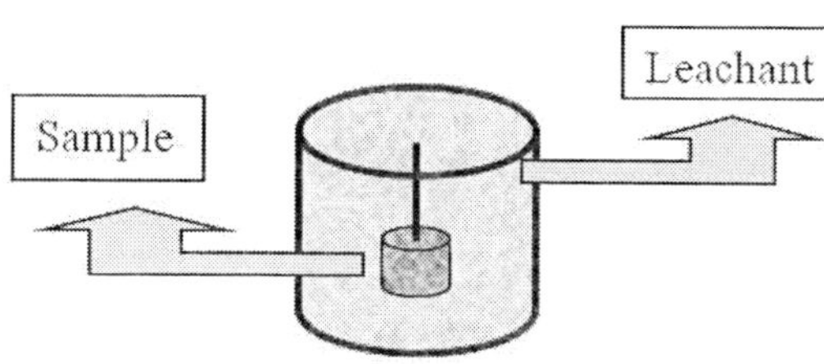

Figure 3. Simplified schematic of a multiple extraction test.

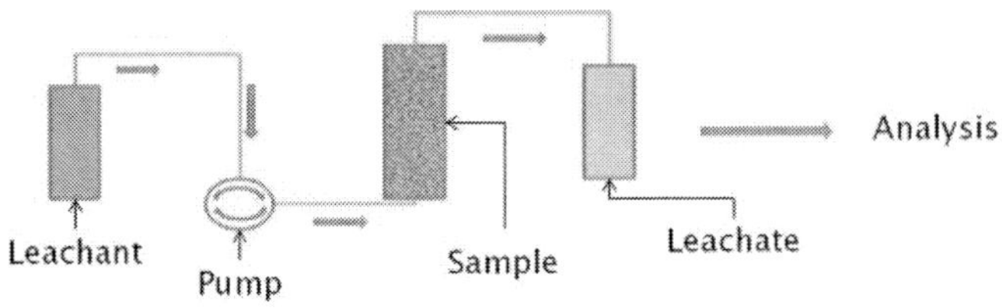

Figure 4. Simplified schematic of a column leaching test.

The second group of tests is characterized by the continuous flow of the leachant through a contained packed with the sample. The leachate is periodically sampled and analyzed for the elements of interest. The column leaching test whose schematic is in Figure 4, is one of the most frequent normalized dynamic test.

However it is very frequent in the literature to find the application of both kinds of leaching tests to study the leaching phenomenon [23,24]. Extraction leaching tests give information about the maximum release and some intrinsic property of the material (e.g. acid-base neutralization capacity) and dynamic tests give information about the leaching mechanism and the controlling parameters that allow the develop of a model for long-term behaviour. For example, Hidalgo et al. [10], apply a neutralization acid test (extraction test) and NEN 7345 (dynamic test) [25], to cement pastes where it was used alternative fuels for cement production (recycled tyres and meat meals) to determine its potential hazard for environment.

Nevertheless, according to the literature, different dynamic leaching tests can lead to different results, even for ranking of concretes, as they can represent different scenarios of natural degradation and could be affected by different factors. Llorente et al. [26], studied the leaching process of three different types of concretes, cast with different binders submitted to two different leaching methods: a tank water test (TWT) based on the standard ANSI/ANS-16.1-1986 [27], typically used in the field of radioactive wastes and a running water test (RWT). Comparisons between these leaching methods are made with respect to the leachability of calcium from the different concretes. The design of the different mixes is detailed in Table 1 and the leachant employed was deionised water.

Table 1. Mixes of the materials used

Mix	1	2	3
Cement	I-42.5 R/SR		
Binder (Kg/m^3)	none	140 (FA)	36 (SF)
w/c	0.45		
Cement (Kg/m^3)	400	260	364
Sand (Kg/m^3)	911		
Gravel (Kg/m^3)	949		
Remarks: FA: Fly ash, SF: Silica Fume			

The standard ANSI is a multiple extraction test that establishes that the leachant must be entirely replaced at designated time intervals: 2, 7, 24, 48, 72, 96, 120, 456, 1128 and 2160 hours from the initiation of the test. Therefore, the duration of the standard test is 90 days. An aliquot of the leachate must be taken at the end of each leaching interval in order to carry out the determination of calcium concentration. Running water test is a non-normalized dynamic test with continuous flow downwards of the leachant (100 mL/h). The test elapsed 6 months, during this time a periodic sampling of the leachate was carried out considering the evolution of Ca^{+2} in the leaching water.

The cumulative fraction leached (CFL) per surface of calcium versus time and the percentage of calcium leached at the end of the tests are plotted in Figure 5.

As shown in this picture, the evolution of Ca^{+2} and the rankings of concretes are strongly dependent of the scenario of degradation.

The ranking of the concretes in the TWT is, in descending order, mix 1 > mix 3 > mix 2 whereas in the RWT is mix 2 > mix 1 > mix 3. These different ranking of the concretes could indicate a different mechanism of degradation. It is worthwhile to mention that there are important differences in the total amount of calcium leached for the different concretes in the used leaching tests which is an indication of its different degree of aggressiveness. For example, in the case of mix 1, the percentage of calcium leached in the ANSI test was about 3% and in the running water test 17% at 90 days of leaching (25% at the end of the test).

Authors use the standard NEN 7345, (revised in 2004 and published as NEN 7375) to establish the leaching mechanisms. According to this standard, the relation between the logarithm of the cumulative release (in mg/m^2) and the logarithm of leaching time (s) is used to identify the controlling leaching mechanisms studying the slope of the linear regression lines through the data points. The following mechanisms can be distinguished: Dissolution (slope >0.65), surface washoff (slope <0.35) and matrix diffusion (0.35 <slope < 0.65). Figure 6 shows the obtained curves for the release of calcium versus time (log-log) for the used tests. As shown in the Figure 6, there are two slopes in all the curves and the first one is always higher than the second one. Table 2 gives both slopes and the r^2 coefficient corresponding to the fitting of the curves in Figure 6.

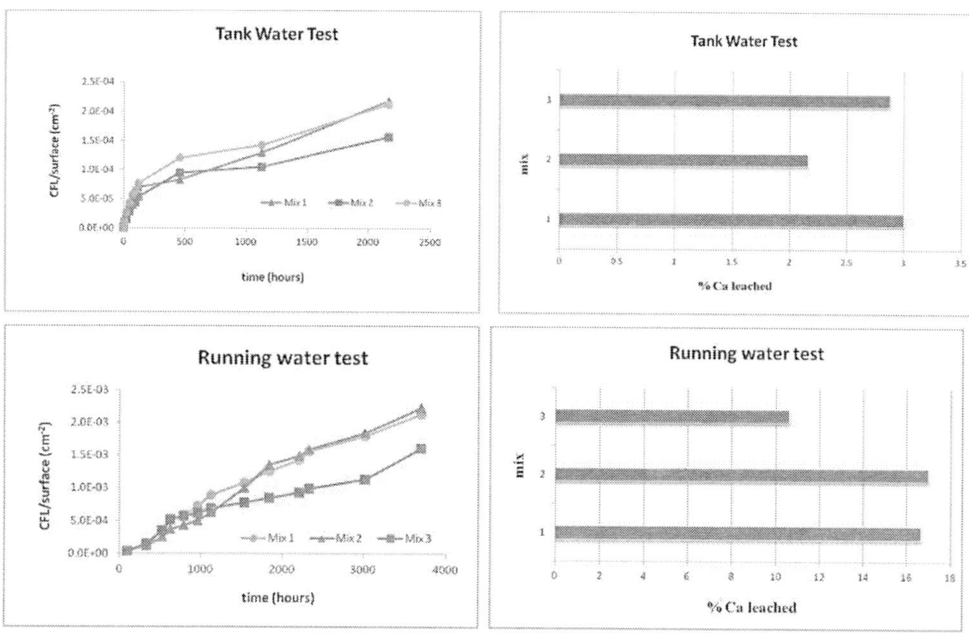

Figure 5. The cumulative fraction leached (CFL) per surface of calcium versus time at the end of the tests and the percentage of calcium leached at 90 days.

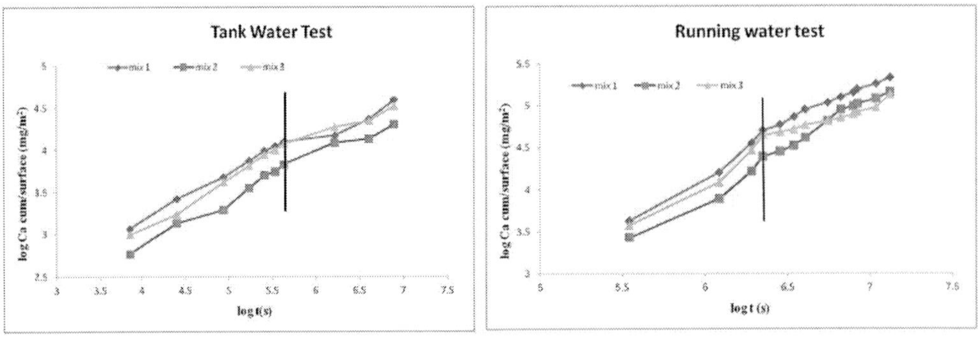

Figure 6. Relation between the logarithm of the cumulative release (in mg/m^2) and the logarithm of leaching time (s) for the used tests.

As can be seen in Table 2, for TWT, the first slope corresponds with the typical one of a diffusion mechanism but the second one is in the limit between wash-off and diffusion mechanism. For RWT, both slopes are higher than 0.65, except for the second slope of mix 3. Thus, the mechanism in this test is dissolution, but there is a general tendency to change it to diffusion. Due to this different mechanism of degradation, the ranking of the concretes in TWT and RWT is different. However, it should be also pointed out that the most important difference between TWT and RWT is about the renewal of the leachant. In fact, in TWT, the leachant is entirely replaced at designated time intervals, but in running water test a continuous flow around the sample is established (see Figure 7).

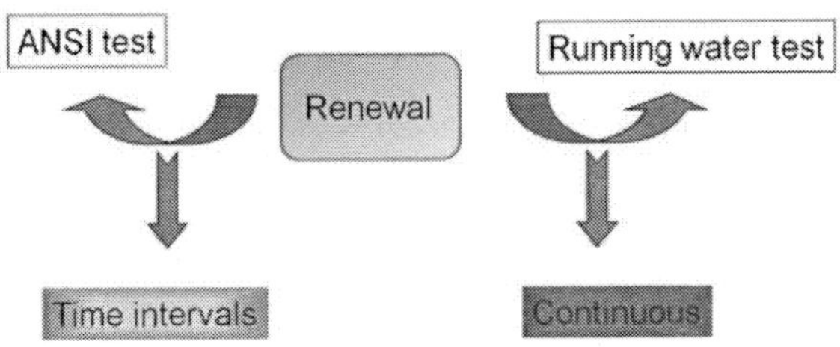

Figure 7. Scheme of the differences between both employed tests.

Table 2. Slopes and r^2 coefficient corresponding to the fitting of the curves shown in Figure 6

Leaching Test	Slope			R^2		
	mix			Mix		
	1	2	3	1	2	3
Tank Water Test	0.573 0.379	0.584 0.349	0.622 0.334	0.998 0.870	0.983 0.956	0.994 0.966
Running Water Test	1.279 0.819	1.128 1.091	1.261 0.485	0.980 0.994	0.962 0.980	0.960 0.995

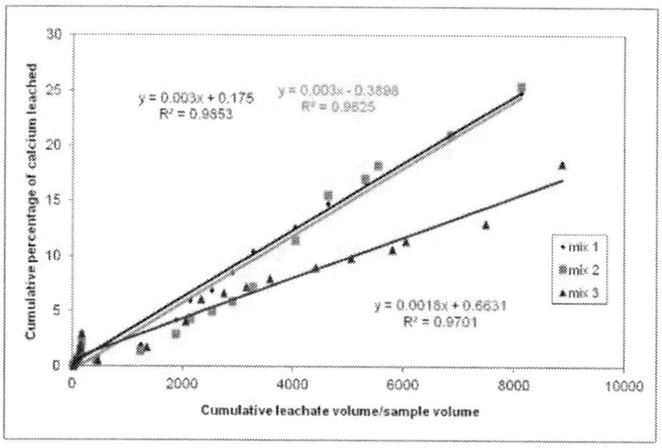

Figure 8. Cumulative percentage of calcium leached at different times in both TWT and RWT versus the cumulative leachate volume to sample volume ratio.

So, in running water test, there is an increase in the frequency of the leachant renewal that produces an acceleration of the process as can be seen in the corresponding obtained slopes. Due to this acceleration the amount of calcium leached is much higher in the running water test than in TWT.

In order to compare both tests and evaluate this acceleration rate a standardization of the conditions is needed. Thus, as the leachant volume in contact with the sample at the same time is different for both tests, to make a comparison between them it is necessary to plot the cumulative percentage of calcium leached at different times in both tank water test and running water test versus the cumulative leachate volume to sample volume ratio (see Figure 8).

As can be seen in Figure 8, there is a linear relationship between the percentage of calcium leached and the cumulative leachate volume to sample volume ratio. Therefore, it can be deduced that the frequency of renewal is the key parameter in treating both scenarios in a global manner and it seems that mixes 1 and 2 behave in the same way (same slope) while mix 3 is more resistant to decalcification (smaller slope).

The conclusion of the authors is that the frequency of renewal of the leaching water is the reason for the differences between the test types, which in spite they induce different leaching mechanisms can be compared by standardizing the Ca^{+2} leached by the volume of leachate with respect to the sample volume ratio.

In this previous work it has been compared the results of two different dynamic leaching tests but using the same kind of concrete but with different binders and deionised water as leachant. But what would be expected if concretes with very different characteristic are submitted to both kinds of leaching tests? In the last decade, the chapter concerning special uses of concrete has gained much relevance. Nowadays, the industry is demanding high resistant concretes with very specific properties which make them to successfully withstand extreme conditions [28]. Thus, the challenge is to produce tailored concretes for long-term storage of radioactive waste materials, confinement of toxic wastes, self compacting, high and ultra high resistance and fire-resistant concrete among other uses. It is of great relevance the study of the leaching behaviour of these concretes. In recent work by Llorente et al. [29], it is compared the leaching behaviour of seven concretes sorted in three different families: traditional concrete, high resistance concrete (HRC) and ultrahigh resistance concrete

(UHRC), using two types of leaching tests (dynamic tests). The leachant employed was granitic water. The difference among these concretes is in their compressive strength: For traditional concretes this value is below 50 N/mm^2 (normalized test), for HRC is between 50-100 N/mm^2 and for UHRC is above 100 N/mm^2. Strength of hydrated cement paste can be improved by paying close attention to the following parameters [30]:

- Porosity: a large number of big pores or voids especially when concentrated in one location, are detrimental to strength;
- Grain size: in general, the strength of a crystalline phase increases with decreasing grain size;
- Inhomogeneities: with multiphase materials, microstructural inhomogeneities are a source of strength loss.

Porosity of the cement paste depends on the water/cement ratio. When this ratio is reduced, the cement particles come close together and there is less capillary porosity. This is possible by means of the use of higher amount of cement in the mix and the use of binders, especially silica fume. However, to give workability to concrete, it is necessary the use of superplasticizers.

Decreasing the water/cement ratio or the water/binder ratio favours the formation of hydration products characterized by a fine texture [28], avoiding the formation of large stacks of hexagonal platelets of lime, long needles of ettringite or long hairy filaments of C–S–H that are characteristic of high water/binder ratio pastes. To reduce inhomogeneities it is necessary a proper consolidation of the mix. In the other hand, the use of superplasticizers play a key role in improving the dispersion of cement particles in freshly mixed paste minimizing entrapped air bubbles. A scheme of the main differences among these families of concrete is given in Figure 9 and a resume of the composition of the different mixes is in Table 3.

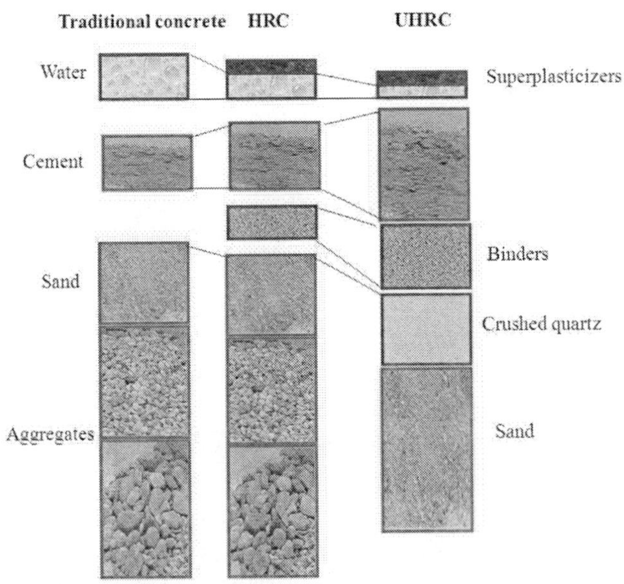

Figure 9. Scheme of the differences in composition among traditional concrete, HRC and UHRC.

Table 3. Composition of the different tested mixes

Mix	Amount of cement (Kg/m^3)	Water/Binder ratio	Binders (%)
Traditional concrete	400	0.45	No
HRC	450-530	0.28	5-20
UHRC	700-760	0.13-0.18	15-33%

It was studied the leaching behaviour of these concretes by using two different dynamic leaching tests. First one, a tank water test (TWT) was based on the standard ANSI/ANS-16.1-1986, and described previously, and a running water test (RWT). Although standard ANSI test lasted 90 days, in order to know the effect of long-term leaching, two additional leaching periods were carried out: the first one, in which the renewal of leaching water was made every 6 weeks up to 12 months and after that a further renewal of leached water was made every 6 months, up to a total leaching period of 30 months. For RWT, a special leaching device was designed for the test whose scheme can be seen in Figure 10. The method works in an open system, similar to a leaching column with a continuous upwards flow at a rate of 7 mL/h. RWT lasted 18 months. Leachant employed in both tests was a natural ground water (granitic water) whose composition is given in Table 4.

As in the previous work, the results are strongly dependent on the scenario of degradation, not only the ranking of the concretes but the degree of aggressiveness. The ranking of concretes was for TWT, in descending order: traditional concrete≈HRC>UHRC, whereas for RWT: traditional concrete>UHRC>HRC. There are important differences in the total amount of calcium leached for the different concretes in the used leaching tests which is an indication of its different degree of aggressiveness. For example, in the case of traditional concrete, the percentage of calcium leached in the TWT was about 4% after 30 months of test and in the RWT 8% after 18 months testing. Again, frequency of leachant renewal is a key parameter and it was necessary to make a standardization taking into account the volume of leachate with respect to the sample volume ratio to compare both scenarios. However, in spite of this standardization, there are some differences that could be only explained taking into account the leachant agent.

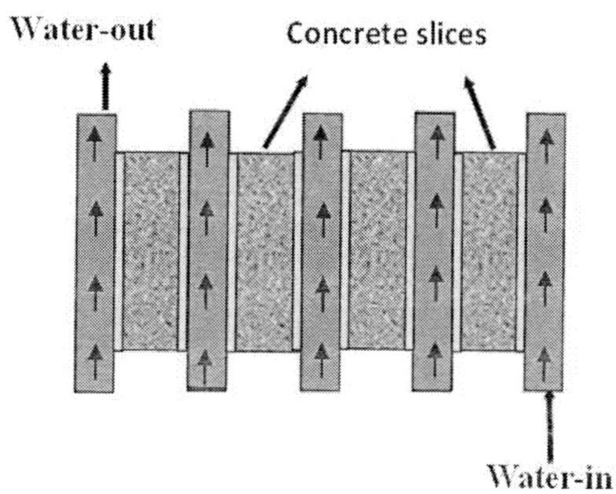

Figure 10. Leaching arrangement for running water test.

Table 4. Chemical composition of natural ground water (GW) employed for leaching tests

Ionic species	Concentration (mmol/l)
Calcium	0.84 ± 0.04
Sodium	0.55 ± 0.03
Magnesium	0.270± 0.005
Potasium	0.023± 0.003
Bicarbonates	2.90 ± 0.05
Sulphates	0.114 ± 0.050
Chlorides	0.39 ± 0.01
Silicon	0.350 ± 0.005
pH	7.91 ± 0.30
Conductivity (mS/cm)	0.337 ± 0.020

The leachant agent employed in both tests was granitic water, which has a [HCO_3^-] given in Table 4 and the systems were not prevent from the external CO_2. Under these conditions, the following reactions could take place:

$$OH^- + HCO_3^- \longleftrightarrow CO_3^{2-} + H_2O \quad (1)$$

$$Ca^{2+} + CO_3^{2-} \longleftrightarrow CaCO_3\downarrow \quad (2)$$

The first reaction will be conditioned by the different alkalinity of the concretes; in this sense UHRC could have more difficulties to produce an external layer of calcite because they do not have portlandite in their composition. On the other hand, the specific characteristics of both tests could introduce some variations in the establishment of these equilibriums. TWT is a static system during the whole period between renovations with no stirring of the system allowing the formation of a more o less compact layer of calcite with the leached calcium on the concrete surface. The compactness of this layer will depend on the capacity of the concrete to produce reaction Eqs. 1 and 2. In this sense, UHRC will produce a less compact layer of calcite than the rest of concretes. In the other hand, RWT is a dynamic leaching test and a constant flow of leachant is established. This flow hinders the calcite formation due to concentration effects, and it will be more evident for UHRC because of their low alkalinity. As a consequence, it is not expected that these concretes could present an external layer of calcite or it would be very weak and incompact. The calcite layer could produce a protective effect and avoid the further attack of the concrete. In the case of the traditional concrete, the positive effect of the calcite could be masked by the relative high porosity and the leaching process would continue. For HRC, however, due to their low porosity, the calcite could difficult the access to the concrete and the leaching could be slowed down. In UHRC the formation of the calcite layer is difficult especially for RWT, thus for these concretes in this test a higher degradation than HRC would be obtained, as can be seen in the ranking of concretes. A comparison in the formation of calcite layer in both tests can be seen in Figure 11, where it is shown SEM images of the surface of the concretes.

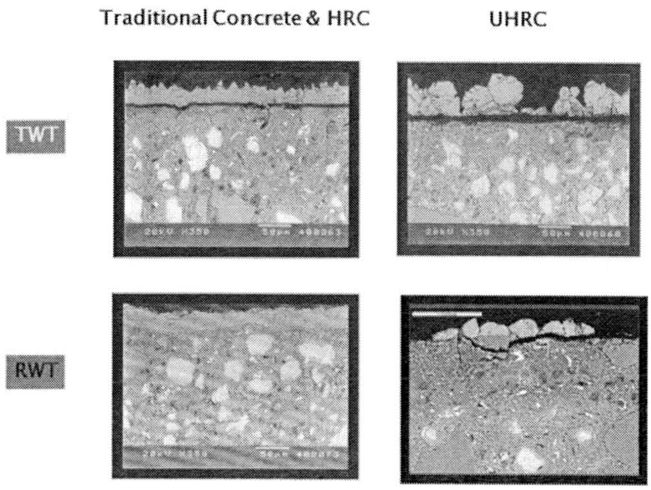

Figure 11. Comparison in the formation of the calcite layer among traditional concrete, HRC and UHRC in the both tests employed.

As can be seen in the previous examples, the variation in only one factor can produce a great variation in the leachability of the material, so for long-term predictions, the question is: How well do the leaching test data can be used as prediction tools? In a real scenario, the leaching behaviour is controlled by various factors that generally are not taken into account in the laboratory tests: e.g.: biological degradation. Besides, there are differences between laboratory and field conditions: heterogeneity of the waste or material under real scenario, difficulties in obtaining representative samples, differences in the liquid-solid ratio. If conditions in the field do not match the parameters of the laboratory tests, the predictions could be inaccurate. However, leaching tests provide a means for evaluating trends and determine the main factors that affect the leachability of an element under site-specific conditions (scenario). In this context, the European standard ENV 12920: "Basic characterization of leaching behaviour in specified conditions", establish the following steps for the environmental behaviour assessment:

- Definition of the problem and description of the environmental scenario by the identification of the significant environmental factors.
- Characterization of the material intrinsic properties: composition and physical characteristics, pollutants solubilisation according to pH.
- Determination of the parameters influence on the leaching behaviour and identification of the main transfer mechanisms.
- Modeling of the leaching behaviour of the material, validation and long-term simulation of the soluble species release.

Following this scheme, Schiopu et al. [7] presented a coupled chemical-transport model for the case of a concrete based construction product (paving slabs) defining a specific scenario: rain water under outdoor exposure conditions. For the development of the model, the authors realised leaching tests at laboratory scale (extraction and dynamic tests) and an experimental study at a field scale under outdoor conditions (product placed in a real exposure scenario). From the extraction tests physico-chemical parameters were obtained and used for the development of the chemical-model. Two dynamic tests were performed (continuous flow and sequential renewal of the leachant, based on NEN 7345). The results of the dynamic leaching tests were used for the development of the transport model. On the other hand, the concrete slabs were exposed horizontally to rain water for one year in order to reproduce two real situations: run-off and stagnation scenarios. For a horizontal exposure of the concrete products to rainfall, run-off scenario corresponds to a quick draining of the rainwater whereas the stagnation scenario stands for the situation where the rain remains in contact with the product for longer periods of time. This distinction is important because the product/leachant conditions (quantity of the leachant in contact with the product, contact time, wet/dry periods) are different, so the leachability of the elements can be different. The chemical-transport model developed a laboratory scale was completed by adding external parameters specific for natural exposure conditions such as rain water balance, water/material specific contact conditions and the atmospheric CO_2 uptake. Authors conclude that the simulation results are considered satisfying given the difficulty to consider and integrate natural conditions.

Novel Cementitious Binders

Since the development of Portland cement over 175 years ago, it has become the dominant binder used in concrete for construction. The very low use-cost of conventional Portland-based cement system is due to the great natural abundance and widespread distribution of the raw materials (calcium carbonates, aluminosilicates and calcium sulfates) coupled with a manufacturing process which has been highly optimized for energy- and cost-efficiency over the last few decades [31]. However, it is not yet optimized in terms of CO_2 emissions, and the current global perspective on sustainability will probably require this to be done over the next few decades, due mainly to the enormous volumes of cement used worldwide. Annual worldwide Portland cement production is approaching 3 Gt in 2010 [32].

The manufacturing of Portland cement consumes $10\text{-}11\text{x}10^{18}$ J annually, approximately 2-3% of global primary energy use. Furthermore, Portland cement production results in approximately 0.87 tones of carbon dioxide for every tone of cement produced [33]. Therefore, the cement industry is regarded to be responsible for 6-7% of all greenhouse gases emitted worldwide. At this pace, by 2025 the cement industry will be emitting CO_2 at a rate of 3.5 billion tones/year, more or less equal to the total emissions in Europe today (including the transport and energy industries) [34].

CO_2 is one of several gases that are known to have a positive effect on the retention of heat by the Earth's atmosphere, and thus have the potential to increase average global surface temperatures. The main problem for the cement industry is that all conventional constructions cements are based on "clinkers" (products of thermal sintering or melt process) containing basic calcium compounds, for which the major raw material is limestone (calcium carbonate). The decarbonation of limestone results in the release of "fossil CO_2" into the atmosphere, which currently accounts for about half of the cement industry's CO_2 emissions; and this problem is even greater for the lime industry [31]. Therefore, the cement industry is under pressure to reduce both energy use and greenhouse gas emissions and is actively seeking alternatives to this familiar and reliable material.

On the other hand, the fact that OPC (ordinary Portland cement) structures, which have been built a few decades ago, are already facing disintegration problems points out the handicaps of OPC binders too. In fact, the number of premature cases of OPC structures disintegration is overwhelming. Beyond the durability problems originated by imperfect concrete placement and curing operations, the real issue about OPC durability is related to the intrinsic properties of the material. OPC structure presents a higher permeability that allows water and other aggressive elements to enter concrete, leading to carbonation and corrosion problems [35]. Moreover, Portland cement can present a serie of shortcomings in certain applications and environments. For instance, rapid-repair applications demand a faster strength gain than Portland cement concrete can provide. Similarly, environmental conditions with high acidity or high sulfate concentrations can cause substantial degradation of Portland cement concrete [32].

So far, research works carried out in developing alkali-activated binders show that this new binder is likely to have enormous potential to become an alternative to Portland cement.

Alkaline Cements as an Alternative to Portland Cement

The reaction of a solid aluminosilicate with a highly concentrated aqueous alkali hydroxide or silicate solution produces a synthetic alkali aluminosilicate material generically called "inorganic polymer" [36].

The development of alkali-activated binders had a major contribution in the 1940s with the work of Purdon [37]. That author used blast furnace slag activated with sodium hydroxide. In 1967 Glukkhovsky made a significant breakthrough in the understanding and development of binders from low calcium or calcium-free aluminosilicate (clay) and alkaline metal activators [38]. He called these binders "soil cements" and the respective concretes "soils silicates". Glukkhovsky classified binder in two groups, depending on the composition of the starting materials: alkaline binding systems Me_2O-Al_2O_3-SiO_2-H_2O and alkaline-alkaline-earth binding systems Me_2O-MO-Al_2O_3-SiO_2-H_2O (where Me = Na, K… and M = Ca, Mg…).

In 1978 Davidovits named "geopolymers" the tridimensional alumosilicates that are formed at low temperature and short time by naturally occurring aluminosilicates [39]. For the chemical designation of the geopolymer Davidovits suggests the name "polysialates", in which sialate is an abbreviation for aluminosilicate oxide. The sialate network is composed of tetrahedral anions $[SiO_4]^{4-}$ and $[Al_2O_3]^{5-}$ sharing the oxygen, which need positive ions such as (Na^+, K^+, Li^+, Ca^{2+}, Ba^{2+}, NH_4^+, H_3O^+) to compensate the electric charge of Al^{3+} in tetrahedral coordination. The polysialate has the following empiric formulae:

$$M_n\{-(SiO_2)_z\text{-}AlO_2\}_n.wH_2O$$

where n is the degree of polymerization, z is 1, 2 or 3, and M is an alkali cation, such as potassium or sodium, generating different types of polysialates.

Krivenko [40] later called these binders "geocements", to highlight the presence of nature mineral analogues in their hydration products and the similarities in the mechanisms governing the formation of these binders and natural geological materials.

The last ten years, have seen exponential growth in research related alkali-activated cements and concretes not only on their benefits in terms of low energy costs and environmental impact, but also their good mechanical performance and long durability [41-49].

Classification of Alkali-Activated Cements

Alkali-activated usually consist of two components: a cementitious component and alkaline activators. A variety of industrial by-products and waste as well as a number of aluminosilicate raw materials has been used as the cementitious components. These materials include granulated blast, furnace slag, granulated phosphorous slag, steel slag, coal fly ash, volcanic glass, zeolite, metakaolin, silica fume and non-ferrous slag. Caustic alkali or alkaline salts are normally used as alkaline activators in alkali-activated cements and concretes [50].

Alkali-activated cements can be classified into five categories using the composition of the cementitious component/-s as a criterion [51,52]:

1) alkali-activated slag-based cements
2) alkali-activated pozzolan cements
3) alkali-activated lime-pozzolan/slag cements
4) alkali-activated calcium aluminate blended cements
5) alkali-activated Portland blended cements (hybrid cements)

Over the past decade, any number of papers has been published on alkali-activated pozzolans, especially alkali-activated fly ash cements and in this chapter of the book we are going to focus on study of these new cements.

Reaction Mechanisms

According to Glukhovsky [38], the stages comprising the alkaline activation reaction are as described bellow.

First Stage "Destruction-Coagulation"

This first disaggregation process entails the severance of the Me-O, Si-O-Si, Al-O-Al and Al-O-Si bonds in the starting material. This disaggregation of the solid phase may be governed by the formation of complex unstable products whose origin lies in the change in the ionic strength of the medium prompted by the addition of electron donor atoms (the alkaline metals). The result is a redistribution of the electronic density around the silicon atom, rendering the Si-O-Si bond more susceptible to rupture.

The presence of alkaline metal cations neutralizes these anions, generating $Si\text{-}O\text{-}Na^+$ bonds, thereby hindering reaction reversibility. Moreover, the conditions created by the Si-O-

Na^+ complexes, which are stable in alkaline media, are suitable for the transport of the reacting structural units and the development of the coagulated structure.

Since the hydroxyl groups have the same effect on the Al-O-Si bond, the aluminates in the alkaline solution form complexes such as $Al(OH)_4^-$ or $Al(OH)_6^{3-}$.

Second Stage "Coagulation-Condensation"

The accumulation enhances contact among the disaggregated products, forming a coagulated structure where polycondensation takes place. The polycondensation rate is determined by the state of the dissolved ions and the existence or otherwise of the conditions necessary for gel precipitation. Silicic acid condensation is therefore favored at pH values in which the acid is slightly dissociated or in a molecular state. At pH>7, for instance, disaggregation of the Si-O-Si bond gives rise to $Si(OH)_4$ like hydroxylated complexes.

Third Stage "Condensation-Crystallization"

The presence of particles from the initial solid phase, together with the appearance of microparticles resulting from condensation, favors product precipitation. The mineralogical composition of the initial phase, the nature of the alkaline component and the hardening conditions determine the qualitative and quantitative composition of the crystallized products.

More recently, different authors have elaborated on and extended the Glukhovsky theories and applied the accumulated knowledge about zeolite synthesis in order to explain the geopolymerization process as a whole [53-57].

Fernández-Jiménez et al. [58,59] proposed the next graphics model for the microstructural development of alkaline aluminosilicate cements (based essentially on MAS-NMR and FTIR findings) depicted in Figure 12 [50].

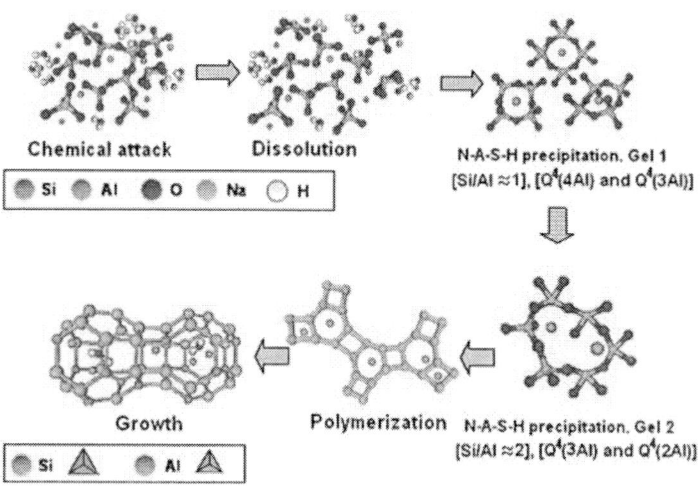

Figure 12. Descriptive model for alkali activation of aluminosilicates.

In the first step, the contact between the solid particles (aluminosilicates) and the alkaline solution (pH>10) induces the dissolution of the vitreous or amorphous component of the former, realising (probably monomeric, see Figure 12, step 2) aluminates or silicates. These monomers inter-react to form dimers, which in turn react with other monomers to form trimers, tetramers and so on. When the solution reaches saturation an alkaline aluminosilicate hydrate gel precipitates. Initially, this gel is an aluminium-rich aluminosilicate gel (Gel 1, Si/Al ratio ≈ 1, according to NMR and FTIR data). The Si and Al tetrahedral initially bond together to form rings containing four secondary (usually alternating aluminium and silicon) tetrahedral units, for example, S4-type secondary building units with a predominance of Q^4(4Al) units [54,58,60]. The cations that neutralize the electrical charge resulting from replacing a silicon with an aluminium tetrahedron are positioned in the gaps left in the structure (see Figure 12, step 3).

Gel 1 formation may be explained by the higher Al^{3+} ion content in the alkaline medium in the early stages of the process (from the first few minutes to the first 4 to 5 hours). This, in turn, can be attributed to the fact that Al-O bonds are weaker and therefore more readily severed than Si-O bonds, making reactive aluminium more readily soluble than silicon [60]. As the reaction progresses, more Si-O groups in the solid particles dissolve, raising the silicate concentration in the medium. This gel gradually becomes richer in silicon, giving rise to an alkaline aluminosilicate hydrate gel, called Gel 2 (Si/Al ratio ≈ 2, see Figure 12, step 4) [54]. The intensity of the NMR signals associated with slightly condensed structures (Q^4(4Al)) declines, while the intensity of the signals associated with the presence of closely condensed structures (Q^4(3Al) and Q^4(2Al)) groups [54,58], giving rise D6R- and D8R-type structures [60].

This model was recently revised by Duxson et al. [36] to include water in the reaction. The water normally consumed during dissolution is released in the condensation process. While water constitutes the reaction medium, it is found inside the gel pores. This type of gel structure is commonly referred to as biphasic, the phases being the aluminosilicate binder and water.

Alkali-Activation of Fly Ash

Thermal power plants using coal as a source of energy also produce significant amounts of fly ash. It is estimated that the world production of fly ash in 2011 will reach 800 million tones. Until now, only a minor part of this material has been recycled (20-30%), while the rest has been disposed of in landfills, thus contributing to the pollution of soil, water and air. The problem of fly ash disposal is far more complicated than disposal of majority of other industrial waste materials and requires finding the possibilities of valorization of this material.

In light of this information, the material proposes to replace to Portland cement in concrete for use in construction is a binder prepared by mixing type F (classification in accordance with ASTM standard C618-03 [61]) fly ash from coal-fired steam power plants with a highly alkaline solution (pH>13) [41].

As this material sets and hardens under moderate thermal conditions, the resulting precipitate is a sodium aluminosilicate hydrate gel (N-A-S-H gel) with cementitious properties. This material could consequently be considered to be a "zeolite precursor". Indeed, small amounts of certain zeolites such as hydroxysodalite, chabazite-Na and zeolite P are often detected in alkali-activated fly ash systems.

The activation rate and chemical composition of the reaction product depend on factors such as particle size and chemical composition of fly ash, type and concentration of the activator, the rheology, curing conditions and so on.

Particle Size and Chemical Composition of Fly Ash

Theoretically, any material composed of silica and aluminum can be alkali-activated. Fernández-Jiménez et al. [62] studied the reactivity of several fly ashes to be alkali-activated. Results obtained in the investigation showed that the main characteristics of a type F fly ash for leading to a material with optimal binding properties by alkali activation were: percentage of unburned material lower than 5%; Fe_2O_3 content not higher than 10%; low content of CaO, content of reactive silica between 40-50%; percentage of particles with size lower than 45 μm between 80-90%; and also high content of vitreous phase.

On the other hand, Van Jaarsveld et al. [63] have reported that source materials used during the synthesis of geopolymers from industrial by-products such as fly ash have an important role in determining the final properties of the geopolymer matrix. In this work, the authors studied the effect of phase composition of fly ash on the dissolution, reactivity, and final physical and mechanical properties of fly ash-based geopolymeric materials. Results showed that a higher aggregation coefficients and the presence of calcium compounds lead to higher mechanical strength.

Type and Concentration of the Activator

Influence of Alkali Cations

It is evident that the type of cation involved in the activation reaction affects the microstructural development of the systems as well as the Si/Al ratio of the N-A-S-H gel. The alkaline metal cation acts as a structure-forming element, balancing the negative framework charge carried by tetrahedral aluminum. In this respect and given that the first stage of the reaction is controlled by the aptitude of the alkaline compound to dissolve the solid fly ash network and to produce small reactive species of silicates and aluminates, it would be reasonable to think that in the case of sodium and potassium hydroxides, KOH should show a greater extent of dissolution due to its higher level of alkalinity [36].

Nevertheless, reality demonstrates that it is NaOH that possesses a greater capacity to liberate silicate and aluminate monomers [64,65]. Therefore, the size of the cation also affects the eventual crystal morphology. Na^+, having a smaller size than K^+, displays strong pair formation with smaller silicate oligomers (such as monomers), while the larger size of K^+ favors the formation of larger silicate oligomers with which $Al(OH)_4^-$ prefers to bind [66]. In general, matrices containing K had higher compressive strength but also higher specific surface area values and proved to have a lower degree of crystallinity as well as lower resistance to attack by HCl [64,65].

The sodium cations have better zeolitization capabilities in geopolymer-forming system [67], possibly because they are smaller than potassium cations and therefore more able to migrate through the moist gel network, or possibly due to their higher charge density.

Finally, Figure 13 ratifies that the type of cation used in the fly ash activation process (sodium or potassium) plays a fundamental role in the Si/Al ratio of the aluminosilicate hydrate gel.

This figure shows the gradual enrichment in silicon that, over time, is suffered by the aluminosilicate gel in the case of the systems activated with sodium. In the case of the systems activated with potassium, this evolution is not demonstrated quite so clearly.

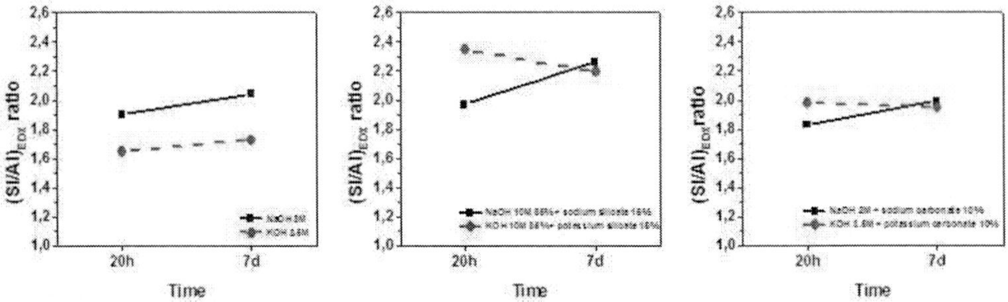

Figure 13. Evolution over time of the Si/Al ratio of the alkaline aluminosilicate hydrate gel. Comparative effect of sodium versus potassium solutions.

Influence of Alkali Anions

The type of anions can induce relevant differences in kinetics reaction and the mineralogical and microstructural characteristics of alkali-activated fly ash systems. The most used alkaline anions are hydroxide, silicate and carbonates. The main reaction product of fly ash alkaline activation is an alkaline aluminosilicate hydrate gel, different microstructures may develop depending on the type of anion in the activator solution.

Fernández-Jiménez et al. [68] have carried out a study about the effect of type of activator on the mechanical strength of alkali activated fly ash mortars and on the nature and microstructure of the reaction product. When the activating solution was an alkaline hydroxide, the OH^- ions acted as a reaction catalyst during the activation process, favoring the dissolution of Si^{4+} and Al^{3+} from coal fly ash. The N-A-S-H gel presented a Si/Al = 1.6-1.8 and Na/Al = 0.46-0.68 ratios and the mechanical strength value development at 8 hours was 42 MPa, see Figure 14.

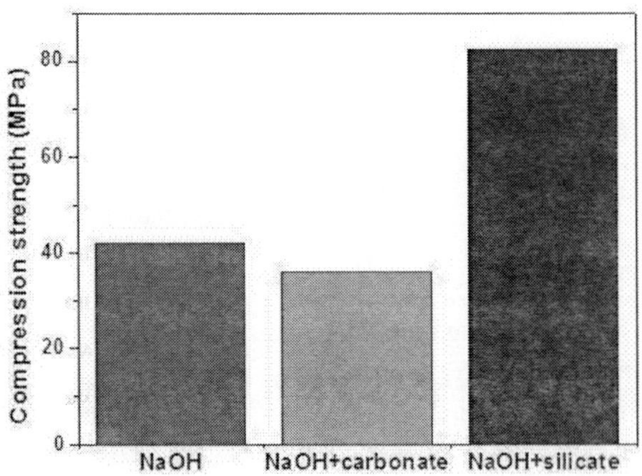

Figure 14. Mechanical strength development at 8 hours for fly ash activated with different alkaline solutions.

When the activating solution includes Na^+ and also Si^{4+} (sodium silicate=waterglass), both elements were incorporated in the reaction products. The gel N-A-S-H had Si/Al = 2.7 and Na/Al = 1.5 ratios. These ratios were clearly higher than those found with the NaOH activator [68]. The addition of waterglass to the activating solution enhanced the polymerization process of the ionic species in the system. Therefore, the condensation degree increased and the mechanical strength increased too (see Figure 14).

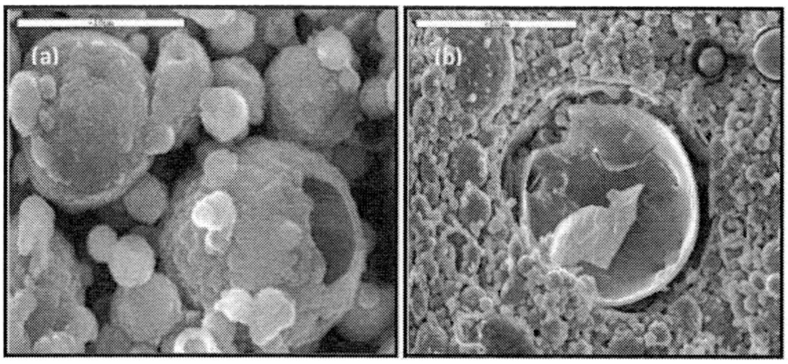

Figure 15. a) 8M KOH and b) 18M KOH activated fly ash.

When in the activating solution there was a certain content of carbonates, the formation of sodium bicarbonate (trona and/or nahcolite) was promoted. It involved an acidification of the medium leading, consequently, to a relatively low amount of Al and Si dissolved from the fly ash. This material showed a very porous microstructure and low mechanical strengths [68-70], see Figure 14.

On the other hand, it is very important to remark that the aluminosilicate hydrate gel formed depends on the concentration of alkaline activator. This parameter has proved to play an extremely important role in designing the dosage in ash-activated mortar and concrete. Figure 15 clearly illustrates the microstructural modifications that can be induced in the final reaction product by varying the alkali concentration in the system [69].

These micrographs show a series of changes, which consist essentially in a rise in matrix density and the formation of a quasi-glassy material (similar to the product obtained with sodium hydroxide + waterglass activators), as the activator concentration increased.

Hardjito et al. [71] have observed that geopolymers prepared with a 14M solution of NaOH showed greater resistances to compression than the samples preprared with a 8M solution of NaOH, regardless of curing temperature and age. Bakharev [72] and Hu et al. [73] have shown that the concentration of the NaOH solution played the most important role in the strength of alkali-activated fly ash-based geopolymers. However, there is an ideal concentration of the alkaline activator which contributes to increase the resistance of geopolymers. Beyond this ideal concentration, there could be loss of the material's mechanical properties, due to the presence of free OH^- in the alkali-activated matrix, which could change the material's geopolymer structure [41].

When the alkaline activator is sodium silicate, the aluminosilicate gel formed depends also on the polymerization degree of the added soluble silica, which depends on the alkalinity of the solution [74-76]. It has been observed that the incorporation of low amounts of soluble silicate to the system favored the tendency of the tectosilicate structure to achieve a high level of ordering in a short time. The kinetics of the transformation of the zeolitic precursor to crystalline phases (aluminosilicates) was notably accelerated. However, when the polymerization degree of soluble silica was increased, the gels formed were apparently more amorphous, and had broader, more featureless peak in ^{29}Si MAS NMR (see Figure 16) [76].

On the other hand, the requirement of a large concentration of silicon in the activating solution to form amorphous geopolymer with a high mechanical strengths and microstructures comprised of small pores has been known for some time in fly ash systems [70,77,78,79].

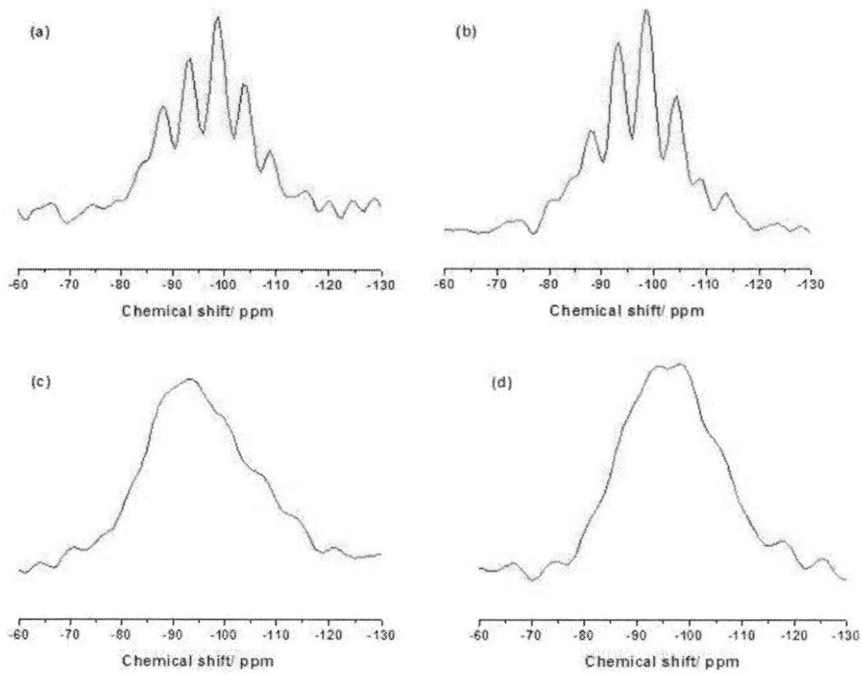

Figure 16. ^{29}Si MAS NMR of N-A-S-H gel from fly ash activated 7 days with a) 8M NaOH, and with sodium silicate solutions, SiO_2/Na_2O= (b) 0.19, (c) 0.69 and (d) 1.17.

Curing Conditions

Curing time, temperature and relative humidity affect the nature of the aluminosilicate gel obtained.

Bakharev [72] believed curing temperature was a crucial aspect about the alkali activation of fly ashes because of the thermal activated barrier, which must be over passed so that the reaction may start.

That stamen was confirmed by the investigations of Katz [80] who noticed an impressive strength increase with curing temperature. Related works carried out with fly ashes [76,81] have demonstrated that time and temperature affected the mechanical development of geopolymer materials, see Figure 17.

As the temperature of reaction increases, the mechanical strength development increases also. However, there is a threshold value for mechanical strength development, above which the strength gaining rate is slow.

The structural transition from amorphous to crystalline of geopolymers synthesized at low to mild temperatures in concentrated medium also implies that the synthesis temperature and aging are critical in determining the structure of the reaction products. XRD patterns for the original fly ash and fly ash after activation with 8M NaOH are show in Figure 18 [74].

The alkali activation of the fly ash also gave rise (by increasing the time of thermal treatment) to the formation of crystalline phases identified as Chabazite-Na and Hydroxysodalite.

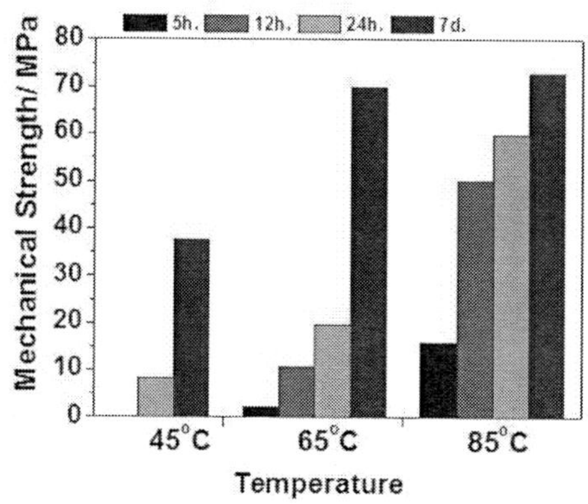

Figure 17. Evolution of the mechanical strength of the activated fly ash mortars as a function of time.

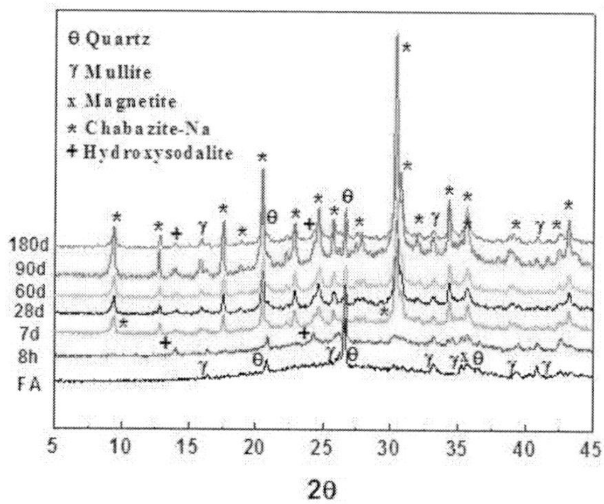

Figure 18. X-ray powder diffraction patterns of fly ash and activated ashes with 8M NaOH solution.

In addition to curing time and temperature, relative humidity (RH) has been shown to play a role in the initial curing of alkali-activated fly ash cement. Unsuitable curing conditions may accelerate carbonation [82-84], lowering pH levels and as a result retarding ash activation with the concomitant water loss and persistence of high aluminium content. Morphological and microstructural characteristics of the two systems cured with Method 1 (RH over 90%) and Method 2 (RH≈40-50%) after 60 days curing are shown in Figure 19.

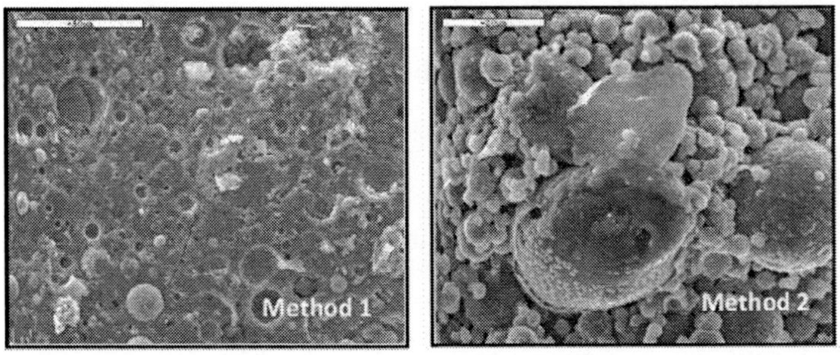

Figure 19. Microstructure of the two systems cured under Method 1 and Method 2 for 60 days.

Curing at low relative humidity (RH≈40-50%) with the pastes in direct contact with the atmosphere, the final product was granular, porous and characterised by low mechanical strength, while curing at a relative humidity of over 90%, in which the pastes were kept in air-tight containers, yielded a dense and compact material, with a good mechanical development over time.

Rheology

Rheology is a very important aspect for the possible technological application of these materials. Working with alkali-activated fly ash systems, Palomo et al. [85] showed that information on their reaction kinetics could be obtained by processing the respective rheological data.

High-range water reducing, admixtures or superplasticizers are used to improve rheology and workability of OPC concretes. Criado et al. [86] have reported that chemical admixtures used do not work the same in the Portland cement systems than in alkali-activated fly ash systems, due to their high alkalinity. The effect of the mixtures was to make the fly ash pastes more fluid and hence less viscous (plastic viscosity and yield stress values were lower than in the reference fly ash paste), see Figure 20.

As a general rule, it seemed that the most efficient admixtures for these new cementitious pastes were those based on polycarboxylates, in which inter-particle electrostatic repulsion prevailed over complex formation.

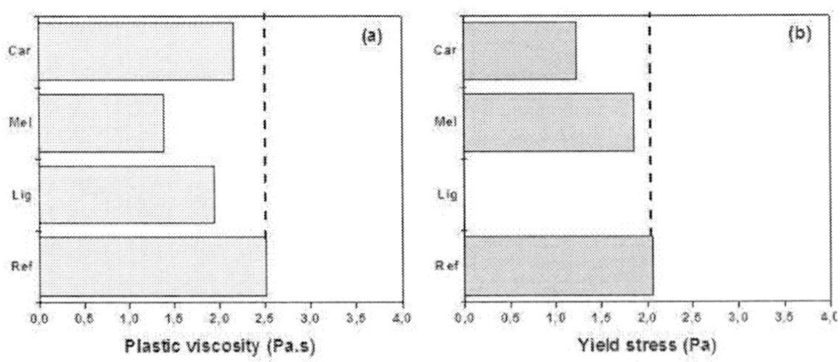

Figure 20. a) Plastic viscosity and b) yield stress of alkali-activated fly ash pastes with and without admixtures with a "liquid/solid" ratio of 0.4. Ref: reference; Lig: Lignosulphonate; Mel: Melamine; Car: Polycarboxylate.

Durability of Alkali-Activated Fly Ash Binders

The deterioration of material may occur through a variety of chemical or physical processes, especially when it is exposed to aggressive environments. The durability of a material has a significant influence on its service behaviour, design life and safety. Concrete made from Portland cement is also subject to certain durability problems such as sulfate attack, seawater attack, acid attack, alkali silica reaction, frost attack and corrosion of steel reinforcement.

Bakharev [87] have studied the evolution of weight, compressive strength, products of degradation and microstructural changes of geopolymer materials manufactured using class F fly ash and alkaline activators when were exposed to a sulfate environment (5% Na_2SO_4, 5% Mg_2SO_4). The best performance in different sulfate solutions was observed in the geopolymer prepared with sodium hydroxide and cured at elevated temperature, see Table 5.

Materials activated with NaOH had 4-12% increase of strength when immersed into sulfate solutions, due to the formation of a more stable cross-linked aluminosilicate polymer structure.

In recent work, Fernández-Jiménez et al. [47] studied the behaviour of alkali activated fly ash completely submerged in HCl solution (0.1 N, pH=1.0). The compressive strength results obtained are given in Figure 21.

In alkali activated fly ash mortars the strength declined approximately 23-25%, whereas in the OPC mortars strength dropped at nearly twice that rate (47%). Similar results were obtained when alkali activated fly ash were exposed to 5% solutions of acetic (pH=2.4) and H_2SO_4 (pH=0.8) [88].

Other aspect related to durability is the alkali silica reaction (ASR). The alkali aggregate reaction is a chemical process involving alkaline oxides generally deriving from the alkali in the cement and certain forms of the reactive silica present in the aggregate. In alkaline environments, these forms of silica are susceptible to attack, ultimately leading to the dissolution of the aggregate. A study carried out by García-Lodeiro et al. [89] with the purpose of evaluate the performance of alkali activated of fly ash in the context of the ASR and compare their behaviour to the finding for traditional Portland cement concrete. Expansion patterns were determined for each and every one of the mortars, see Figure 22..

Table 5. Summary of compressive strength evolution in sulfate environment (%)

Sample	Na_2SO_4 solution (%)	$MgSO_4$ solution (%)	Na_2SO_4+$MgSO_4$ solution (%)
NaOH	+4	+12	+12
NaOH+ KOH	−65	+35	+10
Waterglass	−18	−24	−4.5
OPC	−30	−21	−35
OPC+FA (NaOH)	−14	−4.8	−19

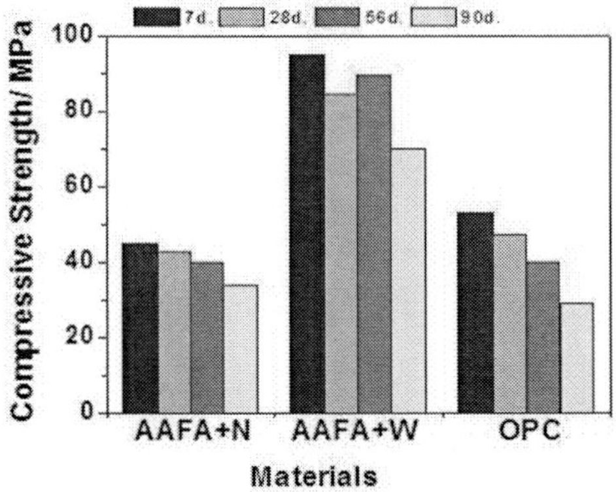

Figure 21. Mechanical strength of alkali activated fly ash (AAFA+N: with NaOH and AAFA+W: with waterglass) and OPC mortars attacked with 0.1N solution of HCl.

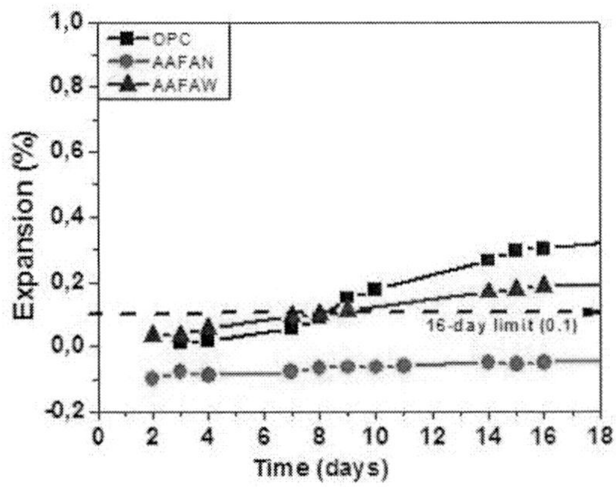

Figure 22. Expansion curves with the time for OPC and fly ash (AAFAN: with NaOH and AAFAW: with waterglass) mortars.

The systems based on activated fly ash were less susceptible to generate expansion by alkali silica reaction than traditional Portland cement systems. AAFA mortars with NaOH expanded less than the 0.1% limit stipulated in the ASTM standard C1260-94 after 16 days. The evidence indicated that the calcium in the materials played an essential role in the expansive nature of gels.

The main reason for the premature failure of reinforced concrete structures (RCS) is corrosion of the reinforcement. Several works [90-92] have demonstrated that alkali activated fly ash mortars passivate steel reinforcement as rapidly and effectively as OPC mortars. Figure 23 shows the variation of i_{corr} values in steel rounds embedded in different blended fly ash-Portland cement mortars and exposed to different chloride containing atmospheres. Mortar specimen composition is shown in Table 6.

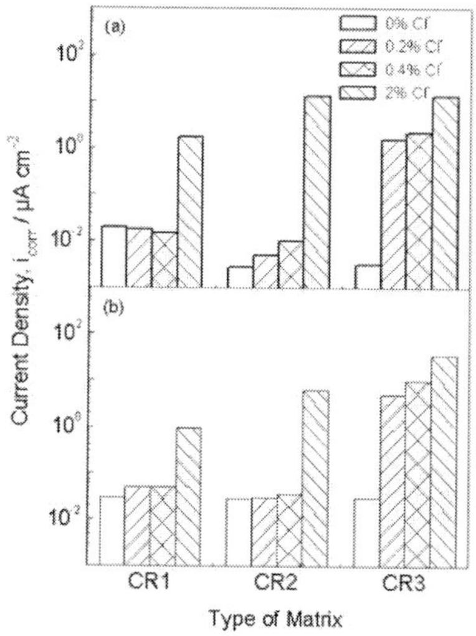

Figure 23.Corrosion current density (i_{corr}) for carbon steel bars embedded in CR1, CR2 and CR3 mortars with 0%, 0.2%, 0.4% and 2% chloride exposed to RH cycles for 450 days, (a) estimated from time constant (τ) values, and (b) from R_p values.

Table 6. Mortar specimen composition

Mortar Matrix	Binder Material	Sand/Binder Ratio	Type of Activator	Liquid/Binder Ratio
[1]CR1	Fly ash	2/1	15% Waterglass, solution 85% (12.5 M NaOH), solution	0.5
[2]CR2	70% Fly ash 30% Portland	2/1	15% Waterglass, solution 85% (12.5 M NaOH), solution	0.5
[2]CR3	58.8% Fly ash 25.2% Portland	2/1	8% Piramid, solid 8% Na_2CO_3 solid Water	0.4

[1]Initial curing: 20 h at 85 °C in a water-vapour-saturated atmosphere.
[2]Initial curing: 20 h in a curing chamber (21±1 °C and 99% RH).

Activated fly ash mortars passivate reinforcing steel as rapidly and effectively as Portland cement mortars. The passivate state stability depends on the activator employed and can be also affected by the environmental conditions. The addition of 2% (by binder weight) chlorides multiplies the steel corrosion rate by a factor of approximately 100.

Regarding the corrosion of steel reinforcement, we would like to emphasize that reinforced concrete is the most widely used the construction material in the world. Nevertheless, the premature failure of reinforcement bars seriously affects the stability of RCS. Concrete deterioration of RCS has become a costly economic issue, an estimated three to four percent of gross national product (GNP) for direct and indirect cost in the developed countries in maintenance and repair operations.

Therefore, use of new corrosion prevention strategies is reported as a palliative method to increase the reinforced concrete service life. For example, use of stainless steel (SS) reinforcements may be a possible solution even though SS reinforcements have proved their ability to prevent corrosion for a very long time, even in very aggressive environments and for this reason, last part of this chapter is devoted to study the corrosion behaviour of stainless steels reinforcement embedded in OPC or in new alkaline cements.

Engineering Properties and Applications

These mortars and concretes made with Portland cement-free activated fly ash develop a high mechanical strength in short periods of time, have a moderate modulus of elasticity and bond better to reinforcing steel and shrink much less than ordinary Portland cement concrete [43].

Since geopolymers are considered as two-component systems (reactive solid components-alkaline activation solution) they can be used as suitable binders in pre-cast industry for the manufacture of reinforced products such as large-diameter pipes, roofing tiles, pre-stressed monoblock sleepers [93-95].

Immobilization techniques are used for the treatment of large amounts of heavy metals and radioactive wastes, thus geopolymerization has received over the years significant attention due to its low cost, flexibility and increased durability versus time. Alkali-activated fly ash matrices have already been used to immobilize and stabilize low-level radioactive wastes as well as heavy metals [96-101].

Corrosion of Reinforced Concrete Structures

Concrete corrosion has serious environmental, societal and economic impact each year, 3-4% gross national product (GNP) for direct and indirect cost in the developed countries in repair, replacement, and environmental impact as shown in Figure 24. The major cause of deterioration is corrosion of steel reinforcing bar in concrete that occurs when concrete pH is reduced by chloride ions penetration and carbonation attack, destroying the passive layer formed on the reinforcing bar.

Reinforced concrete (RC) combines the good compression strength properties of concrete and the excellent mechanical strength properties of steel. Thus, RC materials can be used by designers, architects and civil engineers to meet high mechanical strength, fire resistance, durability, shape adaptability and low cost requirements [102,103]. This explains why Portland cement pastes have been the construction material *par excellence* for decades.

RC is the most widely used construction material in the world, its tonnage surpassing that of all other materials combined. Nevertheless, the premature failure of reinforcement bars seriously affects the stability of reinforced concrete structures (RCS). An idea of the magnitude of the problem can be seen in the fact that losses attributed to RCS corrosion in Spain were of the order of 1.2×10^9 euros in 1995 [102].

Fortunately, if concrete is correctly executed its high pH guarantees the passive state of the rebars. In the passive state, corrosion is insignificant and, if this state is maintained, RCS durability can be practically unlimited; as is demonstrated by the multitude of structures that continue to be in a perfect state of conservation after many decades in service. It is likely that the current situation, with the construction sector mobilising about 10 per cent of the world economy [104], will continue in the short and mid-term.

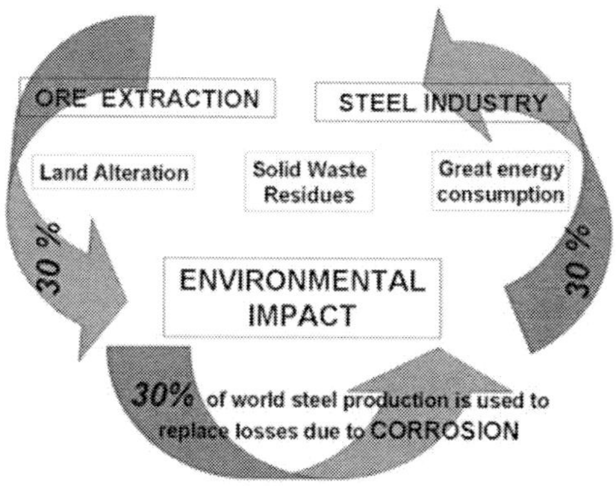

Figure 24. Schematic layout of the environmental impact of corrosion in reinforced concrete structures.

However, in some particular cases, always related with loss of the passive state either locally or in large areas, rebar corrosion becomes significant and at times leads to a dramatic reduction in RCS durability. In this respect, RCS are being constructed today which will require costly repairs or even demolition in very short period of time, resulting in a 10-year to 20-year lifetime. The consequence is that approximately between 40-60% of resources in the construction industry are dedicated to maintenance and repairs [105]. Figure 25 depicts the Tutti model that represents the different stages of the RCS corrosion initiation and propagation, thus governing the lifetime in service.

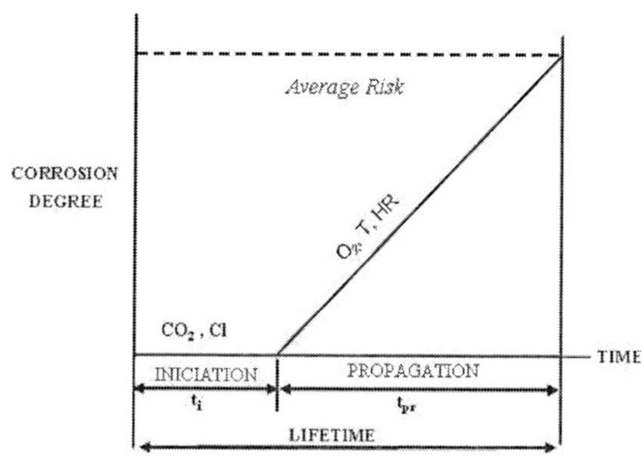

Figure 25. Prediction of the lifetime in service of a RCS by Tutti Model.

These risks are frequently associated with highly aggressive environments or are originated by the incorrect use of RCS know-how. The economic and social importance of the construction sector makes RC failure and specially RCS failure, the principal challenge of civil engineering in developed nations [106].

For instance, almost 50% of the 575,000 bridges in the US inter-state network are in need of some type of repair, due to structural defects and it is estimated that the cost of repair is much higher than that of new construction [107,108].

On the other hand, the very high temperatures (1400-1500 °C) required to manufacture Portland cement, which make it responsible for 40% of all energy consumed, account for the extremely high costs of this process [43]. The environmental impact attributed to the manufacturing of Portland cement is largely due to the energy-intensive processes involved. Indeed, the cement industry is regarded to be responsible for 6-7% of all greenhouse gases emitted world-wide (0.85-1 ton of CO_2 per ton of cement) [89].

Concrete is a composite of cured Portland cement, aggregate, and often various admixtures. The anhydrous compounds in Portland cement react slowly with water and "cure" into a hardened monolithic structure of hydrated compounds having high compressive strength but low tensile strength.

Cement consists of four constituents: tricalcium silicate, dicalcium silicate, tricalcium aluminate, and tetracalcium aluminate ferrite. In the presence of water, these four constituents react to produce concrete gel ($3CaO{\cdot}2SiO_2{\cdot}3H_2O$) and free lime ($Ca(OH)_2$), see Eq.3:

$$3CaO{\cdot}SiO_2 + 6H_2O \rightarrow 3CaO{\cdot}2\,SiO_2{\cdot}3H_2O + 3Ca(OH)_2 \qquad (3)$$

The free lime is somewhat soluble in aqueous solution and it further reacts with various salts to give calcium sulphate and hydroxide ions, see Eq.4:

$$Na_2SO_4 + Ca(OH)_2 \rightarrow CaSO_4 + 2NaOH \qquad (4)$$

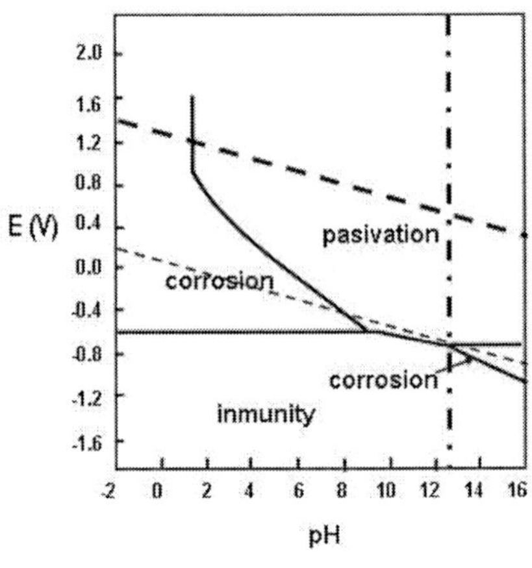

Figure 26. Pourbaix diagram for steel in reinforced concrete.

Hydroxide ions are very soluble in water that contributes to the initial high pH value of the concrete. Steel rebar is protected against corrosion because of this high initial pH value in the concrete matrix as shown in the Pourbaix diagram in Figure 26. At high pH values, the oxide film is thermodynamically stable [109].

Mechanisms for Corrosion in Concrete

For corrosion of steel in concrete to occur, the following conditions must all be satisfied: (a) the provision of an anode-cathode couple with at least part of the steel acting as an anode, (b) the maintenance of an electrical circuit (free flowing ions), (c) the presence of moisture, and (d) the presence of oxygen [110].

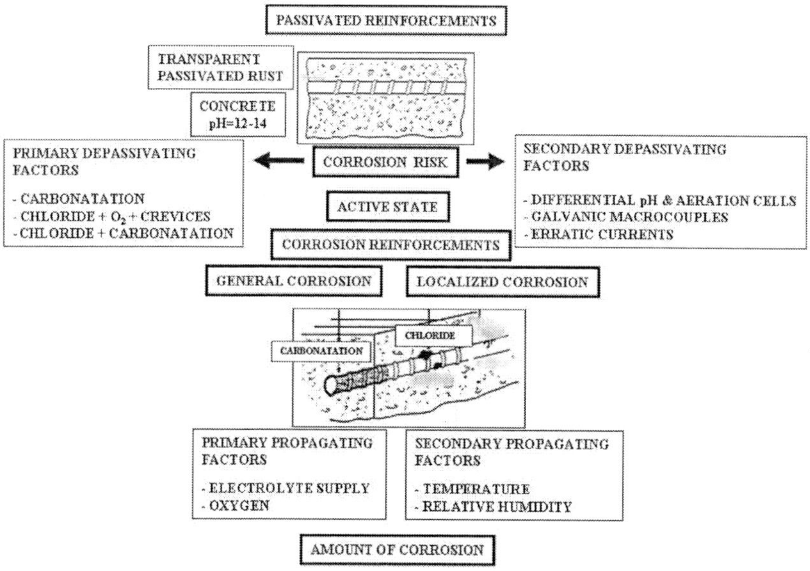

Figure 27. Schematic of reinforced concrete corrosion processes.

There are two major processes that are coupled in corrosion attack on steels in concrete: carbonation reactions and pitting corrosion in presence of chloride ions are shown in Figure 27. Each process is discussed in more detail in the sections below.

Carbonation Reaction

As mentioned already, the passivating oxide film is thermodynamically stable at high pH values. However as the pH decreases, the stability of the film breaks down (with presence of Cl^- ions), exposing metal surface for corrosion. As a result, oxidation and rusting occurs. The decrease in pH in caused by the diffusion of weak acids from the environment. These acids are mainly CO_2 in the air and SO_3 in rain. CO_2 is the more important of the two acids.

Initially, CO_2 gas will not be able to penetrate deeply in the concrete because it reacts with free lime:

$$CO_2 + Ca(OH)_2 \rightarrow CaCO_3 + H_2O \text{ (in presence of } H_2O \text{ and NaOH)} \quad (5)$$

As a result of the above reaction (Eq.5), pH decreases from basic conditions to a more neutral condition. Once the cement is carbonated, the CO_2 gas will be able to diffuse further and further into the cement and eventually the pH near the core, where steel reinforcements reside, becomes neutral. Referring to the Pourbaix diagram in Figure 26, at neutral conditions, the passivating film is no longer stable, therefore, causing the steel to lose its passivity. The penetration depth is known as the "carbonation depth." Because concrete is a macroporous material, the penetration of CO_2 will be determined by the form of the porous structure.

The rate of diffusion of CO_2 is affected by many parameters. One of the most important one is the humidity. If the pore structures are dry, CO_2 diffuses inward readily but the carbonation reaction does not occur because of lack of water. If the pores are filled with water, the carbonation reaction rate is low because of slow CO_2 diffusion in water. Carbonation reaction only occurs when the pores are partially filled with water, which is normally the case at the surface of concrete.

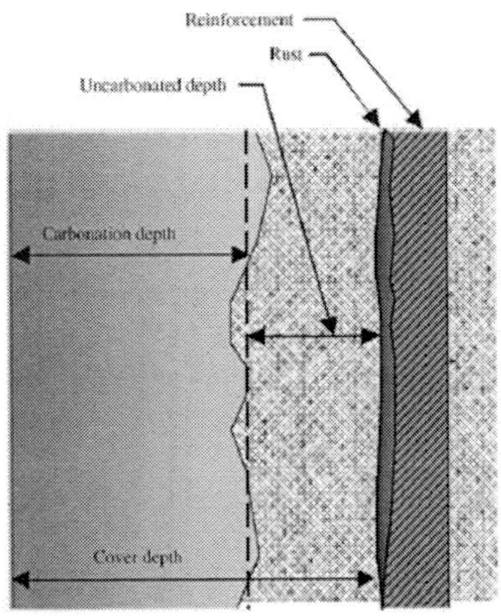

Figure 28. Reinforced concrete corrosion induced by carbonation process, penetration and diffusion throughout cracks.

The environmental conditions determine the moisture content of the concrete surface. The moisture content determines the saturation of the pores and hence the CO_2 permeability. The diffusion length depends on the square root of the time.

With the presence of cracks in the concrete, carbonation penetration is much faster into the interior via the crack. Because of the higher rate of penetration of CO_2 into the interior of concrete, carbonation reaction rate is increased as illustrated in Figure 28. In addition, the water in the crack evaporates more slowly in the cracks than the water at the concrete surface after exposure to water. As a result, the reaction rate is further increased. Once the passive film is broken down, the oxygen acts as oxidizing agent and oxidizes the steel. This corrosion does not have a preferential attack site; steel experiences active corrosion. Carbonation reactions occur in conjunction with pitting corrosion [111].

Pitting Corrosion

Pitting corrosion can be considered as a two-step process: depassivation (chloride attacks) and propagation (pit growth). These steps are discussed further in the sections below. But before pitting corrosion can be discussed, the role of chloride and the transport phenomena in pitting corrosion must be understood.

Role of Chloride Ion

Chloride ions can be introduced into concrete via two ways: i. Contaminates in the original mix, and ii. Exposure to deicing salts, seawater, and other Cl^- ions containing sources. Cl^- ions are incorporated in the lattice of the hydration product.

Unlike the case of CO_2, the diffusion of Cl^- only takes place in presence of water. If the concrete is dry, transport of Cl^- ions is impossible. If the concrete is semi-dry, there will be diffusion of chlorides from the surface to the inner concrete. With presence of cracks, the diffusion of Cl^- ion is enhanced. The diffusion resistance within the crack is negligible to the diffusion resistance of the cover itself.

Depassivation

The maintenance of passivity requires continuing high levels of alkalinity in contact with the steel surface and absence of aggressive ions (i.e. Cl^-) from the concrete. Chloride ions have the special ability to destroy the passive oxide film of steel. Reduction of alkalinity by carbonation reaction in the presence of sufficient Cl^- ion concentrations will lead to passive film breakdown. The initiation of corrosion is attributed to pitting corrosion. Pitting corrosion occurs when the passivating film is exposed to an aggressive chloride-containing environment. The presence of Cl^- ions at local sites enhances the dissolution of metal.

Additionally, O_2 is required to maintain an oxide film over the majority of the metal surface, where the cathodic reactions can be sustained. The minimum amount of chloride required to reduce the passive range is a function of two variables: the ratio of concentration of chloride and hydroxyl ions in the pores.

Propagation

After the initial break down of the passivating film, dissolution and pitting corrosion occurs. The pit represents the anode, while the passive layer is the cathode in the presence of O_2. A large cathode to anode surface area further enhances the corrosion attack. As pitting corrosion continues, the pH in the pit increases due to the hydrolysis of corrosion product, which releases H^+ ions. As a result, the potential in the pit becomes increasingly negative. Correspondingly, the passivating film undergoes a cathodic reaction where hydroxide ions increase the pH. Therefore, pitting corrosion becomes self-supported at these local pitting sites. In order for pitting corrosion to continue, Cl^- ions in the aqueous solution must be replenished. In the process of film breakdown, the Cl^- ions are consumed from the liquid by the formation of intermediate iron-chloride corrosion products. These complexes diffuse away from the zone and will be precipitated out. Replenishment of Cl^- ions is accomplished by the breakdown of chloroaluminate complexes (cement product) and by the diffusion of Cl^- ion from the bulk solution.

In an attempt to overcome these problems, the construction sector is very interested in the development of new cement binder materials as an alternative to conventional Portland cement. In this respect, the most promising emerging approach is based on new raw materials suitable for alkaline activation, essentially fly ash, which originate new binding materials generically known as alkaline cements [67,90,112-116].

These new cements may be used to manufacture concrete well suited to the precasting industry, because when thermally cured they attain compression strength values of up to 50 or

60 MPa just 20 h after mixing [116,117]. Consequently the demoulding and storage of precast elements can be speeded up, thereby raising factory output.

Alkaline cements also show better durability behaviour than conventional Portland cement due to their high resistance to attack by sulphates and the fact that systems based on activated fly ash are less prone to expansion due to alkali-silica reactions (aggregate-alkali reaction, formation of thaumasite, etc.) than Portland cement systems [118]. Furthermore these materials, which adhere extraordinarily well to reinforcing steel, feature high-volume stability, high fire resistance and low thermal conductivity, and would foreseeably afford excellent durability in aggressive environments. It worth noting that fly ash is an industrial waste that is derived in extremely large volumes from a wide range of starting materials. Its generation in the form of dust makes it unnecessary to carry out any type of prior transformation in order for it to be used as a cement material. Therefore it is not surprising that these materials are competitively priced compared to Portland cement.

Many reasons for using fly ash are global, environmental, or social in nature. The production of Portland cement puts about a ton of carbon dioxide (CO_2, a primary greenhouse gas) into the atmosphere for every ton of cement produced-roughly half a ton from the fuel used to cook the raw limestone, and half a ton from the calcination of the limestone. This proposed "green" technology offers a new approach using fly ash, which reduces the cost of electricity generation in coal-fired power plants and save land otherwise needed for filling waste calcium sulfate generated with conventional sulfur removal approaches. The success of the fly ash utilization not only benefits land conservation but also air pollution control and natural resources protection.

Considering that reinforcement corrosion is the main cause of RCS failure [106,119,120], the capacity of activated fly ash mortars and concretes to passivate steel reinforcements is a very important property to guarantee the durability of RCS constructed using these new materials. The passivating capacity and the permanence of the passive state once reached may depend on the nature and the dosage of the binder, on the type of activator used, and on the environmental conditions. Although the results demonstrate that alkaline cements can provide passivation to reinforcements as efficiently and permanently as conventional Portland cement, they also show that this passivity depends on the type and dosage of conglomerate and the type of activators used, so from a scientific, technical and economic viewpoint there is a need for more results on the subject in order to build up a data base and avoid unnecessary risks regarding the durability of reinforced structures manufactured with these new materials.

In recent work, the use of new mortars based on activated fly ash, an alternative to conventional Portland cement was studied by Bastidas et al. [91]. Activated fly ash mortars passivate reinforcing steel as rapidly and effectively as Portland cement mortars. The passive state stability depends on the activator employed and can be also affected by the environmental conditions.

Mortar specimen composition is shown in Table 6.The corrosion current density (i_{corr}) value was estimated from the R_p measurements. Figure 29 shows the current density (i_{corr}) versus time for specimens in a cyclic RH experiment.

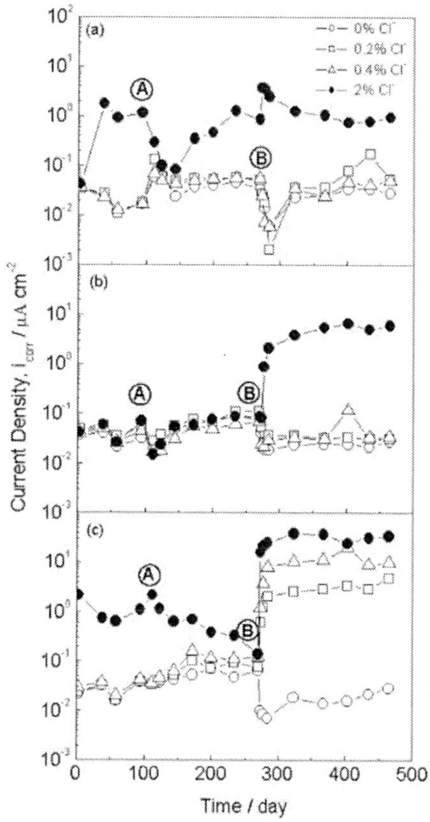

Figure 29. Corrosion current density (i_{corr}) versus time for carbon steel bars embedded in (a) CR1, (b) CR2, and (c) CR3 mortars for different chloride additions and in a cyclic environment: 100 days at a 95% RH, point A, followed by 150 days in the laboratory atmosphere (dry environment, ~30% RH), point B, and 220 days at 95% RH.

Humidity cycles were performed to assess the chemical stability and thus the reliability of the fly ash matrix structure. The specimens were exposed to a cyclic environment: 94 days at 95% RH, point A, followed by 150 days in the laboratory atmosphere (dry environment, ~30% RH), point B, and 220 days at 95% RH. It may be noted that: (i) the CR2 mortar remains in the passive state even with 2% Cl^- in the initial wet period and the intermediate dry period.

An increase of two orders of magnitude in the i_{corr} may be observed with the rewetting of the CR2 mortar (see Figure 29b); and (ii) it is notable that with the CR3 mortar the humidity cycles also cause the loss of the passive state in the specimens with a low chloride content 0.2 or 0.4% Cl^-.

The corrosion potential was monitored vs time as see in Figure 30 for steel bars embedded in the CR1, CR2 and CR3 mortars with different chloride additions, using similar humidity cycles and the same specimens as Figure 29: 94 days at 95% RH, point A, followed by 150 days in the laboratory atmosphere (dry environment, ~30% RH), point B, and 220 days at 95% RH. It may be noted that: (i) the CR2 mortar remains in the passive state even with 2% Cl^- in the initial wet period and the intermediate dry period.

A decrease of 0.4 V (more active) may be observed in the E_{corr} when the specimens are moved from a dry environment to a wet environment (see Figs. 30b and 30c).

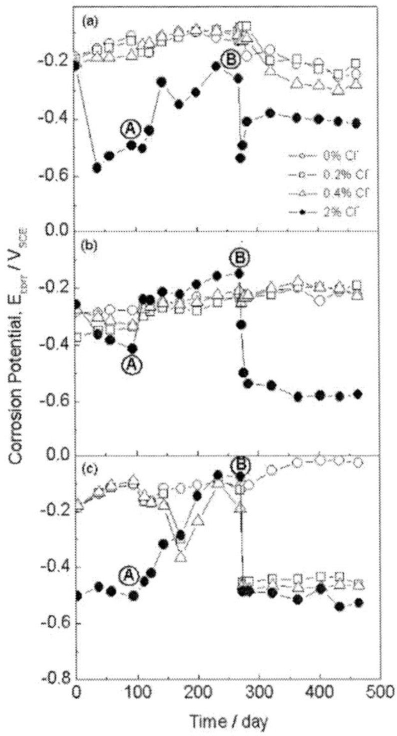

Figure 30. Corrosion potential (E_{corr}) versus time for carbon steel bars embedded in (a) CR1, (b) CR2, and (c) CR3 mortars for different chloride additions and in a cyclic environment: 100 days at 95% RH, point A, followed by 150 days in the laboratory atmosphere (dry environment, ~30% RH), point B, and 220 days at 95% RH.

These changes in the E_{corr} may be attributed to the wetness of the pore network in the mortars. In contrast, the E_{corr} for passive electrodes is of the same order of magnitude in wet and dry conditions. The E_{corr} for specimens kept at a high RH is not included because it does not supply any additional information, only the differences attributed to the passive and active states.

The carbon steel corrosion current density in the experimental conditions yielded similar values for the four tested materials: Portland, CR1, CR2 and CR3 mortars. Similar current density values were also obtained using polarisation resistance and short duration galvanostatic pulses two orders of magnitude higher for activated electrodes than for passivated electrodes.

The addition of 2% (by binder weight) chlorides multiplies the steel corrosion rate by a factor of approximately 100. In this case, the i_{corr} values are slightly higher in fly ash mortars, where the chloride content is likewise higher, due to the higher binder/sand ratio in these mortars.

Corrosion current density values for active electrodes depend on the resistivity of the mortar. An inverse logarithmic relationship was found between i_{corr} and ohmic drop (ΔIR_{Ω}), as has frequently been observed on mortars and concrete manufactured with Portland cement.

For passive steel electrodes there is a smaller dependence between i_{corr} and the resistivity of the construction material.

Absence of stability was observed in the passive state of specimens manufactured with CR3 mortar with 0.2 or 0.4% chloride additions and exposed to HR cycles (~30% RH and ~95% RH). This behaviour was associated with a descending alkaline pH of the CR3 mortar compared to the pH of the CR1 and CR2 mortars and/or a high total porosity.

In a similar way to Portland mortar, in fly ash mortars the corrosion potential (E_{corr}) values for the passive state are hundreds of millivolts higher (more noble) than for the active state in wet environments, but can easily be confused with dry mortars.

Finally, it would also seem appropriate to emphasise that: (a) the observations and conclusions cannot be expected to apply generally to Portland and fly ash cement-based mortars, and (b) the nature and integrity of the layer of solid hydration products formed in close proximity to the embedded steel may have an important role in controlling passivation and depassivation of the metal.

Non-Destructive Monitoring Techniques

The deterioration of reinforced concrete structures (RCS) is one of the greatest civil engineering challenges facing the developed world. Non-destructive and quantitative techniques for measuring corrosion are needed in order to detect the deterioration of structures at an early stage, to predict their residual life, and thus to decide what prevention or repair systems need to be applied. Electrochemical techniques provide virtually the only viable procedure for assessing reinforcement corrosion without removing the concrete cover [121].

The two most common methods for obtaining quantitative information on the corrosion rate of rebars are the DC linear polarization resistance (R_p) ($R_p=\Delta E/\Delta I$) method and the AC impedance method [121-124]. The ideal solution for studying RCS behavior would be to directly measure the corrosion rate of real size structures. The technique most widely used in the field, due to its simplicity, is that of potential measurements, whose most serious limitation is its exclusively qualitative nature [125]. The essential difficulty in real structures stems from the impossibility of knowing with certainty the surface of the rebars affected by the electrical signals in the measuring process.

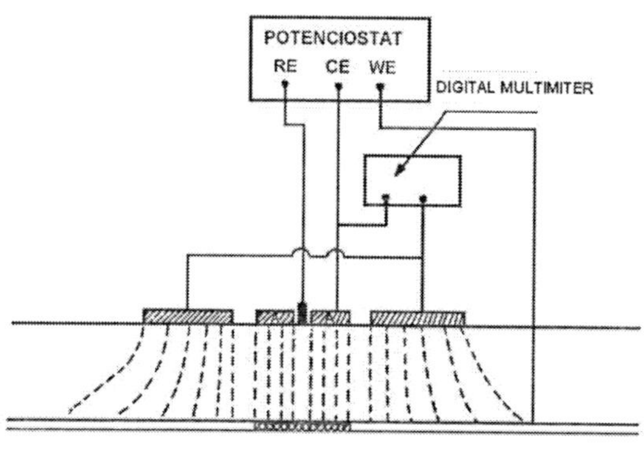

Figure 31. Current distribution confinement using the guard ring model.

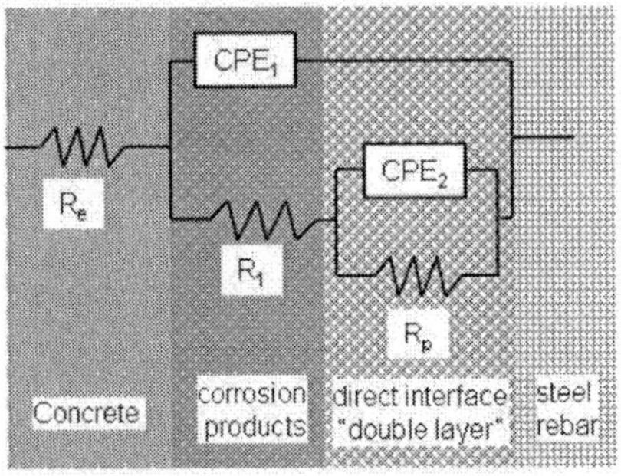

Figure 32. Equivalent electrical circuit (EEC) for steel-concrete system. R_e is resistance of the concrete cover; CPE_1 is the corrosion products capacitance, CPE_2 is double layer capacitance; R_1 is the corrosion products resistance and R_p is the resistance linked to the double layer interface.

The electrical signals are applied with small counter electrodes (CEs) and extend, while progressively decaying, to a critical length (L_{crit}) which is unknown in each case. Using a CE of the same size as the structures, with the aim of achieving a uniform distribution of the current lines, is obviously not possible. Figure 31 show the guard ring geometry used for optimizing current distribution confinement.

The behavior of a RCS may be represented, in a first approximation, by the equivalent electrical circuit (EEC) of Figure 32, though other more complex circuits allow a better simulation of the steel-concrete system [126-129].

Several researchers have explored the possibility of applying to the steel-concrete system corrosion rate measuring methods based on the application of short duration galvanostatic pulses [129-132], or the instantaneous discharging of a capacitor (coulostatic method) [133,134]. Besides their great speed, these methods offer the advantage of causing minimal disturbance of the electrode. Following on from these valuable contributions, the objective pursued is twofold: (i) to demonstrate that it is possible to estimate the corrosion rate of steel in concrete also by means of potentiostatic pulses, and (ii) that by supposing an approximate value of the double layer capacitance (C), this estimation can be made with sufficient exactitude for many practical purposes without the need to know the surface area of the rebars.

Fundaments of the Method

When the polarization of an electrode obtained, either by a potentiostatic pulse, a galvanostatic pulse or an amount of current is interrupted, and assuming that the system is represented by the EEC in Figure 32, the potential decay curve is exponential, after the first moments whend the ohmic drop takes place, according to the equation [133]:

$$\frac{\eta_t}{\eta_0} = e^{-\frac{t}{CR_p}} \qquad (6)$$

where η_t is polarization of working electrode (WE) at time t during relaxation following current interruption; η_0 is the maximum real polarization at moment of current interruption, i.e. for time t equal to zero; and C and R_p have been defined above.

After the current interruption, the increase in the double layer charge is progressively consumed in the corrosion reaction, and taking logarithms of Eq. 6, yields:

$$\ln(\eta_0) - \ln(\eta_t) = \frac{t}{CR_p} \tag{7}$$

Eq. 7 offers two ways for calculating R_P and therefore the corrosion rate through the Stern-Geary expression [135]: $i_{corr}=B/R_p$. One of the two procedures consists of determining the slope of $\ln(\eta_t)$ vs. t plot, as has hitherto been performed [129-134,136]. The second procedure is based on the direct measurement of the time constant ($\tau=CR_p$) attributed to the corrosion process at the steel-concrete interface.

Estimation of R_p from the Slope of $\ln(\eta_t)$ vs. t Plot

Given that η_0 is a constant value for each test, from Eq. 7 it is deduced that the plot of $\ln(\eta_t)$ vs. t is a straight line from whose slope R_p may be estimated if C is known:

$$R_P = \frac{-\Delta t}{C[\Delta \ln(\eta_t)]} \tag{8}$$

Therefore, on the basis of real experience [136,137], it may be assumed without much error that the value of C is, for the system studied, between 10^{-4}-10^{-5} F/cm^2; though much higher values may be determined for rebars that corrode actively [138], perhaps due to the effect of the existence of voluminous corrosion products at the steel-concrete interface.

A disadvantage to be considered is that the overall process mixes several partial processes with different time constants. Previous experience allows this limitation to be overcome, since it is known that R_p is of the order of 10^4 Ω cm^2 for reinforcements that corrode actively and around 10^5-10^6 Ω cm^2 for passive rebars. By selecting an intermediate C, e.g. 5×10^{-5} F/cm^2, it is possible to very approximately estimate the time constant of the corrosion process, which is the magnitude of interest. Having determined τ, the time interval (around τ) is defined in which, in practice, the slope of the straight line for calculating R_p by Eq. 8 must be fitted. In the $\ln(\eta_t)$ vs. t plot it would be necessary to consider tenths of a second in the case of active structures and units or tens for seconds in the case of passive structures.

Direct Measurement of the Time Constant

For a time $\tau=CR_p$, Eq. 6 is reduced to:

$$\frac{\eta_t}{\eta_0} = e^{-1} = 0.37 \tag{9}$$

Thus, it is possible to directly determine the time constant of the corrosion process by using Eq. 9, measuring the time in which η_t is reduced to 37% of its initial value (see Figure 33).

As C is directly proportional to the surface area of the WE and R_p is inversely proportional, τ is independent of this magnitude. Consequently, if C corresponding to the unit of surface area is inserted in the Eqs. 6, 7 and 8, the R_p of the rebars per unit of surface area will be estimated, without their number or their diameter being of importance.

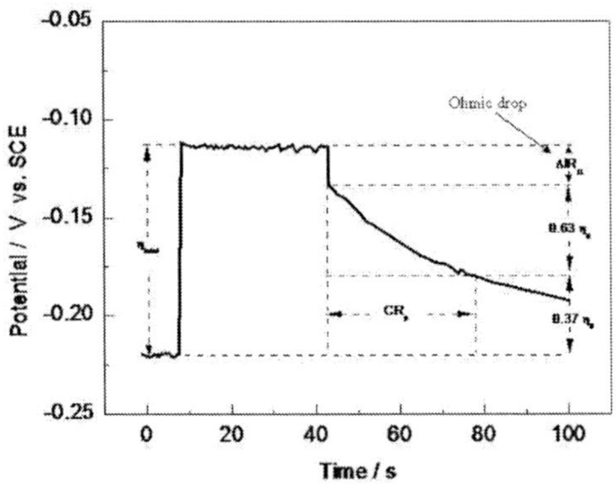

Figure 33. Scheme of direct calculation of the time constant, on the potential decay curve, from the moment of interruption of the potentiostatic pulse applied.

In order for things to happen this way in reality, the possible interference of other partial processes besides corrosion must be of little importance, or must occur in very different times to those used in the measurements, in order not to mask the potential decay due to the corrosion process.

The direct measurement of τ to determine R_p applying galvanostatic pulses was advocated in a highly interesting work by Glass [138]; though previously determining the interfacial capacitance from the charge injected, the potential shift achieved, and the surface area of the steel. The authors of the present communication have shown that it is possible to estimate, with galvanostatic pulses, reliable values of R_p in RCSs without the need to know the surface area of the rebars, assuming an approximate value of the capacitance (C) [139,140].

New Palliative Methods to Prevent Reinforced Concrete Corrosion

Use of new corrosion prevention strategies is reported as a palliative method to increase the reinforced concrete service life, thus contributing to diminish the environmental impact and economic cost.

Reduction of Permeability of the Concrete

The use of high-quality, impermeable concrete with a low water/cement (w/c) ratio and adequate mixing and pouring of concrete cover can do much to alleviate the corrosion problem. Although chloride ions will eventually fully penetrate concrete regularly exposed to de-icing salts or seawater, the rate of penetration can be considerably reduced from the minimized porosity. Concrete with a low w/c ratio has better resistance to cracking and spalling, so the use of partial or full polymer impregnation has been suggested [141].

Protective Coatings on the Concrete

Protective coatings are also commonly used on bridge-deck surfaces. Treatments with water-repellant materials, such as use of overlays of asphalt or polymer concrete, can be very effective. Overlays of this type offer low permeability, good crack resistance, and good bonding properties. A waterproof membrane between the normal concrete and the overlay is needed to serve as an additional barrier to penetration of chloride [142].

Protective Coating on the Steel

Protective coatings can be applied to the reinforcing steel. Metal coatings, such as zinc and cadmium, provide sacrificial anodic protection because they are more prone to corrosion and act as anodes, making the iron cathodic. Although such coating provides protection, they may be insufficient in the presence of chloride ions, which accelerate the corrosion of the coatings. The corrosion products may themselves cause distress in the concrete, and protection ceases when all the metal is used up. Coatings can provide good protection if they provide a continuous film without holes and weak spots; a break in a film may cause severe localized corrosion during fabrication (including bonding to the steel), transportation, handling erection of the structure, and under the service environment [143].

Use of Stainless Steel Reinforcements

Stainless steel (SS) reinforcements were first used many decades ago and have proved their ability to prevent corrosion for a very long time, even in very aggressive environments. Hence the use of SS reinforcements is one of the most reliable methods for ensuring RCS durability. However, their use has been limited due to the high cost of SS compared to carbon steel. For this reason, new SSs, in which the nickel content has been lowered by replacement with other elements, are being evaluated in the literature as possible alternatives to conventional carbon steel [144-146].

Suppression of the Electrochemical Process

Suppression of the electrochemical corrosion of iron is the basis of cathodic protection, which is used to protect steel structures and appears to be another promising method of protecting reinforced concrete structures. If the current supplied to the rebar is sufficiently large, the iron is made cathodic and corrosion is prevented [147]. A sacrificial anode, such as zinc or magnesium bars, can be buried close to the structure to reduce the corrosion of steel rebars.

In addition, various chemicals are known to inhibit corrosion of steel [148-150]. For example, chromates, nitrites, benzoates, phosphate, stannous salts, and ferrous salts may act either to suppress the anodic reaction by stabilizing the passive oxide film, to form a new insoluble coating (chromate, phosphates), or to suppress the cathodic reaction by scavenging oxygen (nitrites, benzoates, stannous and ferrous salts). Inhibitors when compared to the other corrosion protection methods have some advantages such as versatility and cost. Their use in concrete can help to delay the initiation of corrosion of the embedded steel exposed to chloride attack and carbonation.

Conclusion

According to the European Cement Association (CEMBUREAU), the world cement production was above 3 billion tones in 2010. This high value shows the great importance of this material not only for construction but for environmental uses such as stabilization/solidification of hazardous wastes. In this context, the use of reinforced concrete is envisaged for different applications, like low level radioactive waste containers, and more generally nuclear infrastructures containing iron/concrete interfaces. Therefore, is necessary to improve knowledge on the long-term corrosion behaviour of low carbon steels that could be used in concrete to build the substructure of nuclear wastes storage or reversible disposal facilities. This chapter has been focused on some of the most important challenges in cement research, concerning to the production and durability aspects.

Research in the cement production has a great relevance considering the high production of cement above the world and its detrimental environmental effects due to the CO_2 emissions. Thus, researching in this particular field is focused on the reduction of those emissions and the development of alkali-activated binders is likely to have enormous potential to become an alternative to Portland cement. In the other hand, the deterioration of cement-based materials may occur through a variety of chemical or physical processes, especially when it is exposed to aggressive environments. The durability of these materials has a significant influence on its lifetime in service, design life and safety. The design service life is assumed period for which a structure or part of it is to be used for its intended purpose with anticipated maintenance but without major repair being necessary. Thus technical, functional and economic issues will deal with the design and durability performance of reinforced concrete structures. Concrete made from Portland cement is also subject to certain durability problems such as sulfate attack, leaching, alkali silica reaction, frost attack and corrosion of steel reinforcements. Durability research in the field of leaching and corrosion has a great importance due to the environmental and economic consequences.

Acknowledgments

The authors express their gratitude to Project BIA2011-27182 from CICYT, Spain, for finantial support. D.M. Bastidas acknowledges the funding support under Ramón y Cajal Program and M. Criado thanks the Juan de la Cierva Program financed by the Spanish Ministry of Science and Innovation.

REFERENCES

[1] Kanie N, Nishimoto H, Hijioka Y, Kameyama Y, *Int. Environ. Agreements*, 10 (2010) 299–315.

[2] Malhotra V M, 'Making Concrete Greener with Fly Ash', *Concrete International,* 21 (1999) 61–66.

[3] Andac M, Glasser F P, 'Long-term leaching mechanism of Portland cement- stabilized municipal solid waste fly ash in carbonated water', *Cem. Concr. Res*., 129 (8) (1999) 179–186.

[4] Lovera P, Le Bescop P, Adenot F, Li G, Tanaka Y, Owaki E, 'Physicochemical transformations of sulphated compounds during the leaching of highly sulphated cement wastes', *Cem. Concr. Res*., 27 (10) (1997) 1523–1532.

[5] Faucon P, Adenot F, Jaquinot J F, Petit C, Cabrillac R, Jorda M, 'Long-term behaviour of cement pastes used for nuclear waste disposal: review of physico-chemical mechanisms of water degradation', *Cem. Concr. Res*., 28 (6) (1998) 847–857.

[6] Goñi S, Hernandez M S, Guerrero A, Lorenzo M P, 'Effect of temperature on the leaching performance of a simulated cement based immobilisation system. Calcium and hydroxil behaviour', *Constr. Build. Mater*., 10 (3) (1996) 171–177.

[7] Schiopu N, Tiruta-Barna L, Jayr E, Méhu J, Moszkowicz P, "Modeling and simulation of concrete leaching under outdoor exposure conditions", *Science of the Total Environment,* 407 (2009) 1613–1630.

[8] Gerard B, Pijaudier-Cabot G, Le Ballego C, 'Calcium leaching of cement based materials: a chemo-mechanics application. Construction materials theory and application'. Hans-Wolf Reinhart. Zum 60 Geburtstag. (Ed. H von R. Eligehausen, Ibidem-Verlag, Stuttgart, 1999) 313–329.

[9] Taylor H F W, 'Cement Chemistry'. Chap 12, Concr. Chem. 2nd Edn (Thomas Telford, London, 1997) 380–383.

[10] Hidalgo A, Andrade C, Alonso C, 'An accelerated leaching test to evaluate the long term behaviour of concrete in waste disposal', 5th CANMET/ACI Inter. Conf. on Durability of Concrete, Barcelona (June 2000) 373–381.

[11] Gérard B, 'Hydrolysis of cement-based materials: a review', 3th Int Bolomey Workshop. Pore solution in hardened cement paste, (Aedificatio Pub., 2000) 271–304.

[12] Faucon P, Adenot F, Jordá M, Cabrilac R, 'Behaviour of crystallised phases of Portland cement upon water attack', *Mat. Struct*., 30 (8) (1997) 480–485.

[13] Gérard B, Ph.D Thesis, Civil Engineering Department, Laval University, Quebec City, Canada/ENS Cachan, France, (1996) (in French).

[14] Torrenti J M, Didry O, Ollivier J P, Plas F, in: The degradation of the concrete: couplings between cracking and chemical degradation. Hermes Science Press, Paris, (1999).

[15] Heukamp F H, Ulm F J, Germaine J T, Mechanical properties of calcium-leached cement pastes. Triaxial stress states and the influence of the pore pressures. *Cement Concr. Res*., 31 (2001) 767–774.

[16] Faucon P, Ph.D Thesis, Génie Civil, L'Universite de Cergy Pontoise, France, (1997) (in French).

[17] Porteneuve Ch, Ph.D Thesis, Sciences aux Materiaux, L'Universite Paris VI, France, (2001) (in French).

[18] Llorente I, Castellote M, Gonzalez-Arrabal R, Ynsa M D, Muñoz-Martin A, de Viedma P G, Castillo A, Martínez I, Andrade C, Zuloaga P, Ordoñez M, "PIXE/RBS as a tool to study cementitious materials: application to the dynamic leaching of concrete". *Nuclear Instruments and Methods in Physics Research, B* 267 (2009) 3670–3674.

[19] Stone J A, "An overwiew of factors affecting the leachability of nuclear waste forms". *Nuclear and Chemical Waste Management,* 2 (1981) 113–118.

[20] Kylefors K, Andreas L, Lagerkvist A, "A comparison of small-scale, pilot-scale and large-scale tests for predicting leaching behaviour of landfilled wastes". *Waste Management,* 23 (2003) 45–59.

[21] Alonso C, Castellote M, Llorente I, Andrade C, "Ground water leaching resistance of high and ultra high performance concretes in relation to the testing convection regime". *Cement Concr. Res.*, 36 (2006) 1583–1594.

[22] Moszkowicz P, Sanchez F, Barna R, Méhu J, "Pollulants leaching behaviour from solidified wastes": A selection of adapted various models. *Talanta,* 46, (1998), 375–383.

[23] Tiruta-Barna L, Imyim A, Barna R, "Long-term prediction of the leaching behavior of pollutants from solidified wastes". *Advances in Environmental Research,* 8 (2004), 697–711.

[24] Van der Sloot H A, "Comparison of the characteristic leaching behavior of cements using standard (EN 196-1) cement mortar and an assessment of their long-term environmental behavior in construction products during service life and recycling". *Cement and Concrete Research,* 30 (2000), 1079–1096.

[25] NEN 7345, 'Leaching characteristics of soil and stony building and waste materials–leaching tests–Determination of the leaching of inorganic components from building and monolithic waste materials with diffusion tests'. (NNI, Delft, The Netherlands 1994).

[26] Llorente I, Castellote M, Andrade C, "Comparison between several methods for determining the resistance of concrete to leaching". Concrete in Aggressive aqueous environments; Performance, testing and modeling. RILEM Pulications, S.A.R.L., Bagneaux, France, (2009).

[27] ANSI/ANS-16, 'Measurement of the leachability of solidified low level radioactive wastes by short-term test procedure' (1986).

[28] Aïtcin P, in: High Performance Concrete. Modern Concrete Technology 5, EandFN Spon, London (1998).

[29] Llorente I, Ph.D Thesis, Facultad CC. Químicas, Universidad Complutense de Madrid, (2008) (in Spanish).

[30] Nielsen L F, "Strength development in hardened cement paste: Examination of some empirical equations". *Materials and Structures,* 26 (1993) 255–260.

[31] Gartner E M, Machphee D E,'A physico-chemical basis for novel cementitious binders', *Cement and Concrete Research,* 41 (2011) 736–749.

[32] Juenger M C G, Winnefeld F, Provis J L, Ideker J H, 'Advances in alternative cementitious binders', *Cement and Concrete Research,* 41 (2011) 1232–1243.

[33] Damtoft J S, Lukasik D, Herfort D, Sorrentino D, Gartner E M, 'Sustainable development and climate change initiatives', *Cement and Concrete Research,* 38 (2008) 115–127.

[34] http://www.oficemen.com.

[35] Pacheco-Torgal F, Castro-Gomes J, Jalali S, 'Alkali-activated binders: A review. Part 1. Historical background, terminology, reaction mechanisms and hydration products', *Construction and Building Materials,* 22 (2008) 1305–1314.

[36] Duxson P, Fernández-Jiménez A, Provis J L, Lukey G C, Palomo A, Van Deventer J S J, 'Geopolymer technology: the current state of the art", *Journal of Materials Science,* 42 (2007) 2917–2933.

[37] Purdon A O, 'The action of alkalis on blast furnace slag", *Journal of the Society of Chemical Industry,* 59 (1940) 191–202.

[38] Glukhovsky V D, 'Soil silicate articles and structures' Budivelnyk Publisher, Kiev, (1967) p. 156.

[39] Davidovits J, 'Synthesis of new high temperature geo-polymers for reinforced plastics/composites. SPE PACTEC 79 Society of Plastic Engineers, Brookfield Center, (1979) p. 151.

[40] Krivenko P V, 'Alkaline cements: terminology, classification, aspects of durability. Proceedings of the 10th International Congress on the Chemistry of Cement, Göteborg, Sweden, (1997) pp.4iv046–4iv050.

[41] Palomo A, Grutzeck M W, Blanco M T, 'Alkali-activated fly ashes. A cement for the future', *Cement and Concrete Research,* 29 (1999) 1323–1329.

[42] Fernández-Jiménez A, Palomo J G, Puertas F, 'Alkali-activated slag mortars: mechanical strength behaviour', *Cement and Concrete Research,* 29 (1999) 1313–1321.

[43] Fernández-Jiménez A, Palomo A, López-Hombrados C, 'Engineering properties of alkali activated fly ash concrete', *ACI Materials Journal,* 103 (2006) 106–112.

[44] Duxson P, Provis J L, Lukey G C, Mallicoat S W, Kriven W M, Van Deventer J S J, 'Understanding the relationship between geopolymer composition, microstructure and mechanical properties', *Colloids and Surfaces A: Physicochemical and Engineering Aspects,* 269 (2005) 47–58.

[45] Douglas E, Bilodeau A, Malhotra V M, 'Properties and durability of alkali-activated slag concrete', *ACI Materials Journal,* 89 (1992) 509–512.

[46] Puertas F, Amat T, Fernández-Jiménez A, Vazquez T, 'Mechanical and durable behaviour of alkaline cements mortars reinforced with polypropylene fibres', *Cement and Concrete Research,* 23 (2003) 2031–2036.

[47] Fernández-Jiménez A, García-Lodeiro I, Palomo A, 'Durability of alkali-activated fly ash cementitious materials', *Journal of Materials Science,* 42 (2007) 3055–3065.

[48] Allahverdi A, Skvara F, 'Sulfuric acid attack on hardened paste of geopolymer cements. Part 1. Mechanisms of corrosion at relatively high concentrations', *Ceramics-Silikaty,* 49 (2005) 225–229.

[49] Fernández-Jiménez A, Palomo A, Pastor J Y, Martín A, 'New cementitious materials based on alkali-activated fly ash: performance at high temperature', *Journal of the American Ceramic Society,* 91 (2008) 3308–3314.

[50] Shi C, Fernández-Jiménez A, Palomo A, 'New cements for the 21st Century: The pursuit of an alternative to Portland cement', *Cement and Concrete Research,* 41 (2011) 750–763.

[51] Shi C, Roy D M, Krivenko P V, 'Alkali-activated cements and concretes', Ed. Taylor and Francis, (2006) London, United Kingdom.

[52] Shi C, Fernández-Jiménez A, Krivenko P V, Palomo A, 'Classification and characteristics of alkali-activated cements'. First International Conference on Advances in Chemically-Activated Materials (CAM'2010-China), in conjunction with 7th International Symposium on Cement and Concrete (ISCC2010), (2010) Jinan, China.

[53] Fernández-Jiménez A, Palomo A, Criado M, 'Microstructure development of alkali-activated fly ash cement: a descriptive model', *Cement and Concrete Research,* 35 (2005) 1204–1209.

[54] Fernández-Jiménez A, Palomo A, Sobrados I, Sanz J, 'The role played by the reactive alumina content in the alkaline activation of fly ashes', *Microporous and Mesoporous Materials,* 91 (2006) 111–119.

[55] Provis J L, Lukey G C, Van Deventer J S J, 'A statistical thermodynamic model for Si/Al ordering in amorphous aluminosilicates', *Chemistry of Materials,* 17 (2005) 2976–2986.

[56] Provis J L, Duxson P, Van Deventer J S J, Lukey G C, 'The role of mathematical modelling and gel chemistry in advancing geopolymer technology', *Chemical Engineering Research and Design,* 83 (2005) 853–860.

[57] Van Deventer J S J, Provis J L, Duxson P, Lukey G C, 'Reaction mechanisms in the geopolymeric conversion of inorganic waste to useful products', *Journal of Hazardous Materials,* A139 (2007) 506–513.

[58] Fernández-Jiménez A, Palomo A, Alonso M M, 'Alkali Activation of fly ashes: Mechanisms of reaction', 2nd International Symposium on Non traditional cement and concrete, (2005) Brno, Czech Republic.

[59] Palomo A, Fernández-Jiménez A, Criado M, Alonso M M, 'The alkali activation of fly ashes: from macro to nanoscale', 2nd International Symposium on Nanotechnology in Construction; (2005) Bilbao, Spain.

[60] Fernández-Jiménez A, Palomo A, 'A mid-infrared spectroscopic studies of alkali-activated fly ash structure', *Microporous and Mesoporous Materials,* 86 (2005) 207–214.

[61] ASTM C 618-03 Standard, 'Standard specification for coal fly ash and raw or calcined natural pozzolan for use in concrete', Annual book for ASTM Standards, (2003) American for Testing and Materials.

[62] Fernández-Jiménez A, Palomo A,'Characterisation of fly ashes. Potential reactivity as alkaline cements', Fuel 82 (2003) 2259–2265.

[63] Van Jaarsveld J G S, Van Deventer J S J, Lukey G C, 'The characterization of source materials in fly ash-based geopolymers', *Materials Letters,* 57 (2003) 1272–1280.

[64] Duxson P, Lukey G C, Separovic F, Van Deventer J S J, 'Effect of alkali cations on aluminum incorporation in geopolymeric gels', *Industrial and Engineering Chemistry Research,* 44 (2005) 832–839.

[65] Van Jaarsveld J G S, Van Deventer J S J, 'Effect of alkali metal activator on the properties of fly ash-based geopolymeric', *Industrial and Engineering Chemistry Research,* 38 (1999) 3932–3941.

[66] Komnitsas K, Zaharaki D, 'Geopolymerization: A review and prospects for the minerals industry', *Minerals Engineering,* 20 (2007) 1261–1277.

[67] Fernández-Jiménez A, Palomo A, Criado M, 'Alkali activated fly ash binders. A comparative study between sodium and potassium activators', *Materiales de Construcción,* 56 (2006) 51–65.

[68] Fernández-Jiménez A, Palomo A, 'Composition and microstructure of alkali activated fly ash binder: Effect of activator', *Cement and Concrete Research,* 35 (2005) 1984–1992.

[69] Palomo A, Fernández-Jiménez A, Criado M, 'Geopolymers: same basic chemistry, different microstructures', *Materiales de Construcción,* 54 (2004) 77–91.

[70] Komljenovic M, Bascarevic Z, Bradic V, 'Mechanical and microstructural properties of alkali-activated fly ash geopolymers', *Journal of Hazardous Materials,* 181 (2010) 35–42.

[71] Hardjito D, Rangan B V, 'Development and properties of low-calcium fly ash-based geopolymer concrete', Research report GC, 1, (2005) Perth, Australia.

[72] Bakharev T, 'Geopolymeric materials prepared using class F fly ash elevated temperature curing', *Cement and Concrete Research,* 35 (2005) 1224–1232.

[73] Hu M, Zhu X, Long F, 'Alkali-activated fly ash-based geopolymers with zeolite or bentonite as additives', *Cement and Concrete Composites,* 31 (2009) 762–768.

[74] Criado M, Férnandez-Jiménez A, de la Torre A G, Aranda M A G, Palomo A, 'An XRD study of the effect of the SiO_2/Na_2O ratio on the alkali activation of fly ash', *Cement and Concrete Research,* 37 (2007) 671–679.

[75] Criado M, Férnandez-Jiménez A, Palomo A, 'Alkali activation of fly ash: Effect of the SiO_2/Na_2O ratio. Part I: FTIR study', *Microporous and Mesoporous Materials,* 106 (2007) 180–191.

[76] Criado M, Férnandez-Jiménez A, Palomo A, Sobrados I, Sanz J, 'Effect of the SiO_2/Na_2O ratio on the alkali activation of fly ash. Part II: ^{29}Si MAS-NMR survey', *Microporous and Mesoporous Materials,* 109 (2008) 525–534.

[77] Lee W K W, Van Deventer J S J, 'Structural reorganisation of class F fly ash in alkaline silicate solutions', *Colloids and Surfaces A: Physicochemical and Engineering Aspects,* 211 (2002) 49–66.

[78] Lee W K W, Van Deventer J S J, 'Use of Infrared Spectroscopy to Study Geopolymerization of Heterogeneous Amorphous Aluminosilicates', *Langmuir,* 19 (2003) 8726–8734.

[79] Silva de Vargas A, Dal Molin D C C, Vilela A C F, da Silva F J, Pavao B, Veit H, 'The effects of Na_2O/SiO_2 molar ratio, curing temperature and age on compressive strength, morphology and microstructure of alkali-activated fly ash-based geopolymers', *Cement and Concrete Composites,* 33 (2011) 653–660.

[80] Katz A, 'Microscopic study of alkali-activation fly ash', *Cement and Concrete Research,* 28 (1998) 197–208.

[81] Palomo A, Alonso S, Fernández-Jiménez A, 'Alkaline activation of fly ashes: NMR study of the reaction products', *Journal of the American Ceramic Society,* 87 (2004) 1141–1145.

[82] Criado M, Férnandez-Jiménez A, Palomo A, 'Alkali activation of fly ashes. Part I: effect of curing conditions on the carbonation of the reaction products', *Fuel,* 84 (2005) 2048–2054.

[83] Kovalchuk G, Férnandez-Jiménez A, Palomo A, 'Alkali-activated fly ashes: effect of thermal curing conditions on mechanical and microstructural development-part II', *Fuel,* 86 (2007) 315–322.

[84] Criado M, Férnandez-Jiménez A, Palomo A, 'Alkali activation of fly ashes. Part III: effect of curing conditions on the reaction and its graph description', *Fuel,* 89 (2010) 3185–3192.

[85] Palomo A, Banfill P F G, Fernández-Jiménez A, Swift D S, 'Properties of alkali-activated fly ashes determined from rheological measurements', *Advances in Cement Research,* 17 (2005) 143–151.

[86] Criado M, Palomo A, Fernández-Jiménez A, Banfill P F G, 'Alkali activated fly ash: effect of admixtures on paste rheology', *Rheological Acta,* 48 (2009) 447–455.

[87] Bakharev T, 'Durability of geopolymer materials in sodium and magnesium sulfate solutions', *Cement and Concrete Research,* 35 (2005) 1233–1246.

[88] Bakharev T, 'Resistance of geopolymer materials to acid attack', *Cement and Concrete Research,* 35 (2005) 658–670.

[89] García-Lodeiro I, Palomo A, Fernández-Jiménez A, 'Alkali-aggregate reaction in activated fly ash systems', *Cement and Concrete Research,* 37 (2007) 175–183.

[90] Miranda J M, Fernández-Jiménez A, Gónzalez J A, Palomo A, 'Corrosion resistance in activated fly-ash mortars', *Cement and Concrete Research,* 35 (2005) 1210–1217.

[91] Bastidas D M, Fernández-Jiménez A, Palomo A, Gónzalez J A, 'A study on the passive state stability of steel embedded in activated fly ash mortars', *Corrosion Science,* 50 (2008) 1058–1065.

[92] Fernández-Jiménez A, Miranda J M, Gónzalez J A, Palomo A, 'Steel passive state stability in activated fly ash mortars', *Materiales de Construcción,* 60 (2010) 51–65.

[93] Berrryman C, Zhu J, Jensen W, Tadros M, 'High-percentage replacement of cement with fly ash for reinforced concrete pipe', *Cement and Concrete Research,* 35 (2005) 1088–1091.

[94] Sumajouw D M J, Hardjito D, Wallah S E, Rangan B V, 'Fly ash-based geopolymer concrete: study of slender reinforced columns', *Journal of Materials Science,* 42 (2007) 3124–3130.

[95] Palomo A, Fernández-Jiménez A, López-Hombrados C, Lleyda J L,'Railway sleepers made of alkali activated fly ash concrete', *Revista Ingeniería de Construcción,* 22 (2007) 75–80.

[96] Palomo A, López de la Fuente J I, 'Alkali-activated cementitous materials: Alternative matrices for the immobilisation of hazardous wastes: Part I. Stabilisation of boron', *Cement and Concrete Research,* 33 (2003) 281–288.

[97] Palomo A, Palacios M, Alkali-activated cementitious materials: alternative matrices for the immobilisation of hazardous wastes – Part II. Stabilisation of chromium and lead', *Cement and Concrete Research,* 33 (2003) 289–295.

[98] Fernández-Jiménez A, Macphee D E, Lachowski E E, Palomo A, 'Immobilization of cesium in alkaline activated fly ash matrix', *Journal of Nuclear Materials,* 346 (2005) 185–193.

[99] Zhang J, Provis J L, Feng D, Van Deventer J S J, 'The role of sulfide in the immobilization of Cr(VI) in fly ash geopolymers', *Cement and Concrete Research,* 38 (2008) 681–688.

[100] Srivastava S, Chaudhary R, Khale D, 'Influence of pH, curing time and environmental stress on the immobilization of hazardous waste using activated fly ash', *Journal of Hazardous Materials,* 153 (2008) 1103–1109.

[101] Skvara F, Kopecky L, Smilauer V, Bittnar Z, 'Material and structural characterization of alkali activated low-calcium brown coal fly ash', *Journal of Hazardous Materials,* 168 (2009) 711–720.

[102] Grinda E G, in: El Hormigón Armado, Tectónica, ATC Ediciones, Madrid (1995), pp. 4–13.

[103] Jiménez-Montoya P, García-Meseguer A and Morán-Cabre F, *Hormigón Armado*, Ed. Gustavo Gili, Barcelona (1987) pp. 167–186.

[104] Bouzoubaa N, Zhang M H, Malhotra V M, Golden D M, Blended fly ash cement- A review, *ACI Mater. J.*, 96 (1999) 641–650.

[105] Martínez M, Inhibidores de corrosión para hormigón armado, *Hormigón,* 38 (1998) 48–50 (in Spanish).

[106] Flis J, Pickering H W, Osseo-Asare K, "Interpretation of impedance data for reinforcing steel in alkaline solution containing chlorides and acetates" *Electrochimica Acta,* 43 (1998) 1921-1929.

[107] Sehgal A, Kho Y T, Osseo-Asare K, Pickering H W, "Comparison of various corrosion rate measuring devices for determining the corrosion rate of steel-in-concrete systems," *Corrosion,* 48 (1992) 871–880.

[108] Vassie P R, Rubakantha R S, Page C L, Bamfort P B, Figg J W, Editors , *Corrosion of Reinforcement Concrete Constructions*, Springer, London (1996), pp. 156–165.

[109] Glass G K, Buenfeld N R, "Chloride-induced corrosion of steel in concrete", *Prog. Struct. Engng. Mater.*, 2 (2000) 448–458.

[110] Sidney M, Young J F, Concrete. Ed. Prentice-Hall: Englewood Cliffs, NJ, 1981

[111] Schiessl P, Corrosion of Steel in Concrete. Ed. Chapman and Hall: New York, NY, 1988.

[112] Saraswathy V, Muralidharan S, Thangavel K, Srinivasan S, "Influence of activated fly ash on corrosion-resistance and strength concrete", *Cement Concrete Comp.,* 25 (2003) 673–680.

[113] Van Jaarsveld J G S, Van Deventer J S J, Lukey G C, "The effect of composition and temperature on the properties of fly ash and kaolinite-based geopolymers", *Chem. Eng. J.,* 89 (2002) 63–73.

[114] Tommaselli M A G, Mariano N A, Kuri S E, "Effectiveness of corrosion inhibitors in saturated calcium hydroxide solutions acidified by acid rain components", *Constr. Build. Mater.,* 23 (2009) 328–333.

[115] Miletić S, Ilić M, Ranogajec J, Djurić M, "Fly ash-useful material for preventing concrete corrosion" *Studies in Environmental Science,* 71 (1997) 355-364.

[116] Castro-Borges P, de Rincón O T, Moreno E I, Torres-Acosta A A, Martínez-Madrid M, Knudsen A, "Performance of a 60-year-old concrete pier with stainless steel reinforcement", *Mater. Performance,* 41 (2002) 50–55.

[117] Sim J, Park C, "Compressive strength and resistance to chloride ion penetration and carbonation of recycled aggregate concrete with varying amount of fly ash and fine recycled aggregate", *Waste Management,* 31 (2011) 2352–2360.

[118] Choi Y S, Kim J G, Lee K M, "Corrosion behaviour of steel bar embedded in fly ash concrete", *Corros. Sci.,* 48 (2006) 1733–1745.

[119] Page C L, Treadaway K W J , Bamforth P B, *Society of chemical industry*, *Corrosion of reinforcement in concrete*, Elsevier Applied Science, London (1990).

[120] Slater J E, in: Corrosion of Metals in Association with Concrete, ASTM STP 818, Philadelphia, PA (1983).

[121] Millard S G, Gowers K R, Gill R P, "Practical field measurement of reinforcement corrosion in concrete using linear polarization methods", *Eur. J. NDT,* 2 (1992) 17–25.

[122] Dawson J L, John D G, Far M I, Hladky K, Sherwood L, "Electrochemical methods for the inspection and monitoring of corrosion of reinforcing steel in concrete", in *Corrosion of Reinforcement in Concrete*, C.L. Page, K.W.J. Treadaway, P.B. Bamforth, (Eds.), Elsevier Applied Science, (1990), pp. 358–371.

[123] Rodríguez P, Ramírez E, González J A, "Methods for studying corrosion in reinforced concrete", *Mag. Concrete Res.,* 46 (1994) 81–90.

[124] Elsener B, Müller S, Suter S, Böhni H, "Corrosion monitoring of steel in concrete. Theory and practice", in *Corrosion of Reinforcement in Concrete*, C.L. Page, K.W.J. Treadaway, P.B. Bamforth, (Eds.), Elsevier Applied Science, (1990), pp. 348–357.

[125] Escalante E, "Effectiveness of potential measurements for estimating corrosion of steel in concrete", in *Corrosion of Reinforcement in Concrete*, C.L. Page, K.W.J. Treadaway, P.B. Bamforth, (Eds.), Elsevier Applied Science, (1990), pp. 281–292.

[126] Feliu S, González J A, Andrade C, "Multiple-electrode method for estimating the polarization resistance in large structures", *J. Appl. Electrochem.* 26 (1996) 305–309.

[127] Feliu V, González J A, Andrade C, Feliu S, "Equivalent circuit for modelling the steel-concrete interface I. Experimental evidence and theoretical predictions", *Corros. Sci.,* 40 (1998) 975–993.

[128] Feliu V, González J A, Andrade C, Feliu S, "Equivalent circuit for modelling the steel-concrete interface II. Complications in applying the Stern-Geary equation to corrosion rate determinations", *Corros. Sci.,* 40 (1998) 995–1006.

[129] Newton C J, Sykes J M, "A galvanostatic pulse technique for investigation of steel corrosion in concrete", *Corros. Sci.,* 28 (1988) 1051–1073.

[130] Glass G K, "An assessment of the coulostatic method applied to the corrosion of steel in concrete", *Corros. Sci.,* 35 (1995) 597–605.

[131] Rodríguez P, González J A, "Estudio de la interfase acero-mortero por el método coulostático", *Rev. Metal. Madrid,* 29 (1993) 168–175.

[132] Elsener B, Wojtas H, Böhni H, in *Corrosion and Corrosion Protection of Steel in Concrete*, R. Swamy (Ed.), Sheffield Academic Press, Sheffield, (1994), Vol. 1, p. 236B.

[133] Suzuki M, Kano K, Sato Y, "An application of the coulostatic method to corrosion rate measurements", *Werkst. Korros.,* 31 (1980) 347–370.

[134] Sato Y, Kamo K, Suzuki M, "An application of the coulostatic method for rapid evaluation of metal corrosion rate in solution", in *Proc. 11th Int. Congress on Metal. Corros.* Rio de Janeiro, (1978), pp. 1945–1955.

[135] Stern M, Geary A L, "Electrochemical polarization I. A theoretical analysis of the shape of polarization curves", *J. Electrochem. Soc.,* 104 (1957) 56–63.

[136] Rodríguez P, González J A, "Use of the coulostatic method for measuring corrosion rates of embedded metal in concrete", *Mag. Concrete Res.,* 46 (1994) 91–97.

[137] Dawson J L, in *Proc. 1st Int. Symposium on Corros. of Reinforcement in Concrete Construction,* A.P. Crane (Ed.) Ellis Horwood, London, (1983), p. 175.

[138] Glass G K, Page C K, Short N R, Zhang J Z, "The analysis of potentiostatic transients applied to the corrosion of steel in concrete", *Corros. Sci.*, 39 (1997) 1657–1663.

[139] González J A, Cobo A, González M N, Feliu S, "On-site determination of corrosion rate in reinforced concrete structures by use of galvanostatic pulses", *Corros. Sci.*, 43 (2001) 611–625.

[140] Bastidas D M, González J A, Feliu S, Cobo A, Miranda J M. "A quantitative study of concrete-embedded steel corrosion using potentiostatic pulses", *Corrosion*, 63 (2007) 1094–1100.

[141] Johnston C D, "Deicer salt scaling resistance and chloride permeability", *Concrete International*, 16 (1994) 48–55.

[142] Zafeiropoulou T, Rakanta E, Batis G, "Performance evaluation of organic coatings against corrosion in reinforced cement mortars", *Progress in Organic Coatings*, 72 (2011) 175–180.

[143] Lau K, Sagüés A, "Impedance of reinforcing steel with disbonded dual polymer-zinc coating", *Electrochimica Acta*, 56 (2011) 7815–7824.

[144] Fajardo S, Bastidas D M, Criado M, Romero M, Bastidas J M, "Corrosion behaviour of a new low-nickel stainless steel in saturated calcium hydroxide solution", *Constr. Build. Mater.*, 25 (2011) 4190–4196.

[145] Criado M, Bastidas D M, Fajardo S, Fernández-Jiménez A, Bastidas J M, "Corrosion behaviour of a new low-nickel stainless steel embedded in activated fly ash mortars", *Cem. Concr. Compos.*, 33 (2011) 664–652.

[146] Fajardo S, Bastidas D M, Ryan M P, Criado M, McPhail D S, Bastidas J M, "Low-nickel stainless steel passive film in simulated concrete pore solution: A SIMS study", *Appl. Surf. Sci.*, 256 (2010) 6139–6143.

[147] Montoya R, Aperador W, Bastidas D M, "Influence of conductivity on cathodic protection of reinforced alkali-activated slag mortar using the finite element method", *Corros. Sci.*, 51 (2009) 2857–2862.

[148] Söylev T A, Richardson M G, "Corrosion inhibitors for steel in concrete: State-of-the-art report", *Constr. Build. Mater.*, 22 (2008) 609–622.

[149] Dong Z H, Shi W, Zhang G A, Guo X P, "The role of inhibitors on the repasssivation of pitting corrosion of carbon steel in synthetic carbonated concrete pore solution", *Electrochim. Acta*, 56 (2011) 5890–5897.

[150] Ormellese M, Lazzari L, Goidanich S, Fumagalli G, Brenna A, "A study of organic substances as inhibitors for chloride-induced corrosion in concrete", *Corros. Sci.*, 51 (2009) 2959–2968.

In: Brick and Mortar Research
Editors: S. Manuel Rivera and A. L. Pena Diaz

ISBN: 978-1-61942-927-7

Chapter 2

INNOVATIVE USES OF UNFIRED BRICKS AND CLAY PRODUCTS AS SUSTAINABLE BUILDING SOLUTIONS

***Carmen Galán-Marín*[1], *Carlos Rivera-Gómez*[1] *and Fiona Bradley*[2]**

[1] School of Architecture. Department of Building Construction, |University of Seville, Spain

[2] Department of Architecture,University of Strathclyde, UK

INTRODUCTION

Earth is one of the most commonly used building materials on the planet today. Ancient cultures used earth for building houses, fortresses, palaces and religious buildings and it is estimated that at the present time one third of the world's population lives in houses constructed from Earth. Although the majority of its utilisation occurs in developing countries, houses are still built from earthen materials in developed countries. Indeed, there are well known examples in countries such as the "The Tucson residence" [1] in the United States of America, for instance and "The Residence Korbeek" [2] in Belgium. Forms of Earth construction have also been used within religious buildings such as the Chapel of reconciliation in Berlin [3] and within industrial architecture, as illustrated in "The Bodegas La Raia" in Piemonte, Italy.

There is a wide geographical spread of this form of construction across the world. Architectural typologies range from Middle East mosques in Mali and Sedjan Bazar in Iran through to the Great Wall of China. The centre of the Pyramid of the Sun in Teotihuacan (Mexico) and the Alhambra of Granada (Spain) are two other stunning examples of its architectural capability. In each geographical area the indigenous population has, over time, developed different construction techniques and this has given rise to an interesting variations in methods for building with the material.

In ancient times, firstly houses and then cities were built with raw earth. Today, the construction of our homes tends to employ high embodied energy materials which are difficult to recycle and sometimes unfortunately contain toxic elements. Within this contemporary context, it is important that Architects and designers explore every opportunity to construct in a sustainable and ecologically sound manner and therefore there are many advantages in exploring the simple and natural properties of earth.

SUSTAINABLE DESIGN CRITERIA FOR BUILDING CONSTRUCTION MATERIALS

There are several pillars for sustainable architectural design, developed and detailed by multiple authors such as Garrido 2008 [4]. They describe the following main principles for sustainable design:

1) Optimise material resources.
2) Reduce energy consumption
3) Increase the use of renewable energy sources
4) Decrease waste production and carbon emissions.
5) Reduce the maintenance, operation and "in-use" costs of buildings.
6) Increase the quality of life of the occupants within buildings.

Different construction materials can of course be analysed from different perspectives such as strength, stiffness and durability. The key design parameters however when considering the sustainability credentials of a construction material are:

- The amount of energy used in the manufacturing and transportation process (embodied energy)
- The opportunities for recycling or reuse at the end of the process.
- The potential toxicity of any components (the presence of radioactive particles etc.).
- The appropriate use of the material according to geographic location and local skills and knowledge.

To achieve a sustainable form of construction, designers nowadays wish to specify "healthy" materials, without toxicity (declared or suspected) or radioactivity (known or latent) as well as assess whether the material manufacture involves ecological damage or indeed produces significant levels of environmental pollution.

Earth construction complies with all the above design criteria and is therefore an excellent choice for the environment and for more ecologically responsible, "green" construction. There are numerous advantages in utilising earth as a building material and they are listed below:

- Earth is a harmless material which does not contain any toxic substance, provided that it comes from land that has not suffered from pollution.
- It is completely recyclable, if is not mixed with any other manufactured product such as cement. If it is in its natural state, it is possible to fully integrate it back into nature once the building is demolished.

- It is readily available locally. Virtually any kind of soil is suitable for use within earth construction and designers and builders can choose different techniques depending on the soil properties of the available land. Mixtures with other local materials which improve the stability and properties of the composite material can also be carried out and typical added compounds include lime, plaster, and straw.
- Building with raw earth is simple and requires very little energy expenditure. It does not require a significant level of transportation or baking at high temperature and for this reason, it is considered a very low embodied energy material.
- Utilising local earth respects the environment and if the soil is extracted from the site itself, it causes an impact no greater than the one already caused by the construction itself. Furthermore there are no associated problems such as deforestation or extractive mining which can be associated with other building materials.
- Earth has excellent thermal properties and has a great capacity to store heat (known as thermal inertia quality). It mediates fluctuations in outside temperature changes through its inherent thermal mass and therefore creates a pleasant domestic atmosphere which is especially advantageous in arid climates where there are extreme fluctuations in temperature between day and night. If designers include appropriate insulation, it is also ideal in milder climates.
- Earth has good acoustic insulation properties against unwanted noise.
- Earth is an inert material that does not catch fire or rot. It is also not susceptible to insect attack, as long as organic material is avoided in the upper layers of soil.
- Earth is a naturally breathable material which allows a natural regulation of moisture content from the interior of a building thus helping to avoid condensation.
- Finally, it is an affordable (or virtually free) resource that is often in-situ where the house will be erected.

There are many different construction techniques throughout the world that have been adopted for working with raw earth. Most of them are ancient techniques that have existed, with minor changes, through many centuries to the present day, whilst others are modern inputs. Construction methods are often strongly related to local customs, the local climate and the characteristics of the available soil.

The main earthen construction techniques are Rammed Earth or Pisé (de terre), Adobe, Cob, Compacted Earth Blocks (CEB) or Earth Bags. There are also some lesser known techniques which are essentially temporary in nature and of lower quality and these are known as Soil Balls or Cod. Short definitions of the various types of earthen construction are described below:

- Rammed earth is the construction of monolithic walls by compacting earth between a few planks of wood. It has been used in a variety of different building typologies throughout the world.
- Adobe houses are made from unfired bricks of raw soil dried in the heat of the sun, called adobes. Adobe is a technique which originated in the Iberian Peninsula and after the colonization of the Americas was successfully applied into the arid areas of Central America.

- Cob is the construction of earth houses utilising a mixture of soil, water and straw without a specific form. It is common in Britain, although examples exist all over the world. It is especially suitable for rainy areas.
- Compacted earth blocks (CEBs) are bricks of raw soil with a low water content which are manufactured by a simple pressing device which uses mechanical pressure to obtain regular forms and improved loadbearing ability.
- A further form of earth construction, currently under development, is "Superadobe" and this system comprises of earth bags. It was developed by the Iranian architect Nader Khalili [5] and uses polypropylene bags or textiles stuffed with soil to permit solid constructions.

The last few construction systems described above are considered viable options for improving housing problems in more economically disadvantaged countries.

Earth Construction: A Historical Background

Earth as a building material is available in abundance throughout the world and its advantages, which have been previously listed, are numerous and comprehensive. Although primitive forms of houses today are built with raw earth, there are opportunities to construct a wide variety of building typologies. Earthen building techniques are not a thing of the past and interestingly one third to one half of the world's current population lives in houses made from earth.

In places where Earth construction is common, traditional techniques are still prevalent. Traditional earth construction techniques, including Rammed Earth, Mud-block, Wattle and Daub and Cob have had a long and largely successful history butin some developed countries, researchers are now looking at new innovative building systems and investigating a wide range of applications including pre-fabricated forms of construction. The use of traditional vernacular techniques, such as cob, has recently raised the profile of earthen architecture, but a wider impact on modern construction has arisen from compacted earth blocks (CEB) which are used in a variety of countries including Latin America and Africa.

There are many archaeological and historical testimonies for earth construction and "mud" currently abounds in simple construction forms throughout the world. The origins of the use of soil to build shelters dates back to the earliest human settlements. Indeed in Spain, evidence has been found of settlements from the bronze age and, subsequently, of Roman and Iberian structures. However, interestingly, it was Arabian societies who developed and perfected earth building techniques in most of the country.Spain.

Some of the great civilizations including the Persians and Egyptians built entire cities with raw Earth. Some examples include Tobouctou in Mali, Marrakech in Morocco and Shibam in Yemen. In fact, some of these buildings are almost 30 m high and a testimony to the durability and longevity of earth as a building material.

The fact that it is still possible to find many earthen heritage buildings, reflects how durable, in good conditions, these structural forms can be. Soil has been used throughout history to build fortifications, castles, walls, chapels, mosques, barns, and windmills and is popular in places like sub-Saharan Africa, the Maghreb, Central Africa and the East, as well as Latin America and Europe. But it is also found in a number of rainy climates such as Sweden, Norway and Denmark.

It is interesting to note that the countries with the greatest housing need and fewest resources are where earthen structures are most prevalent ie throughoutAfrica, the Middle East and Latin America. In China and India there are more than 50 million homes made from Earth but in Europe, soil construction is virtually ignored in new builds despite being part of the landscape in many rural regions where historical housing conserved. Maybe the 21st Century will be the century where earthen structures start to gain more prominence again due to their many advantageous characteristics – only time will tell.

SOIL STABILIZATION TECHNOLOGIES FOR BUILDING CONSTRUCTION APPLICATIONS

The stabilisation of naturally occurring soils has been practiced for thousands of years. The Mesopotamians and Romans separately discovered that it was possible to improve the load-bearing capacity of roads by the addition of materials such as pulverized limestone and/or calcium thereby improving underlying soil performance(28). Today soil stabilization can also be utilised with regards to earthen forms of building construction such as Cob, Rammed Earth and Adobe to improve strength and durability characteristics by providing a particular soil with mechanical properties which are irreversible in the face of physical constraints.

When architects and engineers specify building materials, they take into account a great many parameters which influence and determine design quality. These include construction techniques, architectural detailing, material quality, the economic aspects of the project and durability issues.

For soil stabilisation to be successful, the chosen process must be compatible with all these design criteria and so when designers are confronted with two basic alternatives they can either utilize a construction system which is dictated by the type of soil available on site or they can specify a pre-determined building system, and this selection dictates the use of a particular type of soil.

In the first instance, the design team evaluates the site context and selects the most appropriate building system/s which will ensure the durability of the building ie the architectural design specifications dictate the most appropriate "stabilizer". This is the preferred approach.

In the second instance, the manufacturing technique, which can often be alien to the site, ensures the durability of the materials, more or less independently of the building system. In this process, additives in the mix act as the "stabilizer".

In this chapter, we deal with the second instance, i.e. the improvement of the soil by the addition of stabilizing materials. Every kind of soil has a corresponding suitable stabilizer. There are more than a hundred products in use today which assist in the stabilization of soils and these admixtures can be used both within the body of the walls and their outer "skins"; for example renders.

Soil stabilization has been practiced for a very long time, but despite its longevity, it is still not an exact science and to date no "miracle" stabilizer has been found amongst the multitude of products available. Indeed some products cannot be considered, either because of their inefficiency or because they are much more expensive than more common alternatives.

Only two characteristics of the soil itself can be treated: either its structure and/or its texture. Stabilization is critical where there is a requirement to achieve a lasting structure from local soil and material properties play a key role in determining the appropriate stabilization method (Montgomery et al., 1991) [6].

There are essentially three ways of treating the structure and the texture of a soil:

- reducing the volume of voids between the particles, i.e. affecting its porosity;
- blocking up the voids which can't be eliminated, i.e. affecting its permeability;
- improving the links binding the particles together, i.e. affecting its mechanical strength.

There are a number of ongoing research initiatives and these are exploring a variety of geotechnical issues such as:

- obtaining better mechanical performances such as increasing dry and wet compressive strength;
- reducing porosity and variations in volume ie trying to improve control of swelling and shrinking with moisture content variations;
- improving the ability to withstand weathering by wind and rain by reducing surface abrasion and increasing waterproofing.

The above geotechnical performance enhancing measures can e essentially be divided into three main categories of Stabilization techniques. These are listed below (see table 1) (Houben and Guillaud, 1995)[7]:

*Mechanical stabilization involves*compacting the soil and changing its density, compressibility, permeability and porosity.

*Physical stabilization essentially*changes the textural properties of the soil. This can be done by controlling the mixture of different grain fractions, drying or freezing, heat treatment and electrical treatment.

Chemical stabilization involves changing the properties of the soil by adding other chemicals or additives. This happens either by creating a matrix, which binds or coats the grains or by a *physio-chemical*reaction between the grains and the additive materials (Gooding and Thomas, 1995)[8].

Many additive materials can be used to stabilize the soil (Hoben and Boubekeur, 1998; Kerali, 2005)[9, 10] and the compressive strength of the soil can be improved multifold by using the right stabilization method. Other benefits will be the improvement of durability by increasing resistance to erosion and water damage.

The common categories of binders used for earth construction are Portland cement, lime, bitumen, natural fibers and chemical solutions such as silicates (Houben and Guillaud 1994)[11], as outlined in the Australian Standard [12] and SAZS 724:2001 Zimbabwe Standard [13].

Table 1. Stabilisation methods

Stabilisation System	Types
Mechanical stabilization	
Physical stabilization	
Chemical stabilization	Soil-cement Soil-lime Soil-bitumen Soil-powerplant wastes (PFA, FGD, GGBFS ?) Soil-polymer Others
Other methods of soil stabilization	Granular stabilization Thermal stabilization Electrokinetic stabilization

CHEMICAL STABILIZATION TYPES

There are a wide variety of additives that can be used to chemically stabilize a soil and the most common materials are listed below:

Soil-cement. Cement is probably one of the best stabilizers for Compressed Earth Blocks (CEBs.) Adding cement before compaction improves the characteristics of the material, particularly its resistance to water, thanks to the irreversible nature of the links it creates between the largest particles. Cement mainly affects sands and gravels, as in concrete or in a sand-cement mortar. This means that it is not necessary, and indeed may be harmful, to use in soils which have too high a clay content (> 20%).

Its use does not require too much water, just an amount necessary to keep the moist compression state of the CEBs. In general, at least 5 to 6% cement will be needed to obtain satisfactory results and compressive strength will be highly dependent on the amount used. With low proportions (2-3%), certain soils will perform less well than when left unstabilized and given similar local conditions, there may be no guarantee that a CEB will use less cement than a cement block (Bahar et al., 2004; Miller and Azad, 2000; Sreekrishnavilasam et al., 2005; Al-Rawas et al., 2005)[14-17].

Soil-lime. Stabilizing soil using non-hydraulic lime (quicklime or slaked lime) is a technique commonly used for roadworks, although mainly for temporary roads. Lime stabilization has the advantage of reacting in a very positive way with clayey soils with a relatively high moisture content, which is often the case for site access roads, for example. Lime forms links with the clays present, but hardly at all with the sands. The use of this stabilizer is therefore on the whole not recommended for the manufacture of CEBs, which require fairly low moisture contents and soils with a relatively high sand content. It should be considered only if cement stabilization is impossible. Results with lime are better than with bitumen or resins etc.

Hydraulic limes, which more closely resemble cement are not considered here. Adding 2 to 3% lime immediately provokes a lowering of the plasticity of the soil. For ordinary stabilization purposes, the amounts generally used range from 6 to 12%, i.e. equivalent to the amounts of cement used. It should be noted that in the case of lime, there is an optimum quantity to be used for each type of soil (Dallas, 2008 and 1996; Jianming et al., 2008; Prusinski and Bhattacharja, 1999)[18-21].

Soil-bitumen. Soil-Bitumen (modified granular bearing skeleton); sand-bitumen (added cohesion); waterproofed granular stabilization and oiled earth surface are all systems in which two or more soil materials are blended to produce a good gradation of particles from coarse to fine. Comparatively small amounts of bitumen are needed and the soil is compacted. (Jenkins, 2008; Browne, 2008; Finberg et al., 2008)[22-24].

Soil-powerplant wastes. Stabilization of cohesive soils can be achieved by adding puzzolanic mineral coal ash, which can be in the form of fly-ash, bottom ash, Ground Granulated Blast Furnace Slag (GGBFS) or FGD (flue gas desulfurization) sludge.

Properties of these siliceous residues are dependent on the type of coal used as a fuel (anthracite, bituminous, subbituminous or lignite) and may or may not require the addition of lime for puzzolanic reaction to take place (Thenoux et al., 2007; Havanagi et al., 2007; Barstis and Metcalf, 2005)[25-27].

Soil-polymer. Waterproofing cohesive soils is achieved by adding to the mixture small amounts of resins or polymers (less than 2% by dry weight of soil) which cement cohesive or non cohesive soils together by means of utilization of artificial or natural resins and polymers. There are a variety of different types of substances including cement-resin mixes such as polymer cements and organic resins which include epoxy, acrylic, polyacrylate, polyurethane and solvinated resins, as well as tomato pulp and alginate.(Galán et al., 2010; Galán, Rivera and Petric, 2010; Daniels, 2006)[28-30].

Others. Other additives includes: salts (chlorides); acids (phosphoric acid); lignin thermoplastic materials for encapsulation (tar, asphalt, polyethylene); sodium and calcium silicates; calcium aluminates; sulfur; sulfates; potassium; iron oxide; hydroxy-aluminum; cement basillius (lime-gypsum-alumina) puzzolanic ash of peat and agricultural waste (rice, peanut, castor bean husk and sugar-cane bagasse). In addition, hydromulching materials such aswood fibre, waste paper, cellulose pulp, ceramic waste and mining wastes can also be utilized (Ahlrich, 2008; Borderick and Daniel 1990)[31, 32]. All these stabilizers can be used to varying degrees and with varying proportions.

OTHER METHODS OF SOIL STABILIZATION

Granular stabilization is a combination of physical and chemical stabilization methods in which granular bearing skeletons are modified by pore-filling and/or cementing natural and extraneous materials such as clay and other concretes and mortars.

Thermal stabilization is a physical method of stabilization that utilizes heating or freezing for long- term or short-term improvement of properties within challenging soils.

Heating soils to high temperatures (typically above 300ºC) by methods that may involve combustion, electricity, microwave or laser beam application causes permanent changes to their physical properties. Furthermore, freezing (refrigeration) is used to achieve temporary ground stability or control of groundwater in soft grounds or excavations below the groundwater table.

Electrokinetic stabilization. Stabilization of soils by the utilisation of an electric field isused in certain circumstances to facilitate the consolidation of fine-grained soils and encourage the movement of stabilizing agents through dense soils and to prevent natural seepage by providing flow barriers.

FIBRES AND SOIL REINFORCEMENT

A standard fibre-reinforced soil is defined as a soil mass that contains randomly distributed, discrete elements (fibers) that provide an improvement in the mechanical behavior of the soil composite [33]. Fibre reinforced soil behaves as a composite material in which fibers of relatively high tensile strength are embedded within a soil matrix. Shear stresses within the soil mobilize tensile resistance in the fibres, which in turn impart increased strength to the soil [34-36].

Different literature reviews show that short fibre soil composites can be divided into two distinct categories. One group comprises a soil with a random inclusion of fibres into the soil mass. The other group consists of oriented fibrous materials, e.g. the Geo-Synthetics family [37, 38]. The former category is not as well-known as the second, not only in terms of optimizing fiber properties, fiber diameter, length, surface texture etc., but also in the reinforcing mechanism [37].

As stated in the introduction early civilizations discovered that it was possible to improve the load bearing capacity of soils through the utilization of a stabilizing agent like pulverized limestone or calcium [39]. An alternative natural strengthening system utilizes the presence of plant roots as a natural means of incorporating randomly oriented fiber inclusions within the soil. Plant fibers serve to improve the soil strength and therefore consequently the stability of natural slopes [38-44].

The concept of fiber reinforcement was discovered more than 5000 years ago. Ancient civilizations used straw and hay to reinforce mud blocks in order to create reinforced building blocks such as those utilized within the Great Wall of China (which demonstrates the earliest example of reinforced earth using branches of trees as a tensile material) and the Ziggurats of Babylon which contain mats of reed.[45].

Modern concepts and principles of soil reinforcement were first developed by Vidal (1969)who demonstrated that the introduction of reinforcement elements within a soil mass increased the shear resistance of the medium [46, 47]. Research activity utilizing various fibrous materials, which had been incorporated in the past, re-started and since this modern development of soil reinforcement, initiated by Vidal in 1969, nearly 4000 structures have been built in more than 37 countries. [48, 49].

Interestingly, randomly distributed fibre-reinforced soils, known as short fibre soil composites, have recently attracted increasing attention in many geotechnical engineering applications, not only in scientific research environment, but also at field applications [50].

Synthetic staple fibres have been used in soil since the late 1980s, when the initial studies using polymeric fibres were conducted [33].

In summary, the concept of reinforcing soil with natural fibres originated in ancient times. However, short natural and synthetic fibre soil composites have recently attracted increasing attention in the geotechnical engineering fraternity for the second time providing a relatively new construction technique within geotechnical projects.

FIBRE PROPERTIES AND CLASSIFICATIONS

Natural Fibres

At the present time, there is great awareness that landfill sites are filling up, non-renewable resources are being depleted and the planet is being polluted.. There is therefore a pressing need to discover environmentally friendly materials and consequently a great deal of recent interest has developed worldwide into the potential applications of natural fibers within soil reinforcement. There have been a number of recent experimental research projects and the term "eco-composite" has emerged within technical literature demonstrating the important role of natural fibers in the modern construction industry [51]. Of course, natural fibers have been used for a long time in many developing countries in cement composites and earth blocks because of their availability and low cost [53-55] so this is not an entirely new construction technology.

There are many factors which affect the performance of natural fibers in a composite natural fiber reinforced soil including the particular part of the plant that the fiber originates from, the age of the plant and the method by which the fiber was isolated. [52]

1. Vegetal Fibres

Coconut fibre (coir). The outer covering of fibrous material of a matured coconut, termed the coconut husk, is the reject of the coconut fruit. Coir fibers are normally 50 to 350 mm long and consist mainly of lignin, tannin, cellulose, pectin and other water soluble substances. However, due to their high lignin content, coir degradation takes place much more slowly than in other natural fibers and this gives it good durability characteristics and an infield life service of 4 to 10 years. The water absorption of coir is about 130 to 180 percent and the diameter is about 0.1 to 0.6 mm.

Sisal. Sisal is a lingo-cellulosed fiber [56] whose traditional use is as a reinforcement for gypsum plaster sheets in the building industry due to its 60 to 70 percent of water absorption and diameter of about 0.06 to 0.4mm. Sisal fibers are extracted from the leaves of the plants, which vary in size between 6-10 cm in width and 50-250 cm in length. In general, Brazil, Indonesia and East African countries are the world´s main producers of sisal fibers [57].

Palm fiber. The palm fibers in date production have filament textures with interesting properties such as low cost, plentitude in the region, durability, light weight, tension capacity and relative strength against deterioration [58]. Fibers extracted from decomposed palm trees are found to be brittle, and exhibit low tensile strength and modulus of elasticity as well as a very high water absorption [59]very advisable for certain types of soil.

Jute. Jute is abundantly grown in Bangladesh, China, India and Thailand. It is extracted from the fibrous bark of jute plants which grow as tall as 2.5 m and contain base stem diameters of approximately 25 mm. There are several different varieties of jute fibers with varying properties [59].

Flax. Flax is probably the oldest textile fiber known to mankind. It has been used since ancient times for the production of linen cloth [60]. Flax is a slender, blue flowered plant grown for its fibers and seeds in many parts of the world [59].

Barley Straw. Barley straw is widely cultivated and harvested once or twice annually in almost all rural areas all over the world and can be used in producing composite soil blocks with better structural characteristics. Unfortunately, relatively little published data is available on its performance as a reinforcement element in either soil or earth blocks. It is however important to note that during Egyptian times, straw or horsehair was added to mud bricks and similarly straw mats were used as a form of reinforcement in early Chinese and Japanese housing construction [61, 62, 63]. From the late 1800s, straw was also used in the United States as a wall bearing element or infill [64]. Interestingly, barley straw is claimed to be the most cost-effective method to retain soil in artificial rainfall tests [65].

Bamboo. Bamboo fiber is a regenerated cellulose fiber. Bamboo can thrive naturally without using any pesticide and is seldom eaten by pests or infected by pathogens. Scientists have found that bamboo contains a unique anti-bacteria and bacteriostatic bio-agent named "Bamboo Kun" [79]. Furthermore, it is important to note that the bamboo roots are excellent soil binders preventing erosion [80, 81].

Cane. Cane or sugarcane belongs to the grass family and grows up to 6 m high with a diameter up to 6 cm. Bagasse is the fibrous residue which is obtained in sugarcane production after extraction of the juice from the cane stalk containing a diameter of up to 0.2 to 0.4 mm. However, waste cane fiber has limited use because of the residual sugars and limited structural properties within the fiber itself. The residual sugars can result in a detrimentally affected finished product due to the fact that a stiffer bonding phase generates in the composite structure.

Therefore, “Cement Board” produced from sugar cane waste has been recently introduced to the market [85]. The authors recommend the application of these fibers in soil reinforcement as a potential area of research.

2. Animal Fibres

α- *keratin fibers.* A review of the existing literature shows that most studies of natural fibers are focused on cellulose-based/vegetal fibers from renewable plant resources. This is due to the fact that natural protein fibers have poor resistance to alkalis and cement (an alkaline product) is present nowadays in many building construction materials. There are therefore very few studies describing composites from protein fibers such as animal hairs. Barone and Schmidt [70] reported on the use of keratin feather fiber as a short-fiber reinforcement in LDPE (Low Density Polyethylene) composites and showed that protein fibers have good resiliency and elastic recovery. The keratin feather fiber for these tests was obtained from chicken feather waste generated by the US poultry industry.

Synthetic Fibres

1. Plastic Fibers

Polypropylene (PP) Fibers. Polypropylene fiber is the most widely used material utilized in laboratory tests of soil reinforcement [86-91]. Currently, PP fibers are used to enhance soil strength properties, reduce shrinkage properties and to overcome chemical and biological degradation [92-94]. Puppala and Musenda (2000) indicated that PP fiber reinforcement enhanced the unconfined compressive strength (UCS) of the soil and reduced both volumetric shrinkage strains and swell pressures of the expansive clays [94].

Polyester (PET) Fibers. Consoli et al. (2002) indicated that due to the inclusion of PET fibers in sand, both peak and ultimate strength characterisitics were improved and this was dependent on fiber content [108]. Kumar et al. (2006) tested highly compressible clay in unconfined compression (UC) test with 0%, 0.5%, 1.0%, 1.5% and 2.0% flat and crimped polyester fibers. Three lengths of 3 mm, 6 mm and 12 mm were chosen for flat fibers, whilst crimped fibers were cut to 3 mm long. The results indicated that as the fiber length and/or fiber content increased, the UC value improved. Crimping of fibers leads to a slight increase in UC [109] and these results are comparable to those found by Tang et al. (2006)[102].

Polyethylene (PE) Fibers. The feasibility of reinforcing soil with polyethylene (PE) strips and/or fibers has also been investigated to a limited extent [91, 115-118] and it has been reported that the presence of a small fraction of high density PE fibers can increase the fracture energy of the soil [119]. Consequently, fibers of this type such as GEOFIBERS® are typically 25-50mm long, discrete fibrillated or taped polypropylene strands that are mixed or blended into sand or clay soils [120,121]. It is important to note that some researchers have applied the term "geofiber" for PP fibers used in soil reinforcement [e.g. 89, 99, 121].

Nylon Fiber. Kumar and Tabor (2003) studied the strength behaviour of silty clay with nylon fibers with varying degrees of compaction. This study indicated that the peak and residual strength of the samples for 93 percent compaction was significantly more than the samples compacted at higher densities [110 and 129]. Gosavi et al. (2004) reported that by mixing nylon fibers and jute fibers, the California Bearing Ratio (CBR) value of the soil was enhanced by about 50% of an unreinforced soil, whereas coconut fiber used in this research increased the value by as much as 96%. The optimum quantity of fiber to be mixed with soil is therefore found to be 0.75%, and any addition of fiber beyond this quantity does not seem to have any significant increase in the CBR value [48 and130].

Polyvinyl Alcohol (PVA) Fibers. Polyvinyl alcohol (PVA) fiber is a synthetic fiber that has recently been used in fiber-reinforced concrete due to its weather resistance, chemical resistance (especially alkaline resistance) and tensile strength being superior to that of PP fiber. PVA fiber has a significantly lower shrinkage from heat than either nylon or polyester. It has a specific gravity of 1.3, a good adhesion property which assists bonding with cement and a high anti -alkali characteristic. For this reason, PVA fibre is suitable for utilization as a soil reinforcing material [138]. The inclusion of PVA fiber therefore seems to produce more effective reinforcement in terms of strength and ductility compared with other fibres under the same cementation. Park et al. (2008) found that the addition of 1% PVA fibre to 4% cemented sand resulted in a two times increase in both the UCS and the axial strain at peak strength when compared with the non-fiber-reinforced specimen [138 and 139]. In addition, Park (2011) reported that a 1% fibre dosage, the values of ductility were much greater, regardless of the cement ratios used.[190].

2. Metal and Glass Fibres

Steel Fibers. Steel fibre reinforcement is found in concrete structures and is also used as reinforcement within soil–cement composites [69,135, 136]. Steel fibres improve soil strength, but the improvement is not as significant as in the other case above mentioned [106 and 137]. Ghazavi and Roustaie have recommended that in cold climates, where soil is affected by freeze–thaw cycles, polypropylene fibers are preferable to steel fibers as polypropylene fibers possess a smaller unit weight than steel fibers. In other words, the PP fibers decrease the level of sample volume increase more than equivalent steel fibers [106].

Glass Fibers. Consoli et al. (1998) indicated that the inclusion of glass fibres in silty sand effectively improved peak strength [23, 110]. In other work, Consoli et al. (2004) examined the effect of PP, PET and glass fibers on the mechanical behavior of fiber-reinforced cemented soils. Their results showed that the inclusion of PP fibres increased the brittle behavior of cemented soils, whereas stresses at failure slightly decreased. Unlike the case of PP fiber, the inclusion of PE and glass fibers slightly increased the stresses at failure and slightly reduced the brittleness [124]. Maher and Ho (1994) studied the behavior of Kaolinite-fiber (PP and glass fibers) composites, and found that the increase in the unconfined compressive strength (UCS) was more pronounced in the glass fiber-reinforced specimens [125].

Table 2. Fibres Classification

NATURAL FIBRES	DESCRIPTION
VEGETAL FIBRES	
	Coconut fibre(coir)
	Sisal
	Palm
	Jute
	Flax
	Barely Straw
	Bamboo
	Cane
ANIMAL FIBRES	
	α-keratin fibres
SYNTHETIC FIBRES	DESCRIPTION
PLASTIC FIBRES	
	Polypropylene (PP)
	Polyester (PET)
	Polyethylene (PE)
	Nylon Fiber
	Polyvinyl Alcohol (PVA)
METAL and GLASS	
	Steel Fibres
	Glass Fibres

UNFIRED BRICKS RESEARCH: GEOGRAPHIC DISTRIBUTION

For more than two thousand years man has been building with unbaked earth. There are examples of its durability from all over the world, but in recent times earth has been abandoned as a building system in developed countries such as France, Germany and Italy. Despite this decline in its use, today more than one third of the world's population still lives in buildings constructed from Earth [104].

Countries such as Canada, the U.S.A and Latin America are now starting to study innovative ways of improving the characteristics of earthen structures due to their excellent sustainability credentials and low embodied energy.

These research projects should help the development of national building regulations which will allow architects and engineers to advance the development of a form of bioclimatic architecture that should never have been lost (Delgado et al., 2005)[105].

The following image 1 and table 3 show graphically the general geographical distribution, throughout the world, of buildings constructed using earthen materials with techniques such as adobe, cob, and rammed earth.

The graph clearly demonstrates a cultural propensity for earth structures in the equatorial and pacific regions of the planet, but there is also an additional cluster of earthen building typologies in the Middle East.

Table 3. Main International associations related to construction materials based on unfired earth

COUNTRIES	Association/Research Group
Germany	
	www.moderner-lehmbau.com
	www.earthbuilding.info
	www.uni-terra.org
	www.gernotminke.de
Spain	
	www.cannabric.com
	www.adoberadelnorte.com
	www.habitat-tierra.org
	www.bioterre.es
	www.ctv.es/USERS/interacc/
	www.ciat.es
France	
	www.lavoutenubienne.org
	www.terre.grenoble.archi.fr
	www.asterre.org
Italy	
	www.emiliocaravatti.it
	www.casediterra.it
	www.terrarossaonline.it
	www.mediterrae.org
	www.mattonesumattone.org
Holland	
	www.artchitecture.nl
	www.tierrafino.nl
Portugal	
	www.centrodaterra.org
	www.construdobe.com
United Kingdom	
	www.rammed-earth.info
	www.dur.ac.uk/p.a.jaquin
	www.eartha.org.uk
	www.historicrammedearth.co.uk

COUNTRIES	Association/Research Group
Switzerland	
	www.iglehm.ch
	www.eccoterra.ch
Turkey	
	www.kerpic.org
EE.UU.	
	www.greenhomebuilding.com
	www.adobealliance.org
	www.jkhomes.com
	www.theearthbuildersguild.org
	www.taoshomesolutions.com
	www.adobeasw.com
	www.wallsofearth.com
	www.calearth.org
	www.adobe-block.com
Brazil	
	www.fatoarquitetura.com.br
	www.abcterra.com.br
Canada	
	www.sirewall.com
	www.terrafirmabuilders.ca
Chile	
	www.marcelocortes.cl
Colombia	
	www.fundaciontierraviva.com
Ecuador	
	www.ecosur.org
COUNTRIES	Association/Research Group
Uruguay	
	www.arquitierra.bitacoras.com
	www.tierraalsur.com
	www.fronterra.org
Australia	
	www.aseg.net/home.htm
	www.mudbricks.com.au
	www.dab.uts.edu.au
	www.ebaa.asn.au
	www.yourhome.gov.au
	www.earthhouse.com.au
New Zealand	
	www.solidearth.co.nz
	www.earthbuilding.org.nz
India	
	www.earth-auroville.com
	www.aureka.com
Nepal	
	www.abari.org
Egypt	
	www.eeca.net
	www.fathyheritage.com

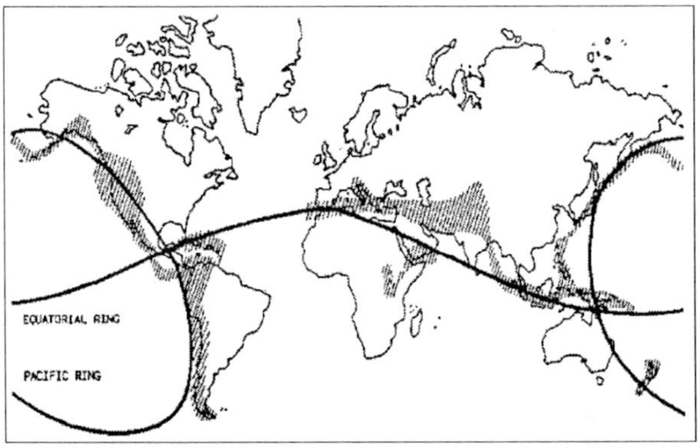

Image 1. Earthquake zones (Houben, Guillaud 1984) [106].

The graph below indicates the relative numbers of researchers throughout the world who are currently studying stabilized soils and unfired bricks.

There is a growing interest in earthen building systems and interestingly Europe and the United States of America appear to be leading the way at present with regards to active research projects in this area.

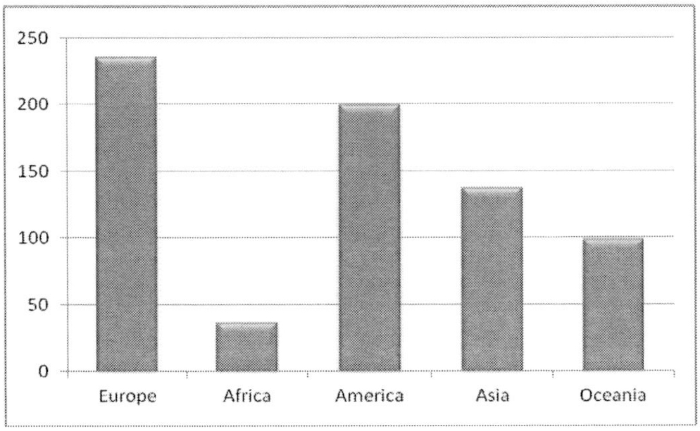

Image 2. Research international distribution on stabilised soil for building proposes.

Appendix 1: Literature Review

The following paragraphs include an updated list of scientific technological and international information on the subjects and themes that embrace research within the field of earthen Architecture.

1. Mortars, Blocks and Soil Reinforced with Fibres

_ACHENZA, M; and FENU, L.: On earth stabilization with natural polymers for earth masonry construction. Materials and Structures, Springer, 2006.

Abstract: In this paper we study the problem of preparing adobe bricks for earth construction as little erodable as possible. We have therefore tested earthen specimens stabilized with both vegetal fibres and a compound of natural polymers, and we have compared their behaviour to that one of specimens stabilised only with fibres. We have studied how the stabilization with these natural polymers modifies porosity and bulk density, and how it improves both the behaviour under the water action and the compression strength. This kind of stabilization appears to give the earth a very good behaviour under the water action; moreover we have found out that the compression strength has been significantly increased, too.

_ADESANYA, D. A.: Evaluation of blended cement mortar, concrete and stabilized earth made from ordinary Portland cement and corn cob ash. Construction and Building Materials, Volume 10, Issue 6, September 1996, Pages 451-456.

Abstract: A study was conducted to determine the effect of using corn cob ash (pozzolanic waste by-product of corn cobs) as a cost-reducing additive in blended cement. Various percentages of the blended cement were used to stabilize laterite and clay. The effects of the blended cement in various concrete mixes, mortar and roofing sheets were also analyzed. The results show that replacing 50.0 and 20.0% respectively of ordinary Portland cement by weight with produces stabilized clay and laterite exhibiting greater strength, lower thermal conductivity and lower water absorption than plain cement stabilized earth. The results also indicate that replacing 20.0% of cement with in concrete mix and mortar improves water absorption and durability of the specimens while there is no significant difference between the strength of concrete produced with 0.0 and 20.0% . Roofing sheets of high quality which comply with BS 690 specifications have been fabricated from clay and laterite stabilized with the blended cement and corn cob fibre. The roofing sheets are superior to asbestos roofing sheets which are commercially available in the Nigerian market.

_AGGARWAL, L. K.: Bagasse-reinforced cement composites. Cement and Concrete Composites, Volume 17, Issue 2, 1995, Pages 107-112.

Abstract: Bagasse is abundantly available in many countries as a by-product from sugar mills and is being mostly used as fuel or disposed of by incineration. An attempt has been made to convert this byproduct into useful eco-friendly cement-bonded composites, which can be used for various internal and external applications in buildings. The investigations include optimization of parameters such as bagasse content, casting pressure and demoulding time for the production of bagassecement composites and the method of their production. The physico-mechanical properties of the composites produced by using the parameters finalised as above were determined. The developed composites were also subjected to accelerated laboratory tests to study the effect of moisture and alternate wetting and drying cycles on their properties. The results obtained from these studies show that the developed composites meet most of the requirements of various standards on cement-bonded particle boards and have high levels of performance even in moist conditions. Therefore, in countries where bagasse is substantially available, it can be used for the production of cementbonded building materials.

_AGOPYAN, V.; SAVASTANO Jr., H.; JOHN, V.M.; and CINCOTTO, M.A.: Developments on vegetable fibre-cement based materials in São Paulo, Brazil: an overview. Cement and Concrete Composites, Volume 27, Issue 5, Natural fibre reinforced cement composites, May 2005, Pages 527-536.

Abstract: Vegetable fibres, which are widely available in most developing countries, can be used as convenient materials for brittle matrix reinforcement, even though they present relatively poor durability performance. Taking into account the fibres mechanical properties, with an adequate mix design, it is possible to develop a material with suitable properties for building purposes. In order to improve the durability of vegetable fibres, this paper presents the approach adopted in the research which is directed towards the development of alternative binders, with controlled free lime, using ground granulated blast furnace slag. Coir fibres demonstrate to be more suitable vegetable fibres for the reinforcement of large components as can be proved by in-use durability performance evaluation of an 11-year old prototype house. More recently, pulp from eucalyptus waste and residual sisal and coir fibres have been studied as a replacement for asbestos in roofing components.

_ANAGNOSTOPOULOS, Costas A.: Cement-clay grouts modified with acrylic resin or methyl methacrylate ester: Physical and mechanical properties. Construction and Building Materials, In Press, Corrected Proof, Available online 30 January 2006.

Abstract: Grouting is a common technical method with many applications, e.g. it is used for soil stabilization and strengthening, for reduction of water ingress to underground facilities or of the water loss through a dam foundation, etc. Grouts comprise several constituents, which are combined in many ways depending on the in situ conditions and the outcome desired. Superplasticizers, accelerators, antifreezers, air-entraining agents and many others are generally used to improve the quality of cement grouts and consequently, their effectiveness on strength (especially bond strength), durability, impermeability and resistance to chemical erosion of the grouted soil or rock mass. A comprehensive laboratory work was carried out in order to study the physical and mechanical properties of grouts prepared by using cement, clay and water in different percentages along with an amount of acrylic resin or methyl methacrylate co-polymer emulsion.

_ASASUTJARIT, C.; HIRUNLABH, J.; KHEDARI, J.; CHAROENVAI, S.; ZEGHMATI, B.; and SHIN, U. Cheul: Development of coconut coir-based lightweight cement board. Construction and Building Materials, In Press, Corrected Proof, Available online 18 October 2005.

Abstract: This paper presents investigation conducted in Thailand on the development of coconut coir-based lightweight cement boards (CCB). These boards were made from coconut coir, cement and water. They are intended to be used as building components for energy conservation. The investigations focused on parameters, mainly, fiber length, coir pretreatment and mixture ratio that affect the properties of boards. The physical, mechanical and thermal properties of the specimens were determined after 28 days of hydration. Results of this study indicated that the best pretreatment of coir fibers was to boil and wash them as it can enhance some of the mechanical properties of coir fiber. The optimum fiber length was 1-6andnbsp;cm fraction, and optimum (cement-fiber-water) mixture ratio by weight was 2:1:2. The produced CCBs satisfied most recommended mechanical standards. In addition, investigation on thermal property of specimens revealed that coconut coir-based lightweight cement board has lower thermal conductivity than commercial flake board composite. That is an important feature to promote the use of CCB's as energy saving material in buildings. Keywords: Internal bond; Modulus of rupture (MOR); Modulus of elasticity (MOE); Thermal conductivity; Water absorption.

_BASHA, E.A.; HASHIM, R.; MAHMUD, H.B.; and MUNTOHAR, A.S.: Stabilization of residual soil with rice husk ash and cement. Construction and Building Materials, Volume 19, Issue 6, July 2005, Pages 448-453.

Abstract: Stabilization of residual soils is studied by chemically using cement and rice husk ash. Investigation includes the evaluation of such properties of the soil as compaction, strength, and X-ray diffraction. Test results show that both cement and rice husk ash reduce the plasticity of soils. In term of compactability, addition of rice husk ash and cement decreases the maximum dry density and increases the optimum moisture content. From the viewpoint of plasticity, compaction and strength characteristics, and economy, addition of 6-8% cement and 10-15% rice husk ash is recommended as an optimum amount.

_BINICI, Hanifi; AKSOGAN, Orhan; BODUR, Mehmet Nuri; AKCA, Erhan; and KAPUR, Selim: Thermal isolation and mechanical properties of fibre reinforced mud bricks as wall materials. Construction and Building Materials, In Press, Corrected Proof, Available online 21 February 2006.

Abstract: Fibre reinforced mud bricks, which are studied in this paper, provide the expected technical performance for the thermal isolation and mechanical properties, according to ASTM and Turkish standards. The mechanical properties of waste materials and some stabilizers were investigated thoroughly and some concrete conclusions were drawn. The fibre reinforced mud bricks fulfill the compressive strength and heat conductivity requirements of the ASTM and Turkish standards. Mud bricks with plastic fibers showed a higher compressive strength than those with straw, polystyrene and without any fibers. Basaltic pumice as an ingredient was found to decrease the thermal conductivity coefficient of fibre reinforced mud bricks. The fibre reinforced mud brick house has been found to be superior to the concrete brick house for keeping indoor temperatures stationary during the summer and winter.

_BLANKENHORN, Paul R.; BLANKENHORN, Brad D.; SILSBEE, Michael R.; and DICOLA, Maria: Effects of fiber surface treatments on mechanical properties of wood fiber-cement composites. Cement and Concrete Research, Volume 31, Issue 7, , July 2001, Pages 1049-1055.

Abstract: The purpose of this research was to determine the effects of treated and untreated hardwood, kraft softwood, and newsprint wood fibers on the 7- and 28-day bending strength, compressive strength, and toughness values for wood fiber-cement composites. Untreated and acrylic- or alkylalkoxysilane-treated hardwood, kraft softwood, and newsprint wood fibers used in wood fiber-cement composites resulted in different bending and compression properties. Fiber characteristics along with different chemical treatments influenced the composite properties.

_BOUHICHA, M.; AOUISSI, F.; and KENAI, S.: Performance of composite soil reinforced with barley straw. Cement and Concrete Composites, Volume 27, Issue 5, Natural fibre reinforced cement composites, May 2005, Pages 617-621.

Abstract: The shortage of low cost and affordable housing in Algeria has led to many investigations into local low cost construction materials. Earth construction is widespread in desert and rural areas but suffers from shrinkage cracking, low strength and lack of durability. The use of natural and vegetable fibres could improve its performance. This paper reports on an experimental study to investigate a composite soil reinforced with chopped barley straw, using four different soils. The effect of fibre length and fibre fraction on shrinkage, compressive strength, flexural strength and shear strength was investigated. Preliminary tests to enhance durability by using different waterproof renders are also briefly reported.

_CARTER, G. R.; and DIXON, J. H.: Oriented polymer grid reinforcement. Construction and Building Materials, Volume 9, Issue 6, Application of Polymeric Materials to the Construction Industry, December 1995, Pages 389-401.

Abstract: Tensar high strength oriented polymer grids were developed in the UK in the late 1970s. Extensive research has been conducted in Europe and North America to investigate the various civil engineering applications for these materials. The availability of these high strength, durable reinforcing grids has led to many innovative and economical developments in geotechnical and highway engineering. This paper describes the major application areas: reinforced soil walls and slopes, reinstatement of slope failures; embankment foundations over soft soil; reinforcement of road bases for paved roads; and asphalt reinforcement.

_CHAND, Navin; TIWARY, R. K.; and ROHATGI, P. K.: Bibliography Resource structure properties of natural cellulosic fibres — an annotated bibliography. Journal of Materials Science, 23, 2, 2/1/1988, Pages 381-387.

Abstract: We present a review of the work on the structure and properties of some natural lignocellulosic fibres with a classified list of references on resource, structure, physico-mechanical properties and on the uses of these fibres. This list of references includes papers published in scientific journals and in the proceedings of conferences.

_COUTTS, R. S. P.: Fibre-matrix interface in air-cured wood-pulp fibre-cement composites. Journal of Materials Science Letters, Volume 6, Issue 2, 1987, Pages 140-142.

Abstract: Presents a study of the fibre-matrix interface in air-cured, wood fibre-cement paste, using scanning electron microscopy (SEM), to determine whether there is a physically observable difference similar to other reinforcements used in cement composites.

_DEMIR, Ismail; BASPINAR, M. Serhat; and ORHAN, Mehmet: Utilization of kraft pulp production residues in clay brick production, Building and Environment, Volume 40, Issue 11, November 2005, Pages 1533-1537.

Abstract: The main objective of this study is to investigate the utilization potential of kraft pulp production residues in clay brick. Kraft pulp production is the primary phase of the paper industry. Long cellulose fibers are produced from wood, straw and reeds in kraft pulp production. Short cellulose fibers are separated as an organic waste material from the production system. This type of residue is only utilized in agricultural purpose or in the production of moulded egg cartons. Kraft pulp production plant of Seka Company (located at Afyon-Çay/Turkey), which has 50,000andnbsp;tpy dried pulp production capacity, generates important amount of organic wastes.

_GALÁN-MARÍN, Carmen; RIVERA-GÓMEZ, Carlos, PETRIC-GRAY, Jelena. Effect of Animal Fibres Reinforcement on Stabilized Earth Mechanical Properties. Journal of Biobased Materials and Bioenergy Vol. 4, 1–8, 2010.

Abstract: In our study we investigate the effect of adding wool—a natural animal fibre, to soil and alginate matrix in order to improve its mechanical strength. All the components of our natural composite were sourced in Scotland, UK. Wool is available in abundance in Scotland, but no longer widely used in local textile industry. Our study has been motivated by a desire to produce a composite material suitable for wet climatic conditions. Natural soil compositions are very variable, so in our experiments we decided to use three different types of soil supplied by brick manufactures in Scotland, in order to analyse the effect of the reinforcement (i.e., addition of wool fibre). Effect of fibre content on mechanical properties (density, tensile strength and flexural strength) was studied. Fibre in the mixture minimizes the shrinkage, reduces the curing time and enhances significantly compressive strength if an optimal reinforcement ratio is used in earth specimens. Addition of animal fibre also increased flexural strength. We additionally observed that more ductile failure was obtained with the reinforced specimens.

_GALÁN-MARÍN, Carmen; RIVERA-GÓMEZ, Carlos, PETRIC-GRAY, Jelena. Clay-based composite stabilized with natural polymer and fibre. Construction and Building Materials 24 (2010) 1462–1468.

Abstract: The research objective is the stabilization of soils with natural polymers and fibres to produce a composite, sustainable, non-toxic and locally sourced building material. Mechanical tests have been conducted with a clay soil supplied by a Scottish brick manufacture. Alginate (a natural polymer from the cell walls of brown algae) has been used as bonding in the composite. Sheep's wool was used as reinforcement.

Tests done showed that the addition of alginate separately increases compression strength from 2.23 to 3.77 MPa and the addition of wool fibre increases compression strength a 37%. The potential benefit of stabilization was found to depend on the combinations of both stabilizer and wool fibre. Adding alginate and reinforcing with wool fibre doubles the soil compression resistance. Better results were obtained with a lower quantity of wool.

_GHAVAMI, Khosrow; TOLEDO Filho, Romildo D.; and BARBOSA, Normando P.: Behaviour of composite soil reinforced with natural fibres. Cement and Concrete Composites, Volume 21, Issue 1, 1999, Pages 39-48.

Abstract: Next to the food shortage, the housing shortage is one of the most crucial problems on earth. To improve this situation and make it possible to build more houses, particularly for low-income families, it is necessary to examine all locally available materials which can be used for construction. Bamboo, sisal and coconut fibres are materials which are available in abundance in Brazil and are not used in civil construction. To increase the amount of information concerning the physical and mechanical behaviour of these materials several research programmers were executed at Pontifical Universidade Católica in Rio de Janeiro (PUC-Rio) and Universidade Federal da Paraiba (UFPb) under the general supervision of the first author. In this paper new results are presented concerning the application of sisal and coconut fibres in conjunction with three types of locally appropriate soil for the production of composite soil blocks reinforced with sisal and coconut fibres.

_GHIASSIAN, H; POOREBRAHIM, G.; and GRAY D. H.: Soil reinforcement with recycled carpet wastes. Waste Management and Research, Vol. 22, No. 2, 108-114 (2004).

Abstract: A root or fibre-reinforced soil behaves as a composite material in which fibres of relatively high tensile strength are embedded in a matrix of relatively plastic soil. Shear stresses in the soil mobilize tensile resistance in the fibres, which in turn impart greater strength to the soil. A research project has been undertaken to study the influence of synthetic fibrous materials for improving the strength characteristics of a fine sandy soil. One of the main objectives of the project is to explore the conversion of fibrous carpet waste into a value added product for soil reinforcement. Drained triaxial tests were conducted on specimens, which were prepared in a cylindrical mould and compacted at their optimum water contents. The main test variables included the aspect ratio and the weight percentage of the fibrous strips. The results clearly show that fibrous inclusions derived from carpet wastes improve the shear strength of silty sands. A model developed to simulate the effect of the fibrous inclusions accurately predicts the influence of strip content, aspect ratio and confining pressure on the shear strength of reinforced sand.

_HATAF, N.; and RAHIMI, M.M.: Experimental investigation of bearing capacity of sand reinforced with randomly distributed tire shreds. Construction and Building Materials, Volume 20, Issue 10, December 2006, pages 910-916.

Abstract: A series of laboratory model tests has been carried out to investigate the using of shredded waste tires as reinforcement to increase the bearing capacity of soil. Shred content and shreds aspect ratio are the main parameters that affect the bearing capacity. Tire shreds with rectangular shape and widths of 2 and 3andnbsp;cm with aspect ratios 2, 3, 4 and 5 are mixed with sand. Five shred contents of 10%, 20%, 30%, 40% and 50% by volume were selected. Addition of tire shreds to sand increases BCR (bearing capacity ratio) from 1.17 to 3.9 with respect to shred content and shreds aspect ratio.

_HATTAMLEH, O. Al; and MUHUNTHAN, B.: Numerical procedures for deformation calculations in the reinforced soil walls. Geotextiles and Geomembranes, Volume 24, Issue 1, February 2006, Pages 52-57.

Abstract: This study presents a membrane analogy method to evaluate the deflection of fabric-reinforced earth walls. The resulting equations were solved using a finite difference scheme to obtain the deflection. The numerical results were compared with a full-scale study. The comparisons show good performance of the model.

_JACOB, Maya; JOSEPH, Seena; POTHAN, Laly A.; and THOMAS, Sabu: A study of advances in characterization of interfaces and fiber surfaces in lignocellulosic fiber-reinforced composites, Composite Interfaces, Volume 12, Issue 1, Jan 2005, Page 95.

Abstract: This article deals with the aspects of interfacial and surface characterization of natural fibers and their composites. Vegetable fibers and their composites have attracted the attention of scientists worldwide because of their favorable properties. The different chemical modifications of natural fibers and characterization aspects have been discussed. The adhesion between fiber and matrix is a major factor in determining the response of the interface and its integrity under stress. Therefore characterization of the interface is of utmost importance. Both fiber surface and polymer matrix surface can be modified to obtain a strong interface. Various treatments being used for the lignocellulosic surfaces and the characterization techniques have been illustrated. The four main techniques of interfacial characterization that are enumerated in this article are the micromechanical techniques, spectroscopic, microscopic and swelling techniques.

_KHEDARI, Joseph: NANKONGNAB, Noppanun; HIRUNLABH, Jongjit; and TEEKASAP, Sombat: New low-cost insulation particleboards from mixture of durian peel and coconut coir, Building and Environment, Volume 39, Issue 1, January 2004, Pages 59-65.

Abstract: The main purpose of this study is to develop low thermal conductivity particleboards with optimized durian peel and coconut coir mixture ratio. To this end, two main parameters were investigated, namely, the mixture ratio of durian peel and coconut coir (by weight) and board density. The particleboards were prepared following common manufacturing technique. It was observed the mixture ratio and board density affect the properties of final particleboards considerably.

_KRIKER, A.; DEBICKI, G.;BALI, A.; KHENFER, M. M.; and CHABANNET, M.: Mechanical properties of date palm fibres and concrete reinforced with date palm fibres in hot-dry climate, Cement and Concrete Composites,

Volume 27, Issue 5, Natural fibre reinforced cement composites, May 2005, Pages 554-564.

Abstract: The study examines four types of date palm surface fibres and determines their mechanical and physical properties. In addition, the properties of date palm fibre-reinforced concrete, such as strength, continuity index, toughness and microstructure, are given as a function of curing in water and in a hot-dry climate. The volume fraction and the length of fibres reinforcement were 2-3% and 15-60andnbsp;mm respectively. Increasing the length and percentage of fibre-reinforcement in both water and hot dry curing, was found to improve the post-crack flexural strength and the toughness coefficients, but decreased the first crack and compressive strengths. In hot-dry climate a decrease of first crack strength with ageing was observed for each concrete type. Water curing decreased the global degree of the voids and cracks with time for each concrete type, but increased it in hot-dry climate.

_KUMAR, Arvind; WALIA, Baljit Singh; and MOHAN, Jatinder: Compressive strength of fiber reinforced highly compressible clay. Construction and Building Materials, Volume 20, Issue 10, December 2006, Pages 1063-1068.

Abstract: Admixtures and geogrids are frequently used in practice to stabilize soils and to improve their load carrying capacity. In this study, polyester fibers were mixed with soft clay soil to investigate the relative strength gain in terms of unconfined compression. Samples were tested in unconfined compression with 0%, 0.5%, 1.0%, 1.5% and 2.0% plain and crimped polyester fibers. Verifications tests were also performed to investigate the repeatability of the test results. The results presented show that the degree of compaction affected the relative benefits of fiber reinforcement for the subject soil. Samples compacted after mixing various proportions of sand into clay (varying from 0% to 12% of clay) were also tested. It was observed that unconfined compressive strength of clay increases with the addition of fibers and it further increases when fibers are mixed in clay sand mixture. Verification tests performed revealed that even though the fibers were randomly oriented, tests results can be reproduced with reasonable accuracy.

_LANGE, D. A.; OUYANG C.; and SHAH, S. P.: Behavior of cement based matrices reinforced by randomly dispersed microfibers. Advanced Cement Based Materials, Volume 3, Issue 1, January 1996, Pages 20-30.

Abstract: A study on behavior of cement based matrices reinforced by randomly distributed microfibers is summarized. Effects of volume fraction, length and type of fiber, and type of cement based matrix were experimentally examined using uniaxial tensile specimens and three-point bend beams. The matrix fracture properties were measured by a RILEM recommended test procedure. A confocal microscopy technique was used to measure fracture surface roughness, a parameter that was shown to correlate with fracture properties. By incorporating the obtained matrix fracture properties, fiber aspect ratio, and fiber-matrix interface bond into a fracture mechanical R-curve approach, mechanical responses of cement based matrices reinforced by fibers can be predicted.

_LEE, Hyun Jong; and ROH, Han Sung: The use of recycled tire chips to minimize dynamic earth pressure during compaction of backfill, Construction and Building Materials, In Press, Corrected Proof, Available online, 17 April 2006.

Abstract: The backfill areas of concrete culverts constructed in roads have been subjected to differential settlement due to poorly compacted soils. Dynamic compaction rollers cannot be fully utilized to compact the backfill soils near the culverts because the high earth pressure induced by dynamic loading frequently results in cracks in the culvert walls. In this study, two cushion materials, recycled tire chips and expanded polystyrene (EPS) boards are applied on the culvert walls in backfill areas to reduce the dynamic earth pressure induced by the compaction loading as well as to improve the characteristics of compacted soils. A numerical analysis is carried out to study the effects of the cushion materials on the stress variation with soil depth in the backfill areas. The numerical analysis shows that a cushion material with low elastic modulus and high damping ratio can effectively reduce the dynamic earth pressure.

_LI, Yan; MAI, Yiu-Wing; and YE, Lin: Sisal fibre and its composites: a review of recent developments. Composites Science and Technology, Volume 60, Issue 11, August 2000, Pages 2037-2055.

Abstract: Sisal fibre is a promising reinforcement for use in composites on account of its low cost, low density, high specific strength and modulus, no health risk, easy availability in some countries and renewability. In recent years, there has been an increasing interest in finding new applications for sisal-fibre-reinforced composites that are traditionally used for making ropes, mats, carpets, fancy articles and others. This review presents a summary of recent developments of sisal fibre and its composites. The properties of sisal fibre itself, interface between sisal fibre and matrix, properties of sisal-fibre-reinforced composites and their hybrid composites have been reviewed. Suggestions for future work are also given.

_MacVICAR R.; MATUANA, L. M.; AND BALATINECZ, J. J.: Aging mechanisms in cellulose fiber reinforced cement composites, Cement and Concrete Composites, Volume 21, Issue 3, 1999, Pages 189-196.

Abstract: This paper examines the effects of laboratory scale accelerated aging exposures on the changes in physicaland mechanical properties of commercially produced cellulose fiber reinforced cement composites. Two different accelerated aging methods were used to simulate the possible aging mechanisms for which the material may experience under service conditions, both methods being compared to material naturally weathered for 5 yr in roofing. The first aging method consisted of different cycles of water immersion, carbonation, and heating exposures whereas in the second method, cycles of water immersion, heating and freeze-thaw exposures were used. The porosity, water absorption, permeability of nitrogen and compressive shear strength of the composites were examined before and after aging exposures. The surface morphologies of the composites fractured in compression shear tests were examined using scanning electron microscope.

_MADAKADZE, I. C.; RADIOTIS, T.; LI, J.; GOEL, K.; and SMITH, D. L.: Kraft pulping characteristics and pulp properties of warm season grasses. Bioresource Technology, Volume 69, Issue 1, July 1999, Pages 75-85.

Abstract: Non-wood fibres are increasingly being used in the pulp and paper industry to help meet the increasing world demand for paper. Their use also helps to reduce demand on declining forest reserves. In this study several warm season grasses, prairie sandreed (Calmovilfa longifolia (Hook.) Scribn.), cordgrass (Spartina pectinata L.), big bluestem (Andropogon gerardii Vitman), and switchgrass (Panicum virgatum L. cv. Pathfinder (PF) and New Jersey 50 (NJ50)), were evaluated as potential raw materials for pulp and paper production. Raw material chemical composition, kraft pulp yield and properties, and fibre characteristics were evaluated. All these grasses were easily pulped under a mild kraft process, with pulp yields ranging from 44 to 51%, highest yields were recorded for NJ50 and big bluestem; and kappa numbers ranging from 10 to 16. The weight-weighted fibre length ranged from 1.29 to 1.43 mm, the highest value being recorded for big bluestem.

_MATTONE, Roberto: Sisal fibre reinforced soil with cement or cactus pulp in bahareque technique. Cement and Concrete Composites, Volume 27, Issue 5, Natural fibre reinforced cement composites, May 2005, Pages 611-616.

Abstract: In order to improve bahareque technique, sisal fibre reinforced soils were stabilised with cement or cactus pulp. Bending, abrasion resistance, water absorption and erosion tests were performed and the results were compared with those obtained on the traditional plasters used in soil technologies. The performance capabilities of sisal fibre reinforced soil stabilized with cement are better than those of cactus pulp stabilised soil. The use of cactus pulp as a stabilising agent to improve the behaviour of the soil, however, is very interesting because this is a natural, ecological material.

_MOHR, B.J.; NANKO, H.; and KURTIS, K.E.: Aligned kraft pulp fiber sheets for reinforcing mortar. Cement and Concrete Composites, Volume 28, Issue 2, February 2006, Pages 161-172.

Abstract: In this research program, aligned pulp fiber sheets were used as reinforcement in cement-based matrices. Flexural testing results validate this reinforcement strategy by demonstrating that: (1) fiber sheet alignment does significantly affect mechanical behavior, indicating that fiber alignment is achieved by the production process, and (2) aligned fiber sheet composite exhibited significantly greater toughness than equivalent volumes of distributed fibers. Additional results indicate that the addition of the wet-strength additive Kymene facilitates the handling of fiber sheets, but has no effect on mechanical performance. Addition of fly ash to the fiber sheets also had no effect on composite behavior. Use of fibrillated (beaten) fibers and the introduction of perforations to the fiber sheets appeared to have no effect on flexural strength, while generally decreasing composite toughness. Increasing the fiber sheet basis weight (thickness) produced increases in toughness, with negligible changes in flexural strength. Reinforcement with multiple fiber sheets (total basis weight increasing) was shown to increase toughness, while layering with total basis weight remaining constant improved strength, but reduced toughness.

_PARK, Taesoon; and TAN, Siew Ann: Enhanced performance of reinforced soil walls by the inclusion of short fiber. Geotextiles and Geomembranes, Volume 23, Issue 4, August 2005, Pages 348-361.

Abstract: This paper presents the effects of the inclusion of short fiber in sandy silt (SM) soil on the performance of reinforced walls. The inclusion of short fiber in soil is expected to increase soil strength and improve stability when it is used as the backfill material. Short fiber of 60andnbsp; mm length was used and the mixing ratio of the fiber was 0.2% by weight of the soil. The finite element method was used to examine the influence of the reinforced short fiber on reinforced walls. The vertical and horizontal earth pressure, displacement and settlement of the wall face were analyzed. These results were compared to the measured results from two full-scale tests. It is shown that use of short fiber reinforced soil increases the stability of the wall and decreases the earth pressures and displacements of the wall. This effect is more significant when short fiber soil is used in combination with geogrid.

_PRABAKAR, J.; and SRIDHAR, R. S.: Effect of random inclusion of sisal fibre on strength behaviour of soil. Construction and Building Materials, Volume 16, Issue 2, March 2002, Pages 123-131.

Abstract: Construction of building and other civil engineering structures on weak or soft soil is highly risky on geotechnical grounds because such soil is susceptible to differential settlements, poor shear strength and high compressibility. Improvement of load bearing capacity of the soil may be undertaken by a variety of ground improvement techniques like stabilisation of soil, adoption of reinforced earth technique etc. Reinforced earth technique is considered as an effective ground improvement method because of its cost effectiveness, easy adaptability and reproducibility. Therefore, in the present investigation, sisal fibre has been chosen as the reinforcement material and it was randomly included in to the soil at four different percentages of fibre content, i.e. 0.25, 0.5, 0.75 and 1% by weight of raw soil. Four different lengths of fibre, i.e. 10, 15, 20 and 25 mm are also considered as one of the parameters of this study. The main objective of this investigation had been focused on the strength behaviour of the soil reinforced with randomly included sisal fibre.

_RAMAKRISHNA, G.; and SUNDARARAJAN, T.: Studies on the durability of natural fibres and the effect of corroded fibres on the strength of mortar, Cement and Concrete Composites, Volume 27, Issue 5, Natural fibre reinforced cement composites, May 2005, Pages 575-582.

Abstract: This paper presents the results of the variation in chemical composition and tensile strength of coir, sisal, jute and Hibiscus cannabinus fibres, when they are subjected to alternate wetting and drying and continuous immersion for 60 days in three mediums (water, saturated lime and sodium hydroxide). Compressive and flexural strengths of cement mortar (1:3) specimens reinforced with dry and corroded fibres were determined after 28 days of normal curing. From the results it is observed that there is substantial reduction in the salient chemical composition of all the four fibres, after exposure in the various mediums. Coir fibres are found to retain higher percentages of their initial strength than all other fibres, after the specified period of exposure in the various mediums. The compressive and flexural strengths of all natural fibre reinforced mortar specimens using corroded fibres are less than the strength of the reference mortar (i.e. without fibres) and fibre reinforced mortar specimens reinforced with dry natural fibres.

_REIS, J.M.L.: Fracture and flexural characterization of natural fiberreinforced polymer concrete. Construction and Building Materials, Volume 20, Issue 9, November 2006, Pages 673-678.

Abstract: Mechanical characterization of epoxy polymer concrete reinforced with natural fibers is investigated in this work to analyze the possibility of substitution by synthetic fibers. These natural fibers studied are coconut, sugar cane bagasse, and banana fibers. All of these fibers come from their specific products after they have been used, i.e. as recycle. As the natural fibers are agriculture waste, manufacturing natural product is, therefore, an economic and interesting option. The main idea is to use the fibers like they come from nature without any kind of preparation. The comparison between epoxy polymer concrete reinforced with natural fibers, unreinforced and reinforced with synthetic fibers is made.

_SAVASTANO Jr. H.; WARDEN, P.G.; and COUTTS, R.S.P.: Brazilian waste fibres as reinforcement for cement-based composites. Cement and Concrete Composites, Volume 22, Issue 5, October 2000, Pages 379-384.

Abstract: Fibre reinforced cement-based composites were prepared using kraft pulps from sisal and banana waste and from Eucalyptus grandis pulp mill residues. The study adapted conventional chemical pulping conditions for the nonwood strands and a slurry vacuum de-watering method for composite preparation followed by air-curing. Plain cement paste and Pinus radiata kraft reinforced cement composites were used as reference materials. Mechanical testing showed that optimum performance of the various waste fibre reinforced composites was obtained at a fibre content of around 12% by mass, with flexural strength values of about 20 MPa and fracture toughness values in the range of 1.0- 1.5 kJ m-2. Experimental results showed that, of the waste fibres studied, E. grandis is the preferred reinforcement for low-cost fibre-cement.

_SAVASTANO Jr., Holmer; and AGOPYAN, Vahan: Transition zone studies of vegetable fibre-cement paste composites. Cement and Concrete Composites, Volume 21, Issue 1, 1999, Pages 49-57.

Abstract: The transition zone of short filament fibres randomly dispersed in a paste of ordinary Portland cement was analysed. Composites with vegetable fibres (malva, sisal and coir) were compared with those with chrysotile asbestos and polypropylene fibres. The composites were prepared for testing at the ages of 7, 28, 90 and 180 days. The water-cement ratio was 0.38; at the age of 28 days specimens with w/c = 0.30 and a w/c = 0.46 were also tested. Mechanical tests evaluated the composite tensile strength and ductility. Backscattered electron imaging (BSEI) and energy dispersive spectroscopy (EDS) were used to identify the major properties of the fibre-matrix interface. Mainly for vegetable fibre composites the transition zone is porous, cracked and rich in calcium hydroxide macrocrystals. These characteristics are directly related to the fibre-matrix bonding and to the composite mechanical performance.

_SHAO, Yixin; Qiu, Jun; and SHAH, Surendra P.: Microstructure of extruded cement-bonded fiberboard, Cement and Concrete Research, Volume 31, Issue 8, August 2001, Pages 1153-1161.

Abstract: The microstructure of cement-bonded fiberboard manufactured by extrusion process was studied using scanning electron microscope (SEM). Comparison between extruded and cast fiberboard revealed that the extruded products were better in strength, stiffness, toughness, fiber distribution, fiber orientation, and bond of fiber with matrix, even in the presence of a higher percent air voids. The dominant component in extruded fiberboard was the type of fiber. Extrusion was capable of incorporating both hydrophilic and hydrophobic fibers into fiberboard production. It was found the sand content had significant effect on toughness. The more sand added, the less the toughness. Fiber dispersion seemed not to be critical. Fiberboard made by non dispersive mixing exhibited satisfactory performance. Accordingly, the mixing time and energy in extrusion production could therefore be reduced.

_TOLEDO Filho, Romildo D.; GHAVAMI, Khosrow; ENGLAND, George L.; and SCRIVENER, Karen: Development of vegetable fibre-mortar composites of improved durability, Cement and Concrete Composites, Volume 25, Issue 2, February 2003, Pages 185-196.

Abstract: The primary concern for vegetable fibre reinforced mortar composites (VFRMC) is the durability of the fibres in the alkaline environment of cement. The composites may undergo a reduction in strength and toughness as a result of weakening of the fibres by a combination of alkali attack and mineralisation through the migration of hydration products to lumens and spaces. This paper presents several approaches used to improve the durability performance of VFRMCs incorporating sisal and coconut fibres. These include carbonation of the matrix in a CO2-rich environment; the immersion of fibres in slurried silica fume prior to incorporation in the ordinary Portland cement (OPC) matrix; partial replacement of OPC matrix by undensified silica fume or blast-furnace slag and a combination of fibre immersion in slurried silica fume and cement replacement. The durability of the modified VFRMC was studied by determining the effects of ageing in water, exposure to cycles of wetting and drying and open air weathering on the microstructures and flexural behaviour of the composites. Immersion of natural fibres in a silica fume slurry before their addition to cementbased composites was found to be an effective means of reducing embrittlement of the composite in the environments studied. Early cure of composites in a CO2-rich environment and the partial replacement of OPC by undensified silica fume were also efficient approaches in obtaining a composite of improved durability. The use of slag as a partial cement replacement had no effect on reducing the embrittlement of the composite.

_WANG, Youjiang ; Zureick, Abdul-Hamid ; CHO, Baik-Soon ; and SCOTT, D. E. : Properties of fiber-reinforced concrete using recycled fibers from carpet industrial-waste. Journal of Materials Science, 29, 16, Pages 4191-4199, Nov 2004.

Abstract A study was carried out to evaluate the use of recycled fibres from carpet industrial waste for reinforcement of concrete at 1 and 2 vol% fractions. Compressive, flexural, splitting tensile and shrinkage tests were performed. Significant increases in shatter resistance, energy absorption and ductility were observed. This paper reports on the experimental programme and compares the effectiveness of such recycled fibres with that of virgin polypropylene fibres specially made for fibre reinforced concrete (FRC). The paper also discusses the benefits of using such FRC for construction applications and possible ways to further enhance the performance of such FRC.

2. Mortars, Blocks and Soil Walls

_ADAM, E. A.; and JONES, P. J.: Thermophysical properties of stabilized soil building blocks. Building and Environment, Volume 30, Issue 2, April 1995, Pages 245-253.

Abstract: The thermal conductivity of lime/cement stabilised hollow and plain earth blocks, produced from Sudanese soil with the Brepak Press, has been measured using the Guarded Hot Box method. The results give thermal conductivity values for oven dry samples at (0.1-0.5%) moisture content by volume of 0.25-0.55 W/mK. For each soil type, the conductivity is highest for cement stabilised soil building blocks. The results compare to existing leading standards for masonry and earthen products.Ingersolls' algorithm was used to calculate the specific heat capacity of the blocks. This gave values around 836 J/kg °C, which again compares favourably to reported results for stabilised soil building blocks.

_BINICI, Hanifi; AKSOGAN, Orhan; and SHAH, Tahir: Investigation of fibre reinforced mud brick as a building material. Construction and Building Materials, Volume 19, Issue 4, May 2005, Pages 313-318.

Abstract: Most of the buildings in the rural areas are made out of limestone, low quality traditional concrete brick and adobe. But these materials do not have sufficiently high compressive strengths. In the present research, an earthquake resistant material with high compressive strength has been sought. To this end, the mechanical properties of certain combinations of fibrous waste materials and some stabilisers were investigated thoroughly and some concrete conclusions were drawn. It was concluded that the interface layers of fibrous materials increased the compressive strength and a certain geometrical shape of these layer materials gave the best results. The mix proposed satisfies the minimum compressive strength requirements of ASTM and Turkish Standards.

_DELGADO, M. Carmen Jiménez; and GUERRERO, Ignacio Cañas: Earth building in Spain, Construction and Building Materials, Volume 20, Issue 9, November 2006, Pages 679-690.

Abstract: This paper is a review of the state of use of the earth building in Spain nowadays. We present researching organizations, modern projects carried out or the existing manufacturers for compressed earth blocks. Besides, we offer an overview of the Spanish general building regulatory system to find that earth construction is not included in it, although there is a pair of non-regulatory guides that could act as national reference documents and whose provisions we examine. Although earth as a construction material is unknown for most people, a growing interest is noticed in two ways, for rescuing the heritage and as a rediscovered environmentally friendly building material. In these areas, we find the problems of how to carry out the conservation works of the great built heritage, usually adobe and rammed earth, as well as the lack of skilled people at all levels, from designer to masons, because it is a forgotten technique.

_GUTOVIC, M.; KLIMESCH, D.S.; and RAY, A.: Strength development in autoclaved blends made with OPC and clay-brick waste, Construction and Building Materials, Volume 19, Issue 5, June 2005, Pages 353-358.

Abstract: This article reports findings on the strength development of autoclaved OPC-clay-brick blends, where a variety of clay-brick types were used. The strength variations are explained in terms of differences in chemical and mineralogical properties of clay-brick as well as the phases formed during autoclaving. An overall increase in strength was characteristic for 30-60 mass% for all clay-brick type additions. The optimum compressive strengths were achieved at 50 mass% clay-brick additions, where highly crystalline 1.1 nm tobermorite coexisted with "fibrous C-S-H".

_HALL, Matthew; and DJERBIB, Youcef: Moisture ingress in rammed earth: Part 1 - The effect of soil particle-size distribution on the rate of capillary suction. Construction and Building Materials, Volume 18, Issue 4, May 2004, Pages 269-280.

Abstract: The novel initial rate of suction (IRS) andlsquo;wickandrsquo; test has been presented, and is suitable for determining the rate of capillary moisture ingress in unstabilised rammed earth that slakes on contact with water. Experimental testing was performed using the andlsquo;wickandrsquo; test to investigate the effect of soil particle-size distribution on moisture ingress in rammed earth. Rammed earth generally absorbs much less water due to capillary suction, and at a slower rate, than conventional masonry building materials such as bricks and concrete. Moisture ingress in rammed earth, due to capillary suction, increases linearly per unit inflow surface area against the square root of elapsed time . The particle-size distribution of the soil is critical in determining the rate at which moisture may ingress. In a suitable soil, the ratio between the total specific surface area (SSA) of the aggregate fraction and the mass of the binder fraction appears to be positively linked with the rate of capillary suction in rammed earth. Experimental data have been included.

_KERALI, A. G.: In-service deterioration of compressed earth blocks. Geotechnical and Geological Engineering, 23, 4, Pages 461-468, 4 2005.

Abstract: With the increasing backlog of shelter in most of the third world, attempts are being made to evolve low-cost but durable walling units. The introduction of compressed earth blocks (CEBs) some 50 years ago was seen as a major milestone. These blocks are made by compressing a damp mix of soil (90–95%), and cement (5–10%) to form strong and dense blocks used for walling. While considerable knowledge is available regarding their initial performance characteristics, little research has so far been conducted on their long-term durability and deterioration due to prolonged exposure to environmental factors. It is now widely recognised that rapid and premature deterioration does take place when the material is used in unrendered walling in most humid tropical environments. Premature defects such as roughening, pitting, erosion, volume reduction, cracking as well as crazing, etc., have all been witnessed within periods ranging from 1 month to 5 years after completion of construction. This paper reports on recent research conducted in Uganda where in-service defects such as pitting and cracking where measured directly from existing exposed walling to quantify their nature and extent. It can be concluded that the greatest deterioration was found to occur on east–west facing facades, and on the lower wall sections. The deterioration was also found to correspond to the age of the structure.

_MEUKAM, P.; JANNOT, Y.; NOUMOWE, A.; and KOFANE, T. C.: Thermo physical characteristics of economical building materials, Construction and Building Materials, Volume 18, Issue 6, July 2004, Pages 437-443.

Abstract: An experimental study was carried out in order to determine the properties of local materials used as construction materials. Cement stabilized compressed bricks were tested. The thermal properties of lateritic soil based materials were determined. The objectives of work reported in this paper are to determine the effect of addition of pozzolan or sawdust in lateritic soil brick on the thermal properties. It was shown that the effect of incorporation of pozzolan or sawdust is the decreasing of the thermal conductivity and density. The moisture content of these materials can modify their thermal performance. Thus a study of the influence of the water content on the thermal conductivity k and the thermal diffusivity [alpha] is presented. The thermal conductivity as a function of water content increases rapidly between 0 and 12% for lateritic soil. The thermal diffusivity curve presents a maximum for values of water content of 15% for lateritic soil and 8% for lateritic soil-pozzolan or lateritic soil-sawdust. However, the composite materials used for building shielding must present sufficient mechanical strength to be suitable for constructions. According to the experimental results the effect of adding cement or pozzolanic stabiliser is expressed in increase of strength of samples studied.

_MOREL, J. C.; and PKLA, A.: A model to measure compressive strength of compressed earth blocks with the "3 points bending test". Construction and Building Materials, Volume 16, Issue 5, July 2002, Pages 303-310.

Abstract: This article proposes a modelling of the andlsquo;3 points bending testandrsquo; on compressed earth blocks (CEB) which will give their compressive strength. This test is already used in situ for CEB quality control. Conventional bending modelling gives the tensile stress at failure using three of the principal assumptions of the Strength of Materials theory, which are in fact not respected here. This is why it is proposed in this paper to model the arch behaviour in compression, which is of primary importance for aspect ratios less than three. Considering a proposed lattice, a behaviour at failure is obtained, close to that of a CEB during a 3 points bending test. The compressive stress at failure is obtained by using this lattice. The validation of the model is given using experimental data. Then the model gives information on the CEB structure such as isotropy, dry density gradient within the CEB, and effect of friction with the experimental direct compression test device, leading to confinement of the sample.

_MOREL, Jean-Claude; PKLA, Abalo; and WALKER, Peter: Compressive strength testing of compressed earth blocks. Construction and Building Materials, In Press, Corrected Proof, Available online 11 October 2005.

Abstract: As with other masonry units, compressive strength is a basic measure of quality for compressed earth blocks. However, as compressed earth blocks are produced in a great variety of sizes the influence of block geometry on measured strength, primarily through platen restraint effects, must be taken into account. The paper outlines current methodologies used to determine compressive strength of compressed earth blocks, including direct testing, the RILEM test and indirect flexural strength testing. The influence of block geometry (aspect ratio), test procedure and basic material parameters (dry density, cement content, moisture content) are also discussed. Proposals for the future development of compressive strength testing of compressed earth blocks are outlined.

_OGUNYE, F. O.; and BOUSSABAINE, H.: Development of a rainfall test rig as an aid in soil block weathering assessment. Construction and Building Materials, Volume 16, Issue 3, April 2002, Pages 173-180.

Abstract: Continual changes in earth technology and the search for new trends in soil block weathering practices have resulted in the development of a rainfall test rig (RTR). By positioning the water-spraying nozzle at a fall-height of 2.0 m and water pressure of 0.5 kg/cm2, the device generates a rainfall intensity of 150 mm/h. On application to the standard flour pellet, the RTR discharges spectra of rainfall in which the range of drop sizes, impact velocities and kinetic energy appropriate to natural rainfall conditions. This was applied to soil block samples arranged on a 0.9-m2 adjustable soil block holder (platform). The results of testing two different soil block samples, stabilised with cement, lime and lime/gypsum at varying proportions and cured at both room temperature (RT) and elevated temperature (ET), under controlled and reproducible artificial rainfall conditions, are briefly discussed.

_OGUNYE, F. O.; and BOUSSABAINE, H.: Diagnosis of assessment methods for weatherability of stabilised compressed soil blocks. Construction and Building Materials, Volume 16, Issue 3, April 2002, Pages 163-172.

Abstract: Stabilised compressed soil blocks (SCSBs) in building fabrics are susceptible to deterioration if their choice conflicts with the required functionality. Various investigators to determine the weatherability potential of these materials have applied different assessment procedures, observed to be fraught with limitations. Such procedures have produced results that often conflict with real life situations. The current inadequately defined weatherability performance measurements require more qualitative descriptions of the field conditions. The underlying premise in this paper is to review the practicality or otherwise of the various testing techniques. The paper highlights some generalised framework essential for SCSBs weatherability assessment method that are rational, explicit enough and applicable to specific and almost all intended uses under a broad range of environmental concerns.

_REDDY, B. V. Venkatarama; and JAGADISH, K. S.: Embodied energy of common and alternative building materials and technologies. Energy and Buildings, Volume 35, Issue 2, February 2003, Pages 129-137.

Abstract: Considerable amount of energy is spent in the manufacturing processes and transportation of various building materials. Conservation of energy becomes important in the context of limiting of green house gases emission into the atmosphere and reducing costs of materials. The paper is focused around some issues pertaining to embodied energy in buildings particularly in the Indian context. Energy consumption in the production of basic building materials (such as cement, steel, etc.) and different types of materials used for construction has been discussed. Energy spent in transportation of various building materials is presented. A comparison of energy in different types of masonry has been made. Energy in different types of alternative roofing systems has been discussed and compared with the energy of conventional reinforced concrete (RC) slab roof. Total embodied energy of a multi-storeyed building, a load bearing brickwork building and a soil-cement block building using alternative building materials has been compared. It has been shown that total embodied energy of load bearing masonry buildings can be reduced by 50% when energy efficient/alternative building materials are used.

_REDDY, B.V. Venkatarama; and LOKRAS, S.S.: Steam-cured stabilised soil blocks for masonry construction, Energy and Buildings, Volume 29, Issue 1, December 1998, Pages 29-33.

Abstract: Energy-efficient, economical and durable building materials are essential for sustainable construction practices. The paper deals with production and properties of energy-efficient steam-cured stabilised soil blocks used for masonry construction. Problems of mixing expansive soil and lime, and production of blocks using soil-lime mixtures have been discussed briefly. Details of steam curing of stabilised soil blocks and properties of such blocks are given. A comparison of energy content of steam-cured soil blocks and burnt bricks is presented. It has been shown that energy efficient steam cured soil blocks (consuming 35% less thermal energy compared to burnt clay bricks) having high compressive strength can be easily produced in a decentralised manner.

_WALKER, P. J.: Strength, durability and shrinkage characteristics of cement stabilised soil blocks. Cement and Concrete Composites, Volume 17, Issue 4, 1995, Pages 301-310.

Abstract: The paper outlines results of a comprehensive investigation undertaken to assess the influence of soil characteristics and cement content on the physical properties of stabilised soil blocks. The dry density, compressive and flexural strength, durability and drying shrinkage of over 1500 block tests are outlined in the paper. Experimental results are compared with current specifications and used to develop empirical guidelines for cement content requirements for a range of soil plasticity characteristics. An empirical relationship between compressive and flexural strength is proposed as a simple means of field assessment.

_ZAVONI, Ernesto A. Heredia; BERNALES, Juan J. Bariola; NEUMANN, Julio Vargas ; and MEHTA, Povindar K. : Improving the moisture resistance of adobe structures. Materials and Structures, 21, 3, Pages 213-221, 8 2006.

Abstract: Mud plasters or soil stuccos are commonly used to protect adobe walls from water erosion. Due to drying shrinkage cracks and high permeability, the commonly used soil stuccos are not durable. Stucco stabilizers such as Portland cement, lime and asphalt are expensive. An experimental study undertaken to evaluate some of the locally available, inexpensive, stucco stabilizers is reported here. Crack-resistant stucco compositions made with a cactus solution are described and techniques to make the stucco surface impervious to water are discussed. The results of durability tests involving simulated rainfall on stucco panels, and stucco applied to adobe walls are presented. Recommendations for field practice are also given.

References

[1] http://www.rammedearth.com/re.html. last visited 20/12/2011.

[2] Minke, Gernot. 2000. Earth construction handbook. The Building Material Earth in Modern Architecture. Southhampton [UK] ; Boston : WIT Press,

[3] http://inhabitat.com/sacred-soil-the-rammed-earth-reconciliation-chapel-in-berlin/. *last visited*, 20/12/2011

[4] De Garrido. Luís. 2008. Análisis de proyectos de arquitectura sostenible. Ed MacGraw-Hill.

[5] http://calearth.org/building-designs/what-is-superadobe.html. last visited 20/12/2011.

[6] Montgomery, D. M., Sollars, C. J., Perry, R., Tarling, S. E., Barner, P., Henderson,E., 1991, Treatment of organic-contaminated industrial wastes using cement-based stabilization/solidification. II. Microstructural analysis of the organophilic clay as a pre-solidification adsorbent. Waste Management and Research, vol. 9, pp.113-125.

[7] Houben, H. and Guillaud, H.. 1995. Traité de Construction en Terre. Editions Parenthèses, Marseille, France, *ISBN* 2-86364-041-0.

[8] Gooding, D.E. and Thomas, T.H. (1995) The Potential of Cement-stabilised Building Blocks as an Urban Building Material in Developing Countries. United Kingdom: University of Warwick, School of Engineering. DTU (Development Technology Unit) Working Paper No. 44. April 2009.

[9] Houben, H. and Boubekeur, S. (1998) *Compressed Earth Blocks: Standards Guide.* Paris, France: CRATerre-EAG Publications.

[10] Kerali, A. G.: In-service deterioration of compressed earth blocks.Geotechnical and Geological Engineering, 23, 4, Pages 461-468, 4 2005.

[11] Houben, H. and Guillaud, H.: Earth Construction - a comprehensive guide, CRATerre-EAG, Intermediate Technology Publications, London, 1994.

[12] STANDARDS AUSTRALIA . 2002. HB 195: The Australian earth building handbook, Sydney, Australia: Standards Australia.

[13] SAZ 2001. Standard Code of Practice for Rammed Earth Structures. Standards Association Zimbabwe Standard SAZS 724. Harare: Standards Association of Zimbawbe (SAZ).

[14] Bahar, R.; Benazzoug, M.; and Kenai, S.: Performance of compacted cement-stabilised soil. Cement and Concrete Composites, *Volume* 26, Issue 7, October 2004, Pages 811-820.

[15] Miller, G. A.; and Azad, S.: Influence of soil type on stabilization with cement kiln dust. Construction and Building Materials, Volume 14, Issue 2, 30 March 2000, Pages 89-97.

[16] Sreekrishnavilasam, A.; Rahardja, S.; Kmetz, R.; and Santagata, M.: Soil treatment using fresh and landfilled cement kiln dust. Construction and Building Materials, In Press, Corrected Proof, Available online 7 October 2005.

[17] Al-Rawas AA, Hago A, Al-Sarmi H. Effect of lime, cement and Sarooj (artificial pozzolan) on the swelling.

[18] potential of an expansive soil from Oman. *Build Environ*, 2005;40:681–7.

[19] Dallas L. June 2008. Pavement Life Cycle Benefits through Stabilization with Lime and/or Cement. New Zealand Institute of Highway Technology Recycling and Stabilisation Conference: Better Roads for a Sustainable Environment,

[20] Dallas L. 1996. Assessment of In Situ Structural Properties of Lime-Stabilized Clay Subgrades. *Transportation Research Record,* 1546, 1996: 13-23.

[21] Jianming L., Huachang X., Runhua G. 2008. Method to Predict Resilient Modulus of Lime and Lime-Cement Stabilized Soils Used in Highway. TRB 87th Annual Meeting Compendium of Papers DVD, Paper #08-2188.

[22] Prusinski, J. R. and Bhattacharja, S. 1999. Effectiveness of Portland Cement and Lime in Stabilizing Clay Soils. *Transportation Research Record* 1652, 1999: 215-227.

[23] Jenkins, K. 2008. WHO and WHAT are Baulking at the Implementation of Cold-Recycling Techniques Using Bitumen. New Zealand Institute of Highway Technology Recycling and Stabilisation Conference: Better Roads for a Sustainable Environment.

[24] Browne, A. 2008. Foamed Bitumen Stabilisation in New Zealand—A Performance Review and Lessons Learnt. New Zealand Institute of Highway Technology Recycling and Stabilisation Conference: Better Roads for a Sustainable Environment.

[25] Finberg, C., Quire, D., Thomas, T. 2008. Granular Base Stabilization with Emulsion in Las Vegas, Nevada. TRB 87th Annual Meeting Compendium of Papers DVD, Paper #08-2343.

[26] Thenoux, G., Halles, F., Vargas, A., Bellolio, J. P., Carrillo, H. 2007. Laboratory and Field Evaluation of Fluid Bed Combustion Fly Ash as Granular Road Stabilizer. *Transportation Research Record*, 1989, 2007: 36-41.

[27] V. G. Havanagi, S. Mathur, P. S. Prasad, C. Kamaraj. 2007. Feasibility of Copper Slag–Fly Ash–Soil Mix as a Road Construction Material. *Transportation Research Record*, 1989, 2007: 13-20.

[28] W. F. Barstis, J. Metcalf. 2005. Practical Approach to Criteria for the Use of Lime–Fly Ash Stabilization in Base Courses. *Transportation Research Record*, 1936, 2005: 20-27.

[29] Galán-Marín, C. ;Rivera-Gómez, C.; Petric, J. 2010. Clay-based composite stabilized with natural polymer and fibre. *Construction and Building Materials*, 24 (2010) 1462–1468.

[30] Galán-Marín, C. ;Rivera-Gómez, C.; Petric-Gray, J. 2010. Effect of Animal Fibres Reinforcement on Stabilized Earth Mechanical Properties. *Journal of Biobased Materials and Bioenergy*, Vol. 4, 1–8, 2010.

[31] Daniels, J. L. 2006. Subgrade Stabilization Alternatives to Lime and Cement. North Carolina Department of Transportation.

[32] Ahlrich, R. 2008. Chemically Stabilized Soils. Mississippi Department of Transportation.

[33] Borderick G.P. and Daniel D. E. Stabilizing compacted clay against chemical attack. *J. Geotech. Eng*., 1990;116(10):1549–67.

[34] Li C. Mechanical response of fiber-reinforced soil, PhD thesis, Faculty of the Graduate School of the University of Texas at Austin, 2005.

[35] Jamshidi R, Towhata I, Ghiassian H, Tabarsa R. Experimental evaluation of dynamic deformation characteristics of sheet pile retaining walls with fiber reinforced backfill, *Soil Dyn. Earth. Eng*., 2010; 30: 438–446.

[36] Ghiassian H, Jamshidi R, Tabarsa A. Dynamic performance of Toyoura sand reinforced with randomly distributed carpet waste strips, 4th dec geol earth eng and soil dyn conf, Sacramento, California, USA, 18–22 May 2008.

[37] Abtahi M, Ebadi F, Hejazi M, Sheikhzadeh M. On The Use Of Textile Fibers to Achieve Mechanical Soil Stabilization, 4th Int Tex Cloth Des Conf, Dubrovnik, Croatia, 5-8 October 2008.

[38] Abtahi M, Sheikhzadeh M, Hejazi M. Fiber-Reinforced Asphalt-Concrete Mixtures- a Review, *Cons. Buil. Mat*., 2010; 24: 871-877.

[39] Baker W, The reinforcement of turf grass areas using plastics and other synthetic materials: a review, *Int. Turf. Grass. Soc. Res. J.*, 1997; 8: 3-13.http://EzineArticles.com/3917867.
[40] Kaniraj R, Gayathri V. Geotechnical behavior of fly ash mixed with randomly oriented fiber inclusions, Geot Geomem 2003; 21: 123–149.
[41] Brown B, Sheu S. Effect of deforestation on slopes. *J. Geotech. Eng., ASCE* 1975; 101: 147–165.
[42] Waldron J. Shear resistance of root-permeated homogeneous and stratified soil, *Soil. Sci. Soc. Ame. J.*, 1977; 41: 843–849.
[43] Wu H, Erb T, Study of soil-root interaction, *J. Geotech. Engng. ASCE,* 1988; 114: 1351–1375.
[44] Wu H, Beal E, Lan C. In-situ shear test of soil-root system. *J. Geotech. Eng. ASCE,* 1988; 114: 1376–1394.
[45] Rao J, Jute Geotextile for improving the performance of Highway Embankment on soft Marine Soil", Proc. Nat. Sem Jute based Geotextiles, New Delhi, India, 1996.
[46] Vidal H, The principle of reinforced earth, *High. Res. Rec.*, 1969; 282: 1–16.
[47] Akbulut S, Arasan S, Kalkan E. Modification of clayey soils using scrap tire rubber and synthetic fibers, i., 2007; 38: 23–32.
[48] Juyol P, Sastry G and Rao M. Rehabilitation of a mined area in Himalaya by Geojute and other measures, Proce 5th Int Conf On Geotextiles, Singapore, 1994.
[49] Azeem A and Ati A. Erosion and Control techniques for Slopes of Banks and Cuttings, Ind. Geotech Conf., Calcutta, 1992.
[50] Leflaive E, Soil reinforced with continuous yarns: Texol, 11th Int Conf on Soil Mech and Found Eng., San Francisco, USA, 1985.
[51] Hanafi I, Few C, Partial Replacement of Silica by white Rice Husk Ash in Natural Rubber Compounds; The Effects of Bond, *Iran. Polym. J.,* 1998; 7: 255-261.
[52] Rowell M, Han S, Rowell S. Characterization and Factors Effecting Fiber Properties, *Nat. Pol. and Agr. Comp.*, 2000; 115-134.
[53] Ghavami K, Filho R, Barbosa P, Behaviour of composite soil reinforced with natural fibers, *Cem. Concr. Comp.,* 1999; 21: 39–48.
[54] Savastano H, Warden G, Coutts P. Brazilian waste fibers as reinforcement for cement-based composites, *Cem. Concr. Comp.,* 2000; 22: 379–384.
[55] Nilsson H, Reinforcement of concrete with sisal and other vegetable fibers, Swed Counc for Build Res, Document DIY, Stockholm, Sweden, 1975.
[56] Mishra S, Mohanty K, Drzal T, Misra M, Hinrichsen G. A Review on Pineapple Leaf Fibers, Sisal Fibers and Their Biocomposites, *Macromol. Mat. Eng.,* 2004; 289: 955–974.
[57] KISHORE J, RAO K. Moisture Absorption Characteristics of Natural Fiber Composites, *J. Reinf. Plast. Comp.*, 1986; 5: 141-150.
[58] YUSOFF M, SALIT M, ISMAIL N, WIRAWAN R. Mechanical Properties of Short Random Oil Palm Fiber Reinforced Epoxy Composites, *Sains. Malay.*, 2010; 39: 87–92.
[59] Swamy N. New reinforced concretes, Surry university press, 1984.
[60] Harriette L, the Potential of Flax Fibers as Reinforcement for Composite Materials, Eindhoven University Press, Eindhoven, the Netherlands, 2004.

[61] Salehan I, Yaacob Z, Properties of Laterite Brick Reinforced with Oil Palm Empty Fruit Bunch Fibers, *Pertanika J. Sci. and Tech.*, 2011; 19: 33 – 43.

[62] Li C. Large volume, high-performance applications of fibers in civil Engineering, *J. Appl. Pol. Sci.*, 2009; 83: 660-686.

[63] Mansour A, Srebric J, Burley J. Development of Straw-cement Composite Sustainable Building Material for Low-cost Housing in Egypt, *J. Appl. Sci. Res.*, 2007; 3: 1571-1580.

[64] Bainbridge B, Athene S. Plastered Straw Bale Construction, The Canelo Project Report, Canelo, Arizona, USA, available from: www.osbbc.ca

[65] Key L, Straw as an erosion control mulch, a technical report from US Agriculture Department, Portland, Oregon, No. 49, 1988.

[66] http://www.swicofil.com/products/015bamboo.html.

[67] Qin Y, Xu J, Zhang Y. Bamboo as a potential material used for Windmill Turbine Blades, a technical report available from www.rudar.ruc.dk, 2009.

[68] Lin D, Huang B, Lin S. 3-D numerical investigations into the shear strength of the soil–root system of Makino bamboo and its effect on slope stability, *Ecol. Eng.*, 2010; 36: 992-1006.

[69] http://www.ati-composites.com/PDF/Bulletin-SugarCaneBuildingBoard.pdf

[70] Barone JR, Schmidt WF. Polyethylene reinforced with keratin fibres obtained from chicken feathers. *Compos. Sci. Tech.*, 2005;65:173–81.

[71] Khattak J, Alrashidi M, Durability and mechanistic characteristics of fiber reinforced soil–cement mixtures, *Int. J. Pav. Eng.,* 2006; 7: 53–62.

[72] Santoni L, Tingle S, Webster L. Engineering properties of sand–fiber mixtures for road construction, *J. Geotechl. and Geoenv. Eng.*, 2001; 127: 258–268.

[73] Tang C, Shi B, Gao W, Chen F, Cai Y. Strength and mechanical behavior of short polypropylene fiber reinforced and cement stabilized clayey soil, *Geotex. Geomem.*, 2007; 25: 194–202.

[74] Viswanadham S, Phanikumar R, Mukherjee V. Swelling behavior of a geofiber-reinforced expansive soil, *Geotex. Geomem.*, 2009; 27: 73–76.

[75] Yetimoglu T, Salbas O. A study on shear strength of sands reinforced with randomly distributed discrete fibers, *Geotex. Geomem.*, 2003; 21: 103–110.

[76] Yetimoglu T, Inanir M, Inanir E. A study on bearing capacity of randomly distributed fiber reinforced sand fills overlying soft clay, *Geotex. Geomem.*, 2005; 23: 174–183.

[77] VASUDEV D. PERFORMANCE STUDIES ON RIGID PAVEMENT SECTIONS BUILT ON STABILIZED SULFATE SOILS, Msc thesis, University of Texas at Arlington, 2007.

[78] Musenda C. Effects of Fiber Reinforcement on Strength and Volume Change Behavior of Expansive Soils, M.S. Thesis, The University of Texas at Arlington, Arlington, Texas, 1999.

[79] Puppala J, Musenda C. Effects of Fiber Reinforcement on Strength and Volume Change Behavior of Expansive Soils, *Trans. Res. Boa.*, 79thAnnual Meeting, Washington, USA, 2000.

[80] Consoli C, Prietto M, Pasa S. Engineering behavior of a sand reinforced with plastic waste, *J. Geotech. and Geoenvir. Eng. ASCE,* 2002; 128: 462-472.

[81] Kumar A, Walia B, Mohan J. Compressive strength of fiber reinforced highly compressible clay, *Cons. Buil. Mat.,* 2006; 20: 1063–1068.

[82] Tang C, Shi B, Chen W. Strength and mechanical behavior of short polypropylene fiber reinforced and cement stabilized clayey soil, Geotex Geomem 2006; 24: 1–9.

[83] Orman E. Interface shear strength properties of roughened HDPE, *J. Geotech. Eng. ASCE*, 1994; 120: 758-761.

[84] Bueno S. The mechanical response of reinforced soils using short randomly distributed plastic strips, in Recent developments in soil and Pavement mechanics, Almeida (ed.) Balkema, Rotterdam, 401-407, 1997.

[85] Dutta K, Sarda K, CBR behavior of waste plastic strip-reinforced stone dust/fly ash overlying saturated clay, *Turk. J. Eng. and Envir. Sci.,* 2007; 31: 171-182.

[86] Sobhan K, Mashnad M. Tensile strength and toughness of soil – cement – fly ash composite reinforced with recycled high density polyethylene strips, *J. Mat. in Civ. Eng. ASCE*, 2002; 14: 177-184.

[87] Miller J, Rifai S. Fiber reinforcement for waste containment soil liners, *ASCE J. Envir. Eng.,* 2004; 130: 891-896.

[88] Tutumluer E, Kim I, Santoni L. Modulus Anisotropy and Shear Stability of Geofiber-Stabilized Sands, *Trans. Res. Rec.*, 2004; 1874: 125-135.

[89] Freilich J, Li C, Zornberg G. Effective Shear Strength of Fiber-Reinforced Clays, 9th *Int. Conf. on Geosyn*, Brazil, 2010.

[90] Jadhao D, Nagarnaik B. Performance Evaluation of Fiber Reinforced Soil- Fly Ash Mixtures, 12th Int Conf of Int Assoc for Comp Meth and Adv in Geomech (IACMAG), Goa, India, 2008.

[91] Kumar S, Tabor E. Strength characteristics of silty clay reinforced with randomly oriented nylon fibers, *EJGE* 2003; 127: 774-782.

[92] Chauhan S, Mittal S, Mohanty B. Performance evaluation of silty sand subgrade reinforced with fly ash and fiber, *Geotex. Geomem.*, 2008; 26: 429–435.

[93] black cotton soil subgrade through synthetic reinforcement, *J. the Inst. Eng.*, (India) 2004; 84: 257–262.

[94] Park S, Effect of fiber reinforcement and distribution on unconfined compressive strength of fiber-reinforced cemented sand, *Geotex. Geomem.*, 2009; 27: 162–166.

[95] Park S. Unconfined compressive strength and ductility of fiber-reinforced cemented sand, *Cons. Build. Mat.*, 2011; 25: 1134-1138.

[96] Segetin M, Jayaraman K, Xu X. Harakeke reinforcement of soil–cement building materials: Manufacturability and properties, *Build. Env.,* 2007; 42: 3066–3079.

[97] Boominathan S, Senathipathi K, Jayaprakasam V, Field studies on dynamic properties of reinforced earth, *Soil. Dyn. and Earth Eng.*, 1991;10: 402–406.

[98] Murray T, Farrar M, Temperature distributions in reinforced soil retaining walls, *Geotex. Geomem.*, 1988; 7: 33–50.

[99] Ghazavi M, Roustaie M. The influence of freeze–thaw cycles on the unconfined compressive strength of fiber-reinforced clay, *Cold. Reg. Sci. Tech.*, 2010; 61: 125–131.

[100] Gray H, Al-Refeai T. Behavior of fabric versus fiber reinforced sand, *J. Geotech. Eng. ASCE,* 1986; 112: 809–820.

[101] Ghazavi M, Roustaie M. The influence of freeze–thaw cycles on the unconfined compressive strength of fiber-reinforced clay, *Cold Reg. Sci. Tech.*, 2010; 61: 125–131.

[102] Consoli C, Montardo P, Donato M, Prietto M. Effect of material properties on the behavior of sand–cement–fiber composites, *Ground. Improv.*, 2004; 8: 77–90.

[103] Maher H, Ho C. Mechanical properties of Kaolinite/fiber soil composite, *J. Geotech. Eng.*, 1994; 120: 1381–93.

[104] Elisabeth, L. and Adams, C eds 2000"Alternative Construction: contemporary natural building methods: New York: *John Wiley and sons*, Inc.

[105] Delgado, M. C.; Jiménez; Guerrero; Cañas, I.: The selection of soils for unstabilised earth building: A normative review. Construction and Building Materials, In Press, Corrected Proof, Available online 28 September 2005.

[106] Guillaud, H., and Houben, H. (1994). Earth construction: A comprehensive guide. U.S.A: Intermediate Technology Publications. University of Michigan.

In: Brick and Mortar Research
Editors: S. Manuel Rivera and A. L. Pena Diaz
ISBN: 978-1-61942-927-7

Chapter 3

IN-PLANE BEHAVIOR OF CFRP RETROFITTED MASONRY: EXPERIMENTAL AND NUMERICAL ASSESSMENT

[1]Viviana Carolina Rougier and [2]Bibiana María Luccioni
[1]Department of Civil Engineering, National Technological University, Fac. Reg. Concepción del Uruguay, Concepcion del Uruguay, Entre Rios- Argentina
[2]Structures Institute, National University of Tucumán, CONICET, Tucumán- Argentina

1. ABSTRACT

Masonry buildings are designed to serve a certain lifetime. However, there are several masonry buildings that have been damaged in a shorter time than it was expected, because of different external actions like earthquakes, impact loads, changes in their use or aggressive agents. Moreover, there are many historical buildings that should be preserved as cultural heritage.

Research activities carried out a few years ago concerning the use of fiber reinforced polymers as external reinforcement of masonry walls have shown that this system considerably improves structural stability with a minor impact over foundations. The use of polymeric fiber composites has also proved to be an efficient repairing technique for historical buildings. However, different aspects of this retrofitting system should still be analyzed. This analysis involves placement techniques, anchorage length, amount and layout of the reinforcement, failure modes of the reinforced element and the behaviour of damaged and even collapsed and then repaired masonry elements.

This chapter is concerned with the in-plane mechanical behaviour of solid clay masonry panels and the same panels but reinforced or repaired with carbon fiber reinforced polymer laminates externally bonded to the wall surface. The experimental program involves; compression normal to bed joints and diagonal compression tests on unreinforced, reinforced and damaged and then repaired panels. Numerical simulation is performed using an existing coupled damaged-plasticity model that is calibrated with the experimental results obtained by the authors. Such model allows simulating the behaviour of masonry elements using the mechanical properties of constitutive materials and their layout.

First a summary of the main characteristics of the mechanical behaviour of unreinforced and retrofitted masonry, subjected to in plane loads is presented. Then the experimental tests and their results are discussed and contrasted with existing design models. After that, the numerical model is summarized and the results of several simulations carried out to verify the efficiency of the reinforcement with carbon fiber reinforced polymers as a retrofitting system for masonry elements are presented. Finally the behaviour of full scale masonry walls externally reinforced with bonded carbon fiber reinforced polymer (CFRP) sheets and subjected to in plane shear load are numerically reproduced using the same numerical model.

2. Introduction

Masonry is a composite material made of units, e.g., clay bricks or concrete blocks, and mortar. A high number of factors affect its in-plane behaviour, among which are the mechanical and geometrical properties of units and mortar, the characteristics of their interfaces, the geometrical arrangement of masonry and the quality of workmanship.

Depending on the type of load, masonry presents a complex behaviour normally accompanied by cracking and brittle and sudden failure. Mechanical properties degradation and structural safety loss make the rehabilitation or reinforcement necessary.

Research into the use of fiber reinforced polymers (FRP) in masonry has shown that this system enhances masonry wall performance, increasing final strength and in some cases ductility and stiffness. In this way, brittle behaviour and sudden failure of unreinforced masonry can be avoided. However, in order to improve these intervention techniques, it is necessary to achieve a better description of the mechanical behaviour of unreinforced and reinforced masonry elements under different load conditions. The present analytical and numerical capacity to quantify this retrofitting system efficiency is still rather limited. It can be noted that for any type of loading condition, the mortar is the first component to fail.

After complete degradation of the mortar, a composite made of blocks connected by the laminate is obtained. Thus, strain concentrations between adjacent blocks occur. On the other hand, as the blocks can support reduced tensile stresses, they fail and produce laminate debonding before it can experiment high deformations.

Numerical simulation of that behaviour is not a simple task, and even though in the last years there has been more research on it (Lucciano and Sacco 1997; Eshani et al. 1997; Giambanco et al. 2001; Marfia y Sacco 2001; ElGawady et al., 2006, Cecchi et al. 2005; Gabor et al. 2006; Grande et al. 2008; Milani 2011; Su et al. 2011, Grande et al. 2011), most of the research carried out is based on experimental work.

3. In Plane Behaviour of Uneinforced Clay Masonry Units

3.1. Axial Compression Normal to the Bed Joints

Masonry transmits compressive loads very effectively. As failure occurs by splitting due to transverse tension in the brick caused by the differential lateral expansion of the stiffer brick and more flexible mortar (Page 1978, Gabor 2006), compression capacity is governed by the tensile properties of the units. In case of low stiffness bricks, they tend to expand laterally faster than high stiffness mortar. But the mortar prevents this lateral expansion and it creates a triaxial compression stress state in bricks with the resulting tensile stresses in the mortar. As soon as the principal tensile stress reaches the tensile strength of the mortar, cracks appear and the failure occurs (Rougier 2007, Prakash and Alagusundaramoorthy 2008).

3.2. Shear Behavior

Masonry walls are primarily designed to resist axial loads. However, they are often subjected to in-plane or out- of- plane forces resulting from lateral loads such as earthquakes. The in-plane shear resistance of load bearing unreinforced masonry (URM) walls is provided by the shear bond strength of the mortar and the friction shear due to the vertical load. Under severe earthquakes loads the shear capacity of the mortar is exceeded, resulting in failure of the wall. (Ehsani et al. 1997).

The typical shear failure modes shown by masonry walls subjected to in-plane forces are: sliding along the bed joints, diagonal tension cracking and shear compression failure (Figure 1).

The first two modes are the most common. Sliding along the bed joints is a failure in which fissure appears along the horizontal joints and it is produced when the unit's strength is higher compared to the mortar bond strength. Consequently, cracking occurs on the weakest element, which in this case is the joint. This causes the sliding of the upper part of the wall onto the lower part (Figure 1a).

The second failure mode is generally characterized by a diagonal tension failure. This type of failure normally occurs when tension strength units is low compared with bond strength mortar.

In general, when there are no compression forces or when they are very small, the failure tends to occur following bed and head joints (staggered cracks) (Figure 1b). When compression is applied cracks can cross over bricks and the failure angle becomes dependent on its magnitude. (Gallegos 1993). The probability that the cracks spread through the units increases with the vertical compression load (ElGawady 2004).Shear compression failure is characterized by the cracking of compressed zones at the wall sides and it makes the wall overturn (Figure 1c).

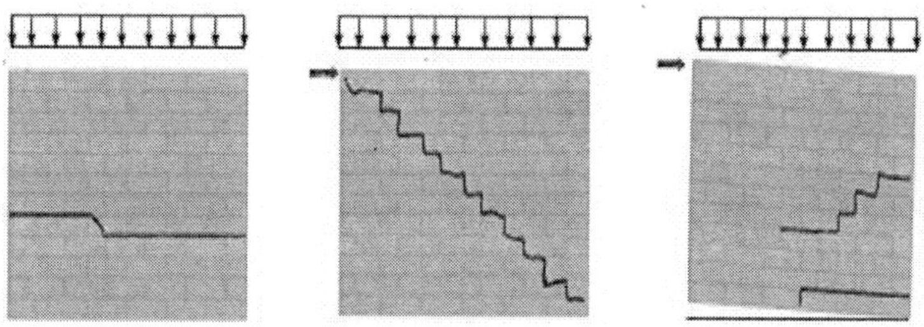

(a) (b) (c)

Figure 1. Modes of failure in URM walls subjected to in plane forces: (a) Sliding along the bed joint. (b) Diagonal tension cracking. (c) Shear compression failure.

4. IN PLANE BEHAVIOUR OF FRP REINFORCED CLAY MASONRY UNITS

The reinforcement and rehabilitation technique with fibre reinforced polymers (FRP) has experimentally proved to be very effective (Saadatmanesh 1997, Valluzzi et al. 2002, Chuang et al. 2003, ElGawady 2004, ElGawady 2005, Gabor et al. 2006, Alcaino and Santa María 2008, Prakash and Alagusundaramoorthy 2008, Papanicolaou 2011, Luccioni y Rougier 2011). There are three common types of FRPs, which are glass (GFRP), aramid (AFRP), and carbon (CFRP).

The properties of FRP's have been summarized in various publications (Hollaway 2010, Shrive 2006), and are continually improving. Recent introductions involve combinations of fibres to provide non-linear responses attempting to achieve some "ductility" in the material. FRP's not only have the advantage of very high strength over ''conventional'' materials, but are also light weight and highly durable in many environments.

In the following sections compression normal to the bed joints and shear behaviour of FRP retrofitting clay masonry units is briefly described.

4.1. Axial Compression Normal to the Bed Joints

Recent researches carried out over masonry specimens subjected to axial compression normal to the bed joints showed different results, depending on material properties of masonry components, retrofitting schemes, amount and layout of the reinforcement.

Prakash and Alagusundaramoorthy (2008) studied the effectiveness of GFRP composite retrofitting on the behaviour of masonry wallettes constructed with high stiffness cement mortar and low stiffness brick, using two layers of GFRP composites.

They found enhancing in load resistance up to 20 %, increasing in stiffness of more than 100 % and reducing up to 40 % in average ultimate strain due to GFRP reinforcing. However, Rougier and Luccioni (2011) tested small solid clay masonry panels retrofitted with one layer of CFRP sheets and strips and they found that FRP reinforcement did not increase strength and stiffness substantially, but it improved ductility and modified the failure mode.

4.2. Shear Behavior

Some researchers have been focused on enhancing the in plane shear capacity of URM walls by means of external reinforcement of composite materials (CFRP and GFRP. Under shear loads and in absence of normal stresses, failure of unreinforced clay masonry panels generally occurs due to mortar joints sliding which causes a very brittle and sudden failure. Depending on the dimensions and configuration chosen, FRP reinforcement modifies this brittle behaviour, avoiding the sliding of the joints and increasing ultimate strength, stiffness and, in some cases, ductility (Valluzzi et al. 2002, Chuang et al. 2003, ElGawady 2004, Gabor et al. 2006, Rougier 2007, Santa María et al. 2008, Papanicolau 2011). The URM retrofitting with FRPs is one method that attempts to improve a structures load carrying capacity and integrity during an earthquake event (Zhuge 2010).

FRP reinforcement is generally made covering the entire wall with fabrics of carbon or glass impregnated in epoxy resin and then applied over the surface previously primed, or through strips applied in the same way (wet process) (Valluzzi et al. 2002, Chuang et al. 2003, ElGawady 2004, ElGawady 2005, Gabor et al. 2006, Alcaino and Santa María 2008, Rougier 2007, Prakash and Alagusundaramoorthy 2008). Some FRP in plane retrofitting schemes for masonry walls are presented in Figure 2.

A quite remarkable gain in strength, an important deformation capability and negligible increasing in-plane stiffness was obtained by Gabor et al. (2006) on hollow brick masonry panels entirely reinforced with bidirectional glass fiber composite. Under dynamic tests, in general, bi-directional surface type materials (fabrics and grids) bonded to the entire wall surface and correctly anchored can help to postpone the three classic failure modes of masonry walls: rocking ("flexural failure"), step cracking and sliding ("shear failures"). Additionally, in some situations, this retrofitting scheme can postpone in-plane collapse by "keeping the bricks together" under large seismic deformations (ElGawady 2004).

Unidirectional strips reinforcement arranged along diagonal direction, improves shear strength and preserves specimen integrity after cracking. The length of FRP bands' anchorage constitutes a very important design variable. The longer it is, the higher the ultimate strength reached by the masonry element will be (Valluzzi et al. 2002, Gabor et al. 2006, Rougier 2007). Santa María et al. (2006) carried out shear tests of full scale masonry walls made of hollow clay bricks and externally reinforcement with CFRP diagonally and horizontally strips.

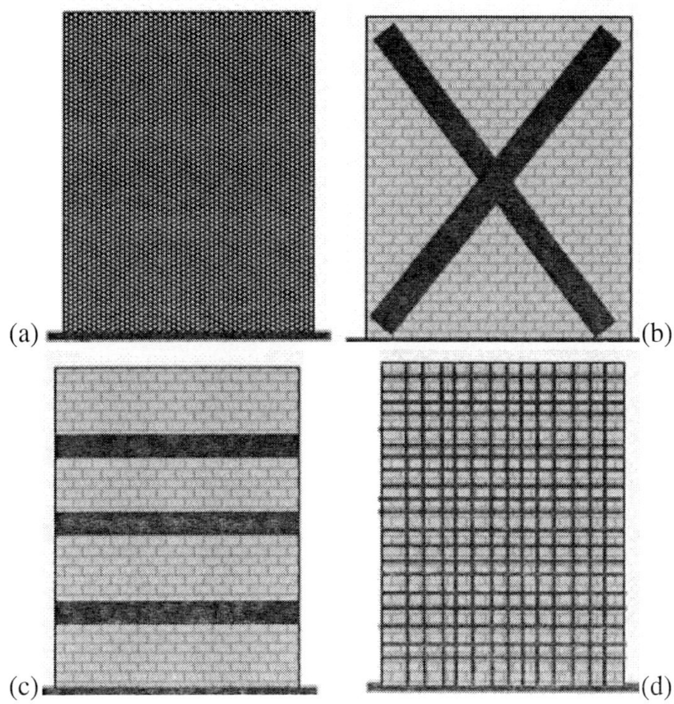

Figure 2. URM in plane FRP retrofitting schemes (a). Composite sheets bonded to entire wall surface. (b) Cross arrangements of FRP plates or laminates. (c) Horizontal strips of FRP laminates. (d) grid disposition of FRP.

They found that both external reinforcement configurations increase the shear strength of the walls, the maximum displacement before failure and the displacement and load of first major crack. The diagonally reinforced wall had a brittle failure with sudden loss of strength, while the horizontally reinforced one showed a less brittle failure.

Carbon plates or fabric strips used in a diagonal pattern by ElGawady (2004) on two hollow clay brick masonry walls subjected to dynamic loads were less successful than fabrics applied on the entire surface of the wall, because premature failure developed (anchorage once and shear-flexure another). Significant increases of the in plane strength, ductility and energy dissipation capacity of retrofitted full scale solid clay brick masonry walls with two and four FRP diagonal strips strengthening were obtained by Chuang et al. (Chuang et al. 2003). The specimens were tested under combined compression and racking cyclic loads.

5. Experimental Study

5.1. General

Two groups of similar panels were constructed and tested according to INPRES-CIRSOC 103 recommendations and trying to reflect material and workmanship qualities similar to those used in actual masonry construction in Argentina: Group I (580x 610x130 mm panels) and Group II (560 x 550 x 125 mm panels).The specimens dimensions differ the dimensions of the solid clay bricks used were not the same. All panels had a 15 mm thick mortar joint.

The strength and elastic properties of bricks, mortar and brick masonry panels were evaluated through testing. Mechanical properties of CFRP composite materials were supplied by manufacture.

The specimens were tested under uniaxial compression normal to the bed joints and diagonal compression. The latter is an in-plane loading condition designed to simulate the resultant of axial load combined with in plane shear load.

For comparison, some of the specimens without reinforcement (control specimens) were tested up to failure, others were reinforced with CFRP and tested up to failure and others specimens without reinforcement were tested until a predefined degree of damage (specified latter for each kind of test), then they were unloaded and repaired with woven carbon fabric, laminated and bonded on site and, finally, they were reloaded up to failure. As no guideline for the design of the FRP strengthening of the panels was available when tests were carried out , a finite element analysis with different retrofitting configurations was performed. The objective of that numerical analysis was both assessing the influence of the different reinforcement schemes on masonry and obtaining the best retrofitting arrangement (Rougier 2007).

5.2. Tests on Clay Bricks and Mortar

The clay bricks were tested under uniaxial compression following IRAM standards (IRAM 12586) to find the material parameters such as modulus of elasticity, Poisson ratio and compressive strength, (Figure 3).

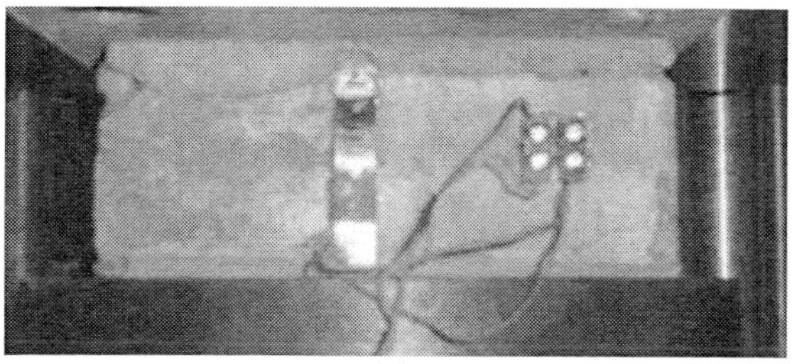

(a)

(b)

Figure 3. Bricks mechanical parameters determination: (a) Uniaxial compression test. (b) Failure mode.

Slightly different types of clay bricks were used for Groups I and II of panels. The dimensions of the bricks for Groups I and II are included in Table 1. The strength, elasticity modulus and Poisson ratio mean values of the two types of bricks used in the tests are also presented in Table 1. They are the average of 20 tests in each case (Rougier 2007).

Mortar used in panels construction was designed according to INPRES-CIRSOC 103 prescriptions, in order to obtain a characteristic compressive strength of 5 MPa at an age of 28 days, which corresponds to a Type N mortar (mix design of 1:1:5 ratio, cement: lime: sand, by volume). The water quantity added to the dry mixture was determined to ensure a good workability of the fresh mortar. Following IRAM standards (IRAM 1622) flexure and compression tests (Figure 4) were performed to obtain the mechanical properties of mortar presented in Table 2. Among the specimens of Group I, two similar mortars can be distinguished: (a) and (b). Mortar (a) was used for specimens of Group I tested under compression normal to bed joints, while mortar (b) was used for specimens of Group I subjected to diagonal compression. The ultimate strength values, elasticity modulus and Poisson ratio in Table 2 correspond to the average of 20 half test specimen of 40 x 40 x 160 mm for each type of mortar. For panels of Group I, the average mortar modulus of elasticity was between 103 % and 160 % higher than that of the bricks under compression. For panels of Group II, the modulus of elasticity of the mortar and bricks was similar. In general, brick masonry built with mortar of higher stiffness than bricks tends to develop lateral tensile stresses in mortar, which produces its early cracking (Rougier 2007).

Table 1. Bricks mechanical properties obtained from tests

Brick	Dimensions [mm]			Elasticity modulus E (MPa)	Comp. ult. strength, $\square_{uc}$ (MPa)	Poisson ratio, $\square$
	Length	Width	Height			
Group I	280	130	50	1662	11.82	0.16
Group II	260	125	55	1400	8.28	0.15

Figure 4. Mortar mechanical parameters determination: (a) Uniaxial compression test. (b) Failure mode.

Table 2. Mortars mechanical properties obtained from tests

Properties	Mortero (Group I)		Mortero (Group II)
	(a)	(b)	
Elasticity modulus E (MPa)	3380	4312	1528
Comp. ult. strength, $\square_{uc}$ (MPa)	6.73	7.72	4.00
Characteristic. Flex. Rupture Modulus (MPa)	2.83	2.83	2.65
Poisson ratio, $\square$	0.21	0.21	0.21

5.3. Carbon Fibre Fabric

Unidirectional carbon fibre fabric with high content of carbon and high modulus and strength, saturated in situ with an epoxy system was used for the reinforcement and retrofitting.

Table 3. Composite mechanical properties

Volume fraction of fibres, k_f	0.3
Longitudinal elasticity modulus, E_l (MPa)	72500
Transversal elasticity modulus, E_t (MPa)	6200
Longitudinal-transversal Poisson's ratio , $\square_{lt}$	0.08
Transversal-longitudinal Poisson's ratio, $\square_{tl}$	0.017
Transversal-transversal Poisson's ratio, $\square_{tt}$	0.20
Longitudinal tensile strength, $\square^{u}_{long}$ (MPa)	960
Transverse tensile strength, $\square^{u}_{t}$ (MPa)	51

The properties of the lamina were obtained from the manufacturer (SIKA) specifications and were numerically validated (Rougier 2007) using a constitutive model for unidirectional fibre reinforced lamina. They are presented in Table 3.

In order to remove mortar remains, the surfaces of the panel were carefully polished with a steel brush and high-pressure air. The carbon fibre fabric soaked up with the epoxy resin was applied to both panel surfaces previously impregnated with the same resin. In all cases, one CFRP lamina of 1 mm final thickness was applied to each surface of the panel.

5.4. Uniaxial Compression Tests

5.4.1. Masonry Panels Description and Experimental Set up

The test setup for compression normal to the bed joint is shown in Figure 5. To ensure uniform pressure and to reduce the lateral displacement, 20 mm thick steel plates with a rubber layer of approximately 8 mm thickness were placed on the top and bottom of the specimens. Vertical and horizontal relative displacement between two fixed points were measured on both sides of the panels and then extrapolated to the total width and height of the panel. Stress–strain curves were derived from load–displacements measurements. In most cases, the measurement equipment had to be removed before failure for safety reasons. Demec gauges with a gauge length of 120 mm and capable of registering displacements of 0.001 mm were used in all measurements.

All the tests were performed with Instron 8204, 600 kN testing machine under increasing load with displacement control. The loading rate was 0.01 mm/s. and it was chosen in order to have a quasistatic test with a total duration of 15 min. Ten masonry panels were cast and tested under compression normal to bed joints. Seven unreinforced specimens were used as control ones and three were externally reinforced with CFRP laminates following two types of retrofitting schemes: total reinforcement and reinforcement with 75 mm width strips (Figure 6). The strip width was chosen based on numerical results (Rougier, 2007). In all cases, both the front and back sides of the panels were reinforced with one CFRP layer and fibres were laid out horizontally, that is, parallel to the bed joints and normal to the load direction. The specimens tested are described in Table 4.

In order to assess the efficiency of the repairing with CFRP, two unreinforced specimens (see Table 1) were tested under uniaxial compression up to approximately 70% of the failure load, when the first cracks could be observed and then unloaded.

Figure 5. Experimental setup for uniaxial compression test on masonry panels.

Table 4. Panels tested

Compression normal to bed joint			
Specimen	Group	Dimension *(l x b x t)*[1] [mm]	Retrofitting / Repairing scheme
MN1	Ia	580 x 610 x 130	-
MN2	Ia	580 x 610 x 130	-
MN3	Ia	580 x 610 x 130	-
MN6	II	560 x 550 x 125	-
MN7	II	560 x 550 x 125	-
MN8	II	560 x 550 x 125	-
MN10	II	560 x 550 x 125	-
MN4Ret	Ia	580 x 610 x 130	Totally retrofitted
MN5Ret	Ia	580 x 610 x 130	Ret 75 mm x 580 mm strips normal to load direction
MN8Rep	II	560 x 550 x 125	Rep 75 mm x 560 mm strips normal to load direction
MN9Ret	II	560 x 550 x 125	Ret 75 mm x 560 mm strips normal to load direction
MN10Rep	II	560 x 550 x 125	Rep 75 mm x 560 mm strips normal to load direction

Ret: retrofitted; Rep: repaired; N: compression normal to bed joint
[1] l = length, b = wide, t = thickness.

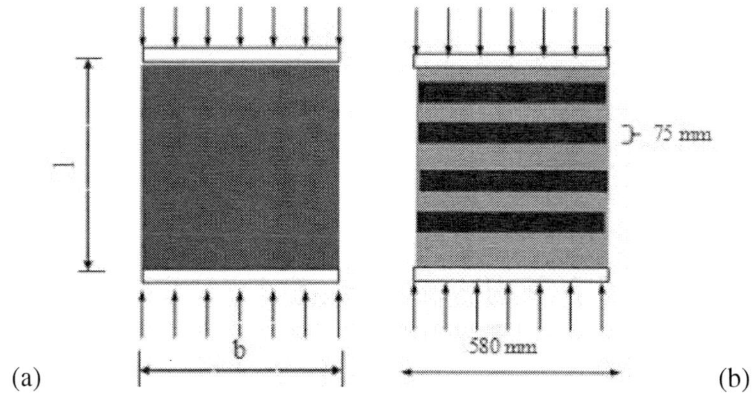

Figure 6. CFRP retrofitted and repaired specimens tested under compression. CFRP. layout. (a) Completely reinforced. (b) Band reinforced.

The panels were repaired with 75 mm width CFRP bands and they were tested again up to failure (MN8Rep and MN11Rep). In all cases, both front and back faces of the panels were repaired with one layer of CFRP bands and fibres were laid out parallel to bed joints and normal to load direction. The layout of the CFRP strips and the number of layers is the same than that used for the retrofitting scheme, see Figure 6b.

5.4.2. Discussion of Test Results

5.4.2.1. Control Unretrofitted Panels

The control specimens tested under compression normal to bed joint failed through vertical cracks along both sides of the panels. Failure was not sudden but ductile even though crushing of the bricks near the loading plates was observed in some cases. The failure pattern is presented in Figure 7.

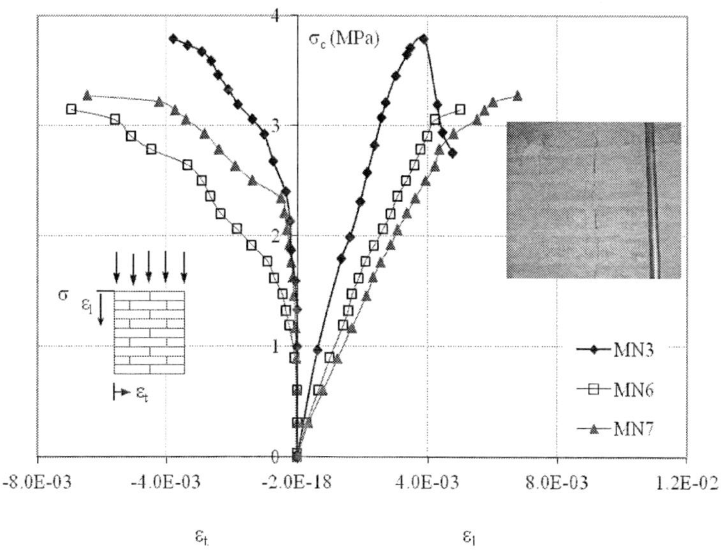

Figure 7. Stress–strain curves for unreinforced specimens under compression and failure pattern.

The stress-strain curves obtained from applied load and the extrapolation of measured displacements are shown in Figure 7. A relatively linear relationship between compressive stress and axial strain, with low values of the lateral strain is observed up to about 50% of the peak strength. Then, a sudden increase in lateral strain and stiffness degradation due to vertical cracks is observed. The integrity of the panels was preserved during the softening part of the stress– strain curve up to the complete failure (Luccioni and Rougier 2011).

Table 5. Panels tested under compression

Specimen	P_{rot} (kN)	[1]□$_u$ (MPa)	[2]E (MPa)	[3]AE/l (kN/mm)	Failure mode
MN1 (Ia)	192	2.55	-		Vertical cracks on front and back sides
MN2 (Ia)	262	3.47	-		Vertical cracks on front and back sides
MN3 (Ia)	286	3.61	1066	145.70	Vertical cracks on front and back sides
Average	246.70	3.21	1066	145.70	
MN6 (II)	216	3.14	821	100.79	Vertical cracks on front and back sides and crushing near supports
MN7 (II)	226	3.29	761	93.43	Vertical cracks on front and back sides and crushing near supports
MN8 (II)	151.20	-	790	97.00	(*)
MN10 (II)	160.00	-	1045	128.30	(*)
Average		3.25	854	105.00	
MN4Ret (Ia)	308	4.08	1126	156.31	Vertical cracks all along the lateral vertical sides
MN5Ret (Ia)	282	3.74	1170	162.42	Small vertical cracks in front, back and vertical sides
MN8Rep (II)	210	3.05	832	104.00	Failure of bricks with pull out of the composite lamina.
MN9Ret (II)	238	3.46	752	94.00	Small vertical cracks in front, back and vertical sides.
MN10Rep (II)	260	3.78	1602	199.82	Failure of bricks with pull out of the composite lamina

Ret: retrofitted; Rep: repaired; N: compression normal to bed joint.
[1]Ultimate strength; [2] Young's modulus; [3]Axial secant stiffness at failure.
(*) No failure mode because they were unloaded before failure and repaired.

The influence of mortar properties in the specimen's response is also clear in the stress-strain curves shown in Figure 7. The curve corresponding to MN3 shows greater stiffness and strength and this difference is mainly due to the mortar´s greater stiffness and strength (see Table 2). As a counterpart, it may be observed that increasing mortar strength leads to a more brittle behavior. For specimens MN1 and MN2 only the ultimate load was recorded (Luccioni and Rougier 2011).

The failure load, ultimate strength, Young's modulus obtained from the tests and the failure mode are shown in Table 5. The average modulus of elasticity was found to be 1066 MPa and 854 MPa for panels of Groups I and II, respectively. The low modulus of elasticity of masonry is mainly due to the low strength of bricks which were procured from the local suppliers.

5.4.2.2. CFRP Composite Retrofitted and Repaired Panels

The failure patterns obtained for the different specimens tested are shown in Figure 8 and they are described in Table 5. For total reinforcement, no crack can be observed in front and back sides, while vertical cracks appear in lateral sides.

The failure is more ductile than for unreinforced specimens and the integrity of the panel was preserved even after it is removed from the test machine. The panels that were reinforced with strips showed small vertical cracks in the front, back and lateral sides. The failure of the superficial layers of the bricks produced the pulling out of the CFRP lamina as it can be observed in Figure 8b. The failure was less ductile than for the case of total reinforcement and similar to that of unreinforced panels. In no case, failure of the CFRP laminate was observed (Luccioni and Rougier 2011).

Stress-axial and transverse strain ($\sigma-\ \varepsilon_l$ and ε_t) curves under compression normal to the bed joint for retrofitted specimens are presented in Figure 9. Additionally, the curves for specimens of the same groups but without reinforcement and the ultimate load values are included in aforementioned figure.

In the case of the completely reinforced panel, only the axial displacements were recorded. Although the retrofitted specimens tested did not show an important increase in strength under compression normal to the bed joints, in all cases a more ductile behavior was obtained for retrofitted panels when compared to unreinforced panels. The ultimate strength **and the Young's modulus** obtained from the tests are shown in Table 5. The strength increase was about 22% for the full retrofitted specimen, 12% for specimen MN5Ret and 5% for panel MN9Ret, both retrofitted with CFRP strips (Luccioni and Rougier, 2011).

(a) 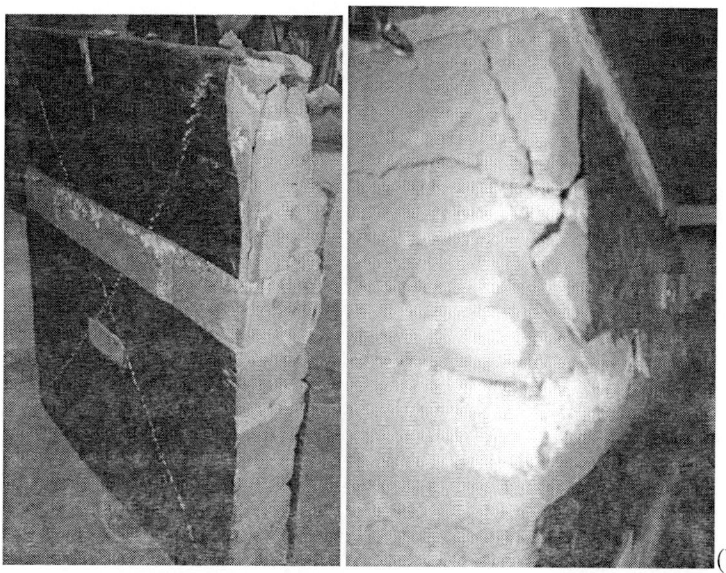 (b)

Figure 8. CFRP retrofitted specimens tested under compression. Failure mode. (a) Completely reinforced (MN4Ret). (b) Band reinforced (MN5Ret).

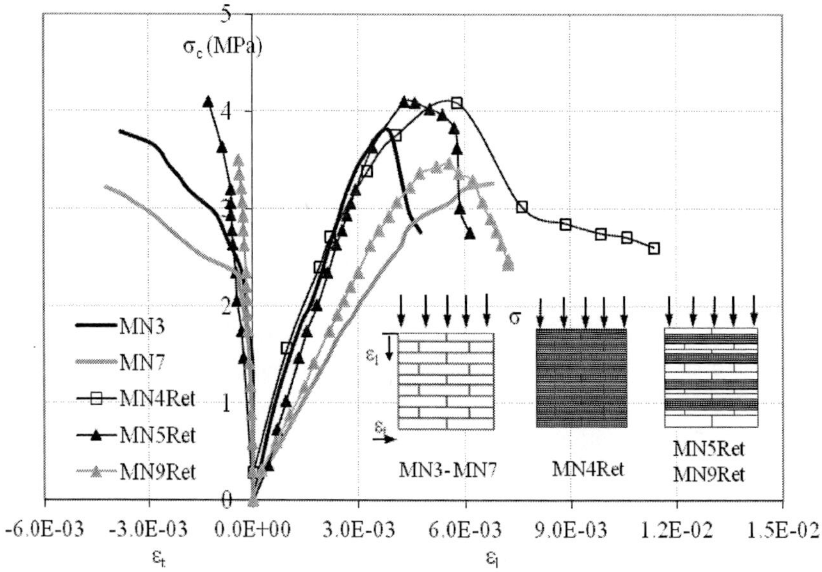

Figure 9. Stress–strain curves for CFRP retrofitted specimens under compression. 580x 610x130 mm panels) and Group II (560 x 550 x 125 mm).

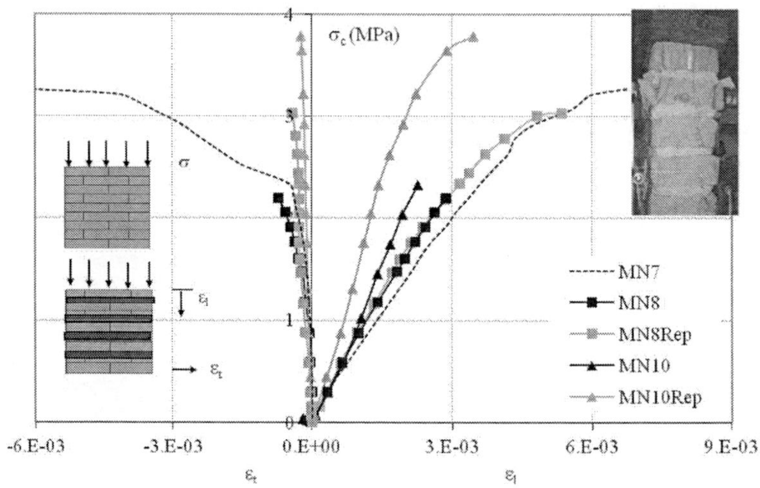

Figure 10. Stress–strain curves for CFRP repaired specimens under compression.

The axial secant stiffness at the failure, AE/l, (A= cross-sectional area of panel, E = Young's modulus and l = length of panel) of the retrofitted specimens, MN4Ret and MN5Ret was slightly higher than the control specimens (Table 5) and the increase in axial secant stiffness of the panels due to retrofitting was up to 7% and 11.5%, respectively. The axial secant stiffness of specimen MN9Ret is 94 kN/mm while average axial secant stiffness of control specimens of the same group is 105 kN/mm.

An increase in axial deformation capability of about 240% was obtained for the totally reinforced panel while this increase decays to 12% and 22% for the specimens reinforced with strips, (see Figure 9). Sudden increase in transverse strains, corresponding to the opening of the mortar joints was prevented with CFRP strips retrofitting (Luccioni and Rougier, 2011). The failure of the repaired specimens was due to the crushing in the upper support zone that produced the pulling out of the composite lamina including the surface layers of the bricks. No cracks in the front and back sides of the panels could be observed (Figure 10).

The stress-axial and transverse strain ($\sigma - \varepsilon_l$ and ε_t) curves for repaired specimens are also presented in Figure 10. An increase in vertical stiffness due to the repairing can be observed. In the case of MN8Rep the increase in stiffness was almost negligible in comparison with MN11Rep.

These results could be attributed to the variability of properties in masonry itself. More tests should be conducted in order to explain the differences observed. The initial strength was just recovered. Only a small strength increase (10%) was achieved for the specimen MN10. Unlike initially reinforced specimens, in this case, no increase in the deformation capacity was observed (Luccioni and Rougier 2011).

5.5. Diagonal Compression Tests

5.5.1. Masonry Panels Description and Experimental Set up

The set of unreinforced, retrofitted and repaired masonry panels tested under diagonal compression are described in Table 6.

Six specimens were used as control ones and three were externally reinforced with CFRP laminates. Three different possible retrofitting schemes presented in Figure 11 were studied: total reinforcement with fibres laid out normal to load direction and diagonal strips bonded orthogonally to the compressive diagonal and strips parallel to the bed joint. The strip width was chosen based on numerical results (Rougier 2007). One layer of CFRP was used in each face of the specimens in all cases. The different specimens tested are specified in Table 6.

In order to study the effect of bond length on the retrofitting effectiveness, four additional specimens of Group II, retrofitted with CFRP diagonal strips of variable length, were tested under diagonal compression. Only the length of the central band was varied (640 mm, 400 mm, 320 mm). The description of the specimens is presented in Table 6.

Additionally four damaged panels (see Table 6) were repaired with CFRP laminas and then tested again.

The test specimens were built following the specifications in INPRES-CIRSOC 103 for the evaluation of masonry shear strength. According to these prescriptions, test panels should be square and with l greater than 550 mm. In order to apply the compression loading, the corners of the panels should be embedded in two metallic supports with embedding length r greater or equal to 200 mm. The test setup and measurement devices arrangement used for diagonal compression tests are shown in Figure 12. A rubber layer was also inserted between the specimens and the metallic supports to reduce friction between them and the specimens. Nevertheless, in some cases it has been observed that this type of support has an undesirable effect because it can cause a localized failure in coincidence with one of its ends, which spreads producing an anticipated collapse (Luccioni and Rougier 2011).

Relative displacements along the compressed diagonal and tensioned diagonal were recorded between two fixed points located along the diagonal and were later extrapolated to the total length of the diagonals to obtain the total diagonal displacements. All the tests were performed under increasing load with displacement control. The loading rate was decreased to 0.008 mm/s for diagonal compression tests because a sudden failure was expected to occur.

Table 6. Panels tested

Diagonal compression			
Specimen	Group	Dimension *(l x b x t)*[1] [mm]	Retrofitting / Repairing scheme
MD1	Ib	580 x 610 x 130	-
MD2	Ib	580 x 610 x 130	-
MD3	Ib	580 x 610 x 130	-
MD7	II	560 x 550 x 125	-
MD8	II	560 x 550 x 125	-
MD12	II	560 x 550 x 125	-
MD3Rep	Ib	580 x 610 x 130	Rep 70 mm strips normal to load. Central band length: 640 mm
MD4Ret	Ib	580 x 610 x 130	Completely retrofitted
MD5Ret	Ib	580 x 610 x 130	Ret 50 mm strips normal to load. Central band length: 640 mm
MD6Ret	Ib	580 x 610 x 130	Ret 50–60 mm strips / bed joints. Central band length: 640 mm
MD7Rep	II	560 x 550 x 125	Rep 70 mm strips normal to load. Central band length: 640 mm
MD8Rep	II	560 x 550 x 125	Rep 70 mm strips normal to load. Central band length: 640 mm
MD9Ret	II	560 x 550 x 125	Ret 70 mm strips normal to load. Central band length: 640 mm
MD10Ret	II	560 x 550 x 125	Ret 70 mm strips normal to load. Central band length: 640 mm
MD11Ret	II	560 x 550 x 125	Ret 70 mm strips normal to load. Central band length: 320 mm
MD12Rep	II	560 x 550 x 125	Rep 70 mm strips normal to load. Central band length: 640 mm
MD13Ret	II	560 x 550 x 125	Ret 70 mm strips normal to load. Central band length: 400 mm

Ret: retrofitted; Rep: repaired; N: compression normal to bed joint

[1] l = length, b = wide, t = thickness

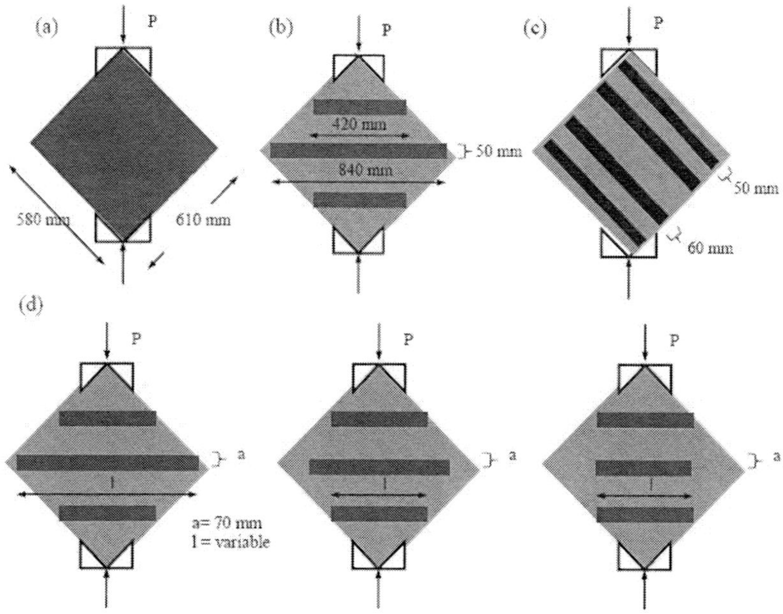

Figure 11. Configuration of strengthening for masonry panels under diagonal compression. (a) Completely retrofitted. (b) Retrofitted with diagonal bands. (c) Retrofitted with.bands parallel to mortar joints. d) Retrofitted with diagonal bands of variable length.

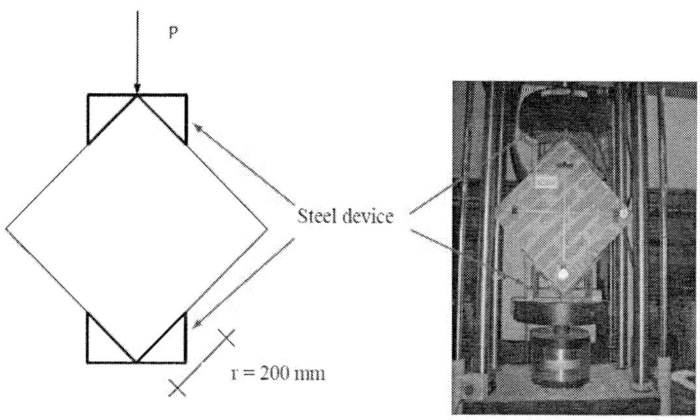

Figure 12. Diagonal compression test setup and measurement devices arrangement.

5.5.2. Discussion of Test Results

5.5.2.1. Control Unretrofitted Panels

The failure patterns of specimens MD1, MD2, MD3 and MD7 (see Table 6) are presented in Figure 13. In general, the type of failure is strongly dependent on the bond strength between mortar and bricks (Luccioni and Rougier 2011). In all cases, a brittle failure with bricks breakage and sliding of mortar joints was achieved. Panel MD1 failed due to the sliding of mortar joint at a load of 50.50 kN. The collapse of panel MD2 at a load of 62.9 kN, was produced by sliding of mortar joint and a localized failure in coincidence with the lower support. In the case of panel MD3, it is clear that failure was initiated in the upper support at a load of 82.90 kN.

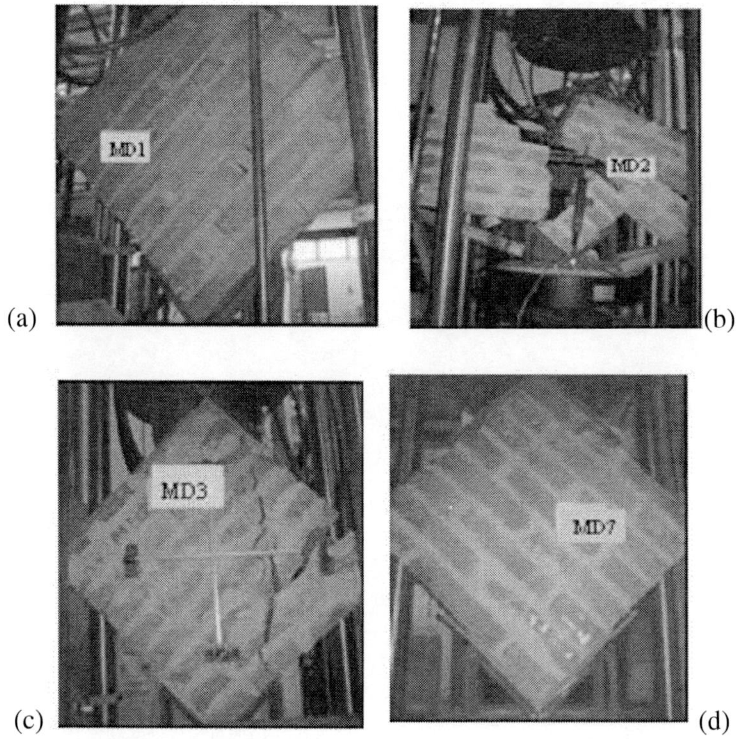

Figure 13. Diagonal compression test. (a) MD1 failure mode. b) MD2 failure mode. (c) MD3 failure mode. (d) MD7 failure mode.

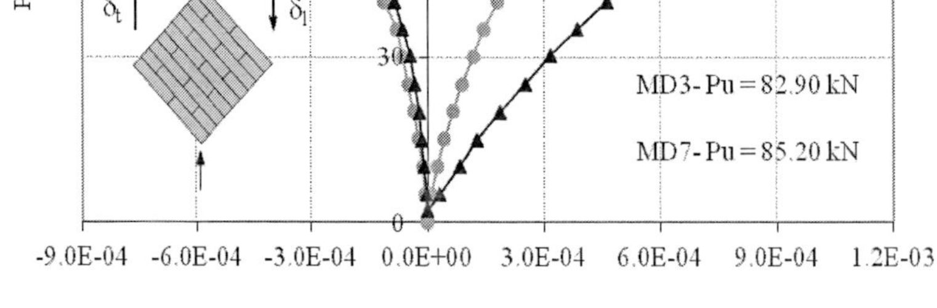

Figure 14. Load–displacement curves for unreinforced specimens under diagonal compression.

In the case of MD7 failure was produced by the formation of a crack along the compressed diagonal and it was less brittle than for the other cases. The ultimate load attained was 85.20 kN. The load–displacement (P - δ_l and δ_t) curves for the unreinforced specimens MD3 and MD7 are presented in Figure 14. Specimen MD7 shows greater deformation capacity mainly due to the differences in mortar properties (see Table 2). The global behavior described by curve representing the applied load as a function of the displacement along the compressed diagonal, is quasi- elastic up to failure load. Ultimate load values are also included in Figure 14 (Luccioni and Rougier 2011).

5.5.2.2. CFRP Composite Retrofitted and Repaired Panels

Different failure modes shown in Figure 15 are obtained for the different retrofitting schemes. In the case of total reinforcement, the integrity of the specimen was preserved without evidence of cracks near to failure load. Failure was produced by bricks crushing near the lower support. Specimen retrofitted with diagonal strips failed due to detachment of the superficial layers of the bricks. This failure was initiated near the upper support and propagates to the lower support.

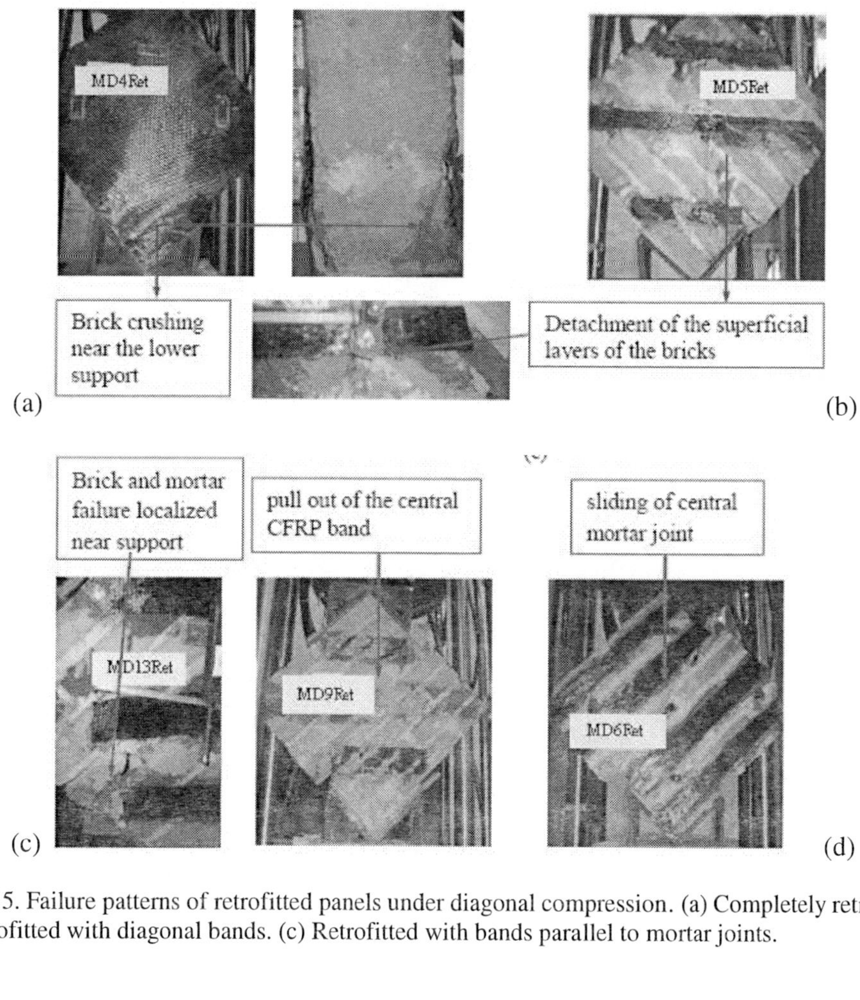

Figure 15. Failure patterns of retrofitted panels under diagonal compression. (a) Completely retrofitted. (b) Retrofitted with diagonal bands. (c) Retrofitted with bands parallel to mortar joints.

In the case of shorter strips, the failure of superficial layers of the bricks that produces the pull out of the CFRP laminas was also observed.

All panels preserved the monolithism after failure and they presented a less brittle failure than the unreinforced specimens, depending on the anchorage length of the central band. (Luccioni and Rougier 2011). A brittle failure initiated in the support and followed by the sliding of central mortar joint was observed in the case of the specimen MD6Ret retrofitted with strips parallel to the bed joints. The stress concentration in the support produced the separation of the superficial layers of the bricks with consequent pull out of the composite lamina. Consequently, the specimen presented an abrupt failure and the expected strength increase was not attained.

Specimen MD9Ret failed abruptly for a load significantly lower than that expected. In this case, failure was produced by the sliding of central mortar joint resulting from the pull out of the central CFRP band evidencing a poor adhesion with masonry (Luccioni and Rougier 2011).

The load–displacement (P – δ_l and δ_t) curves for retrofitted panels under diagonal compression are included in Figure 16 together with those for unreinforced specimens for comparison. In all cases, the measurement equipment was removed before ultimate load was attained to preserve it from breakage due to sudden failure. Figure 16 shows that the total reinforcement significantly increases the stiffness and the strength of the panel. This effect should be considered or modelled when assessing, for example, seismic behaviour of the retrofitted structure. No appreciable increase in stiffness was obtained with diagonal strips but, as it can be observed in Figure 16, the ultimate load was practically duplicated with a significant saving of CFRP material.

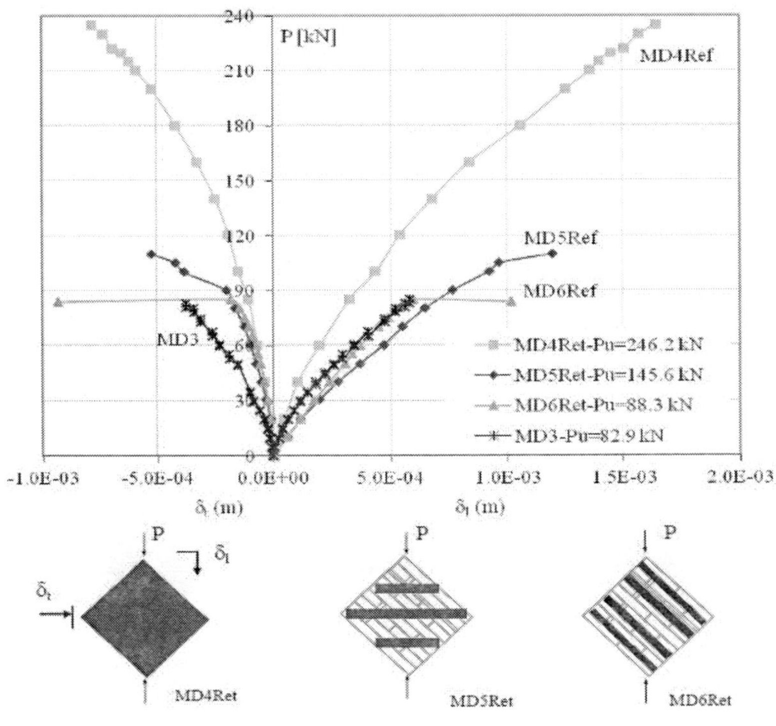

Figure 16. Load–displacement curves under diagonal compression.

Thus the increase of strength is quite remarkable while stiffness is kept practically unmodified. No improvement was achieved with CFRP strips parallel to mortar joints (Luccioni and Rougier 2011). The load–displacement curves for the specimens retrofitted with different CFRP strip lengths and the comparison with the unreinforced panel are presented in Figure 17. In general, strength increases with the increase in the central band length but the improvement is hardly appreciable. All the specimens presented a moderate increase of stiffness and increase of deformation capacity with strip length (Luccioni and Rougier 2011). In order to assess the efficiency of CFRP laminas as a repairing material, four damaged panels (MD3, MD7, MD8 and MD12) were repaired and then tested again. Specimens MD3 and MD7 were loaded under diagonal compression up to failure while specimens MD8 and MD12 were loaded up to 50% of the expected failure load and no damage could be visually observed in this case (Rougier 2007). The cracks were first filled in with cement paste and then the CFRP laminas were adhered to the panels. All the specimens were repaired with diagonal CFRP strips of 70 mm width on the front and back faces of the panels with a layout similar to that presented in Figure 11b (Rougier 2007).

The repaired specimens were reloaded up to failure under diagonal compression. The load–displacement curves obtained for the repaired panels and the comparison with those corresponding to the unreinforced specimens are presented in Figure 18 (a) and (b).

The post-peak response of MD8Rep and MD7Rep show their important deformation capability (Luccioni and Rougier 2011). Actually, all repaired specimens increased their deformation capacity, but the post-peak response is only recorded for MD8Rep and MD7Rep that kept their monolithism almost up to failure. In all the other cases, the measurement equipment was removed before failure to prevent it from damage caused by the collapse of the masonry specimens. Specimens totally damaged and then repaired with CFRP, not only recovered initial strength, but also exhibited a strength increase of about 30%. Brittle failure was initiated near the lower support and produced bricks failure followed by the pull out of the CFRP laminas.

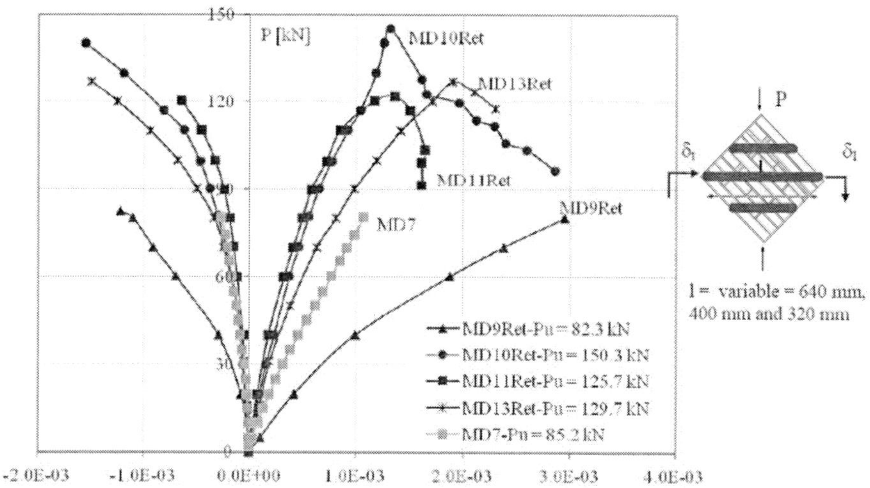

Figure 17. Load–displacement curves for panels retrofitted with CFRP diagonal bands with variable length under diagonal compression.

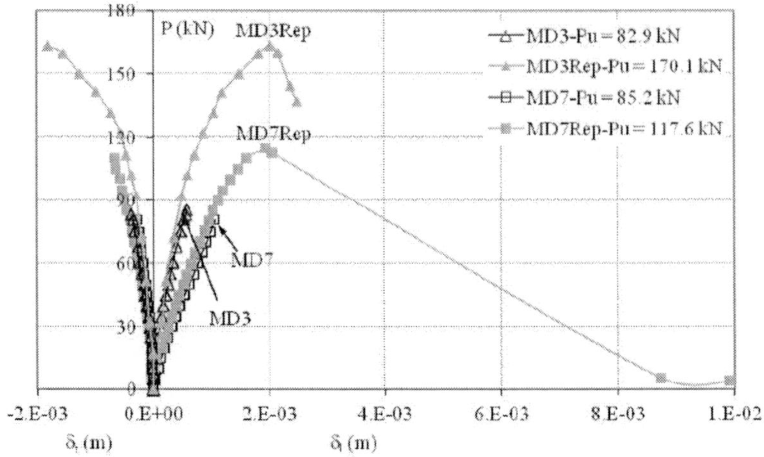

(a)

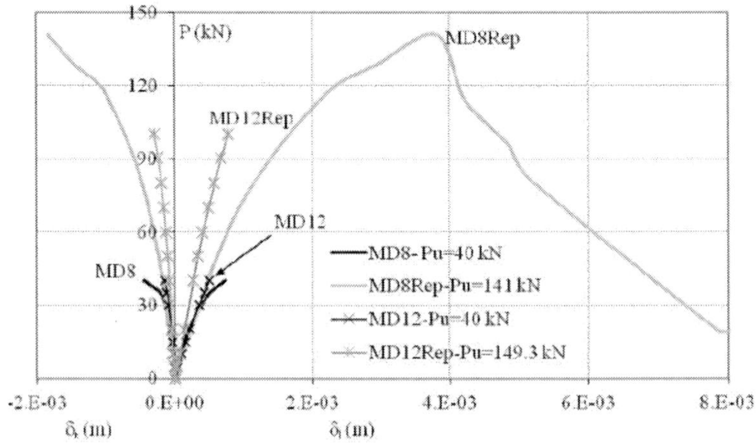

(b)

Figure 18. Load–displacement curves for repaired specimens under diagonal compression. (a) MD3Rep and MD7Rep. (b) MD8Rep and MD12Rep.

Specimens preloaded to 50% of the failure load and then repaired presented an increase of about 70% in ultimate strength. Failure was similar to that presented by the other repaired specimens (Luccioni and Rougier 2007).

5.5.2.3. Comparison between Experimental and Analytically Predicted Shear Strength for FRP-Strengthened Panels

Limited design models have been developed for masonry retrofitted with externally bonded FRP, (Triantafillou 1998; Triantafillou and Antonopoulos 2000; Nanni and Tumialan 2003; ACI 125 2007). Similarly to the case of reinforced concrete, all methods are based on the assumption that the final shear capacity is the sum of two contributions. The first term, F_m, accounts primarily for the contribution of uncracked masonry, whereas the second term, F_{FRP} accounts for the effect of shear reinforcement (Zhuge 2010):

$$F = F_m + F_{FRP} \tag{1}$$

F_m may be calculated according to provisions in existing design codes of each country, so the major difference among different available models is attributed to FRP contribution F_{FRP} (Zhuge 2010). The existing design models to estimate shear capacity of unreinforced masonry retrofitted with FRP can be classified into two categories based on how the models were derived: effective strain-based models and truss analogy-based models. In the case of the effective strain-based models F_{FRP} is determinate by effective strain (□$_{eff}$) of FRP. Triantafillou's model, Triantafillou and Antonopoulos's model and ACI 125 (Zhuge 2010) are base on the effective strain.

The common feature of truss analogy-based models is that F_{FRP} is determined similarly to shear resisted by the stirrups of reinforced concrete beams. None of the models in this category consider the walls strengthened with FRP continuous sheets or in the diagonal directions. Nanni and Tumialan's model (Zhuge 2010) is based on the truss analogy.

Triantafillou model (Zhuge 2010) (hereinafter called the Tr model) is only suitable for the case where FRP laminates are in the form of narrow straps. It is assumed that the contribution of vertical FRP is negligible and the only shear resistance mechanism is associated with the action of horizontal laminates. This action is modeled in analogy to the action of stirrups in reinforced concrete beams. F_{FRP} is calculated as follows:

$$F_{FRP} = \frac{0.7}{\gamma_{FRP}} \rho_h \mathrm{E}_{FRP} \varepsilon_{eff} tL \tag{2}$$

$$\rho_h = \frac{A_{FRP} \cos\alpha}{Lt} \tag{3}$$

where A_{FRP} ρ_h = reinforcement ratio of FRP in horizontal direction; E_{FRP} = modulus of elasticity of FRP in GPa; ε_{eff} = effective FRP strain; A_{FRP} = transverse area of FRP, t = thickness of masonry wall in mm, L= length of masonry wall in mm, □$_{FRP}$ = partial safety factor for FRP in uniaxial tension (1.15, 1.2, and 1.25 for CFRP, AFRP, and GFRP, respectively) and s = angle of inclination of diagonal strips.

The effective FRP strain □$_{eff}$ in Eq.(2) can be determined from Eq.(4), which was developed by the Triantafillou through regression of experimental data for reinforced concrete beams strengthened with FRP in shear.

$$\varepsilon_{eff} = 0.0119 - 0.0205(\rho_h E_{FRP}) + 0.0104(\rho_h E_{FRP})^2$$

$$0 \le \rho_h E_{FRP} \le 1GPa \tag{4}$$

$$\varepsilon_{eff} = 0.0245 - 0.0065(\rho_h E_{FRP}) \qquad \rho_h E_{FRP} \ge 1GPa \tag{5}$$

This model has not been validated by any experimental data and it does not take into consideration the differences between the two cases; wrapped retrofitting in which FRP rupture is the most probable mode of failure and unwrapped retrofitting in which debonding dominates the behavior (ElGawady 2006).

Later on, Triantafillou and Antonopoulos (Zhuge 2010) (hereinafter called the TA model) proposed an improved model in which different axial strain expressions were proposed for different types of FRP and failure modes. The effective strains in FRP are the minimum of the following expressions:

Side or U-shaped CFRP jackets-debonding

$$\varepsilon_{eff} = \min\left[0.00065\left(\frac{f_c^{2/3}}{E_{FRP}\rho_h}\right)^{0.56},\ 0.17\left(\frac{f_c^{2/3}}{E_{FRP}\rho_h}\right)^{0.3}\varepsilon_{FRP,u}\right] \tag{6}$$

where fc is the compressive strength of masonry and $\square_{FRP,u} = 0.015$.

Another model that can be used to calculate the F_{FRP} is the AC125 (Zhuge 2010) that proposes the following expression for F_{FRP}:

$$F_{FRP} = 2\rho_h f_j t_{FRP} H \sin^2\theta$$
$$f_j = 0.004E_{FRP} \le 0.75 f_{FRP,u} \tag{7a}$$

$$F_{FRP} = 0.75\rho_h f_j t_{FRP} H \sin^2\theta$$
$$f_j = 0.0015E_{FRP} \le 0.75 f_{FRP,u} \tag{7b}$$

where f_j = axial force in FRP; $f_{FRP,u}$ = ultimate tensile strength of FRP; t_{FRP} = FRP thickness; H = depth of the wall; and θ=fiber's orientation.

Eqs. (7a) and (7b) have been proposed for rectangular wall sections where fiber bonded on either both sides or one side only. When FRP bonded on both sides of the wall (completed wrapped on all four sides) at an angle θ to the member's axis Eq. (7a) is used. When FRP only bonded to one side at an angle $\theta \ge 75°$ to the member's axis Eq (7b) (Zhuge 2010). Nanni and Tumialan's model (Zhuge 2010) is also based on the truss analogy and in the same form of reinforced concrete (Zhuge 2010):

$$F_{FRP} = K_f\left(\frac{A_{FRP}}{s}\right) f_{fu} d \tag{8}$$

where A_{FRP} = cross-sectional area of FRP shear reinforcement; s = spacing of reinforcement; d = actual depth of masonry in direction of shear considered; and f_{fu} = tensile strength of the FRP, kf = 0.5 (effective stress in the FRP is 0.5 of the ultimate strength).

The comparison among the experimental and the predicted values of the total shear strength obtanined from specimens MD4Ret, MD10Ret, MD11Ret and MD13Ret (see Table 6) is reported in Table 7. Each row containing F is followed by a row containing $Diff_{\text{total}}$ defined as follows (ElGawady et al. 2006):

$$Diff_{total} = \frac{F_{estimated\,according\,to\,the\,proposed\,model}}{F_{Experimental}} \% \quad (9)$$

Table 7. Comparison of Results

Specimens	$\square_h$ (%)	F_{total}. Exp (kN)	Tr	TA	ACI 125
MD4Ret	1.53	174.00	148.50	109	61.26
$Diff_{total}$ (%)			***85***	***63***	***35***
MD10Ret	0.59	106.30	151.50	90.93	61.27
$Diff_{total}$ (%)			***143***	***86***	***58***
MD11Ret	0.59	88.90	151.50	90.93	61.27
$Diff_{total}$ (%)			***170***	***102***	***69***
MD13Ret	0.59	91.70	151.50	90.93	61.27
$Diff_{total}$ (%)			***165***	***99.16***	***67***

[a] **Nanni and Tumialan's model (Zhuge 2010).**

None of the design models distinguish the differences of retrofitting schemes. In fact, Triantafillou (Tr) and Triantafillou and Antonopoulos (TA) (Zughe 2010) are based on FRP strips where F_{FRP} only considers the area fraction of FRP in horizontal direction. However, experimental results showed that when different lengths of the central strip (MD10Ret, MD11Ret and MD13Ret, see Table 6) are used, different ultimate shear strength of panels is obtained.

Triantafillou and TA models (Zughe 2010) use the same formula to calculate the contribution of shear strength from FRP, except the expression for effective strain. Although both adopt formulas from concrete, separate formulas were proposed by TA model for fully wrapped FRP (not a case for masonry retrofitting) and side jacketing which explain why the model gives better results, in the case of diagonal strip retrofitting scheme. For ACI 125 model (Zughe 2010) which is based on FRP continuous sheet, the FRP orientations are considered and a constant strain is used in the model. From Table 7 it can be observed that the model does not lead to a good prediction for the FRP contribution in walls retrofitted with FRP sheets and strips. Nanni and Tumialan's model (Zughe 2010) was excluded as well as the results obtained with it were extremely unconservative.

6. NUMERICAL STUDY

This part of the chapter is devoted to the presentation of the constitutive models used to analyze the behavior of the unreinforced and the strengthened masonry panels and full scale walls. The models have different complexity levels:

1) Detailed modelling, which considers the real configuration of the masonry panels (constituted from bricks and mortar) and the composite reinforcement. This model is applied in both cases (unreinforced and strengthened panels).
2) Simplified modelling, considering the experimentally measured global mechanical parameters of the masonry panels.

In both cases continuous models in which fracture is regarded in a distributed manner were used. The models that were implemented in a non linear plane finite element program are briefly described in the next sections.

6.1. Detailed Modelling of the Unreinforced and Retrofitted Masonry

According to this approach masonry was modelled analyzing the bricks and the mortar separately. For both materials, orthotropic plastic damage models were used (Luccioni and Rougier 2005). In the case of retrofitted or repaired specimens modelled as plane stress states, the sets of brick and CFRP lamina or mortar and CFRP lamina, were modelled with classical ''mixture theory'' (Voigt model) (Luccioni 2006, Toledo et al. 2008).

In this simplified composite model all the components are assumed to have the same strains and the composite stress is obtained as the sum of the component stresses multiplied by their respective volume fractions. The orthotropic model used is based on the assumption that two spaces can be defined (Betten 1988, Luccioni et al. 1995): (a) a real anisotropic space and (b) a fictitious isotropic space.

The problem is solved in the fictitious isotropic space allowing the use of elastoplastic models originally developed for isotropic materials. The isotropic elasto- plastic model used in this paper includes energy-based criteria to make it suitable for brittle materials (Luccioni et al. 1996, Luccioni and Rougier 2005)

6.1.1. Plastic Process

The plastic process is described by a generalization of classical plasticity theory that takes into account many aspects of geomaterials behaviour (Luccioni et al. 1996). The elastic threshold is described by a yield function,

$$F(\sigma_{ij};\alpha_k)=f^p(\sigma_{ij})-\mathrm{K}(\sigma_{ij};\kappa^p)\leq 0 \tag{10}$$

where $f^p(\sigma_{ij})$ is the equivalent stress defined in the tension space and that can take up the form of any of the yielding functions of classic plasticity (Tresca, Von Mises, Mohr Coulomb, Drucker Prager, etc). If this model is used for mortars or bricks a suitable approach for frictional materials as Mohr Coulomb or Drucker Prager must be adopted. $K(\sigma_{ij};\kappa^{p})$ is the yielding threshold and κ^p is the plastic hardening variable.

The following rules are used for the evolution of plastic strains:

$$\dot{\varepsilon}_{ij}^{p} = \dot{\lambda}\frac{\partial G(\sigma_{mn};\kappa^{p})}{\partial \sigma_{ij}} \tag{11}$$

where $\dot{\lambda}$ is the plastic consistency factor and G is the plastic potential function.

The plastic hardening variable κ^p is obtained normalizing energy dissipated by the plastic process to unity and varies from 0, for the virgin material, to 1 when the maximum energy is plastically dissipated (Luccioni and Rougier 2005, Luccioni et al. 1996, Rougier and Luccioni 2007).

6.1.2. Damage Process

The damage threshold is described by a damage function in the following way:

$$F^{d} = f^{d}(\sigma_{ij}) - K^{d}(\sigma_{ij};\kappa^{d}) \leq 0 \tag{12}$$

where $f^{d}(\sigma_{ij})$ is the equivalent tension, $K^{d}(\sigma_{ij},\kappa^{\mathrm{d}})$ is the equivalent damage threshold and κ^d is the hardening variable (Luccioni et al. 1996).

The equivalent tension $f^{d}(\sigma_{ij})$ may be evaluated using known yielding functions (Tresca, Von Mises, Mohr–Coulomb or Drucker–Prager) or any function specially developed for damage (Luccioni et al. 1996, Luccioni and Rougier 2005). The scalar damage variable d varies from 0 to d_c, that is $0 \leq d \leq d_c$ where $0 < d_c \leq 1$ is the level of damage correspondent to the material failure.

6.1.3. Consistency Conditions

The evolution of permanent strains and damage is obtained from the simultaneous solution of the following equations called the consistency conditions of the problem (Luccioni et al., 1996),

$$\begin{cases} \dot{F}^{p} = 0 \\ \dot{F}^{d} = 0 \end{cases} \tag{13}$$

Eq. (13) are two linear equations in $\dot{\lambda}$ y $\dot{d}$ that can be easily solved.

6.2. Simplified Modelling

The detailed modeling of the geometrical structure of the masonry is characterized by great computational effort which limits its applicability to the analysis of small elements (e.g. small laboratory specimens) (Sacco 2008).

A different approach is needed when large actual structures have to be studied, a In particular the equivalent homogenized macro models constitute effective methods to analyze the global response of masonry structures in terms of load and displacements. For this purpose, the use of a simplified model that considers an equivalent orthotropic material, with the overall mechanical properties experimentally obtained over small specimens and without considering the internal geometry of the masonry is proposed. The model is similar to that described in section 6.1 but it is used to analyze the behavior of full scale unreinforced and retrofitted with CFRP laminates masonry walls.

6.3. Composite Materials Modeling

The CFRP laminas applied on both faces of the masonry panels were not modelled independently but together with brick or mortar. This gives place to a composite material consisting of brick or mortar and two sheets of composite. Mixtures theory can be used to model the in plane behaviour of this type of composite where the strains are the same for both component materials.The reinforcement material made up of polymeric matrix and carbon fibres is itself a composite material formed by a matrix with embedded long fibres. To simplify the numerical simulation and to reduce calculus volume, it was modelled with an equivalent homogeneous model. An orthotropic elasto plastic model with the composite properties was used for that purpose. As the properties provided by the manufacture were not enough to model this material, a generalization of mixture theory (Luccioni 2006, Toledo et al. 2008) was used to obtain all the reminding mechanical properties. In this calculus, the properties of the composite were obtained from the properties of the fibres and the epoxy matrix and the fibres volume ratio. In this way, the lamina properties already given by the manufacture were also verified (Rougier 2007).

6.4. Influence of the Composite Strips Configuration on the Global Behaviour of Masonry Panels

The results of a numerical study carried out in order to evaluate the capacity of the different retrofitting schemes to improve the in plane behaviour of solid clay masonry panels, are shown in this section. Compression normal to bed joints tests and diagonal compression tests were simulated with a non linear plane finite element program in which the models described in sections 6.1 and 6.3 were implemented. The results of these simulations are hereafter presented. All the specimens were modelled with three node triangular plane stress finite elements. Masonry elements were meshed distinguishing bricks and mortar elements. The mechanical properties of mortar and bricks were experimentally obtained and are summarized in Table 8 (Rougier 2007, Luccioni et al. 1996). A 1mm thick CRFP retrofitting lamina was applied on each face of the panels in all cases.

Table 8. Bricks and mortar mechanical properties

Specimens	580 x 610 x 130 [mm^3]			560 x 550 x 125 [mm^3]	
Properties	Mortar (a)	Mortar (b)	Brick	Mortar	Brick
Elasticity Modulus, E (MPa)	3380	4312	1662	1528	1400
Poisson's ratio, □	0.21	0.21	0.16	0.21	0.15
Tension ultimate strength, $□_{ut}$ (MPa)	0.673	0.772	0.591	0.54	0.414
Compression ultimate strength, $□_{uc}$ (MPa)	6.73	7.72	10.60	4	8.28
Uniaxial compression elastic threshold, $□_{fc}$ (MPa)	5.60	6.4	-	3.5	-
INITIAL COMPRESSION/TENSION STRENGTH RATIO, R^P_0	10	10	20	10	20
Plastic damage variable for the peak stress, κ^P_{comp}	0.20	0.20	-	0.20	-
Fracture energy, G_f^P (MPa m)	6.01-5	4.01-5	3.0E-5	1.01-5	3.0E-5
Crushing energy, G_c^P (MPa m)	6.01-3	4.01-3	2.0E-3	1.01-3	2.0E-3
Yield criterion	[1]M C	M C	[2]D P	M C	D P
Plastic flow	M C	M C	D P	M C	D P
Damage Criteria	Drucker-Prager	Drucker-Prager	Drucker-Prager	Drucker-Prager	Drucker-Prager
Friction angle for damage function (°)	7°	7°	7°	7°	7°
UNIAXIAL COMPRESSION DAMAGE THRESHOLD, $□^D_C$ *(MPA)*	5.9	7.0	10	3.7	7.5
DAMAGE FRACTURE ENERGY, G_D *(MPA M)*	6.0E-3	6.0E-3	5.0E-2	6.0E-3	5.0E-2

[1]Mohr Coulomb; [2] Drucker-Prager.

Loading and boundary conditions and typical finite element meshes used for a retrofitted panel under compression normal to bed joint and diagonal compression are presented in Figure 19a and b respectively. In the first case, taking advantage of the specimen symmetry only a quarter of them was modelled. The whole panel was modeled in the case of diagonal compression tests.

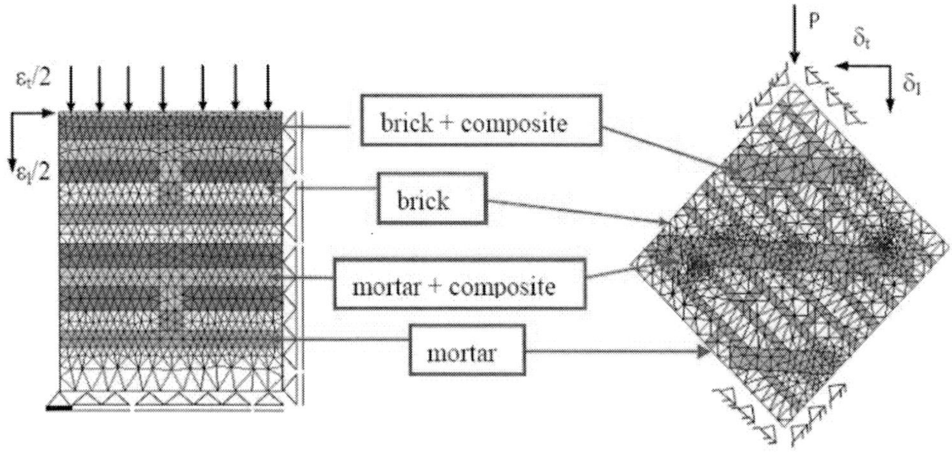

Figure 19. CFRP retrofitting panels loading and boundary conditions and FE mesh. (a) Compression normal to the bed joints, (b) CFRP strips retrofitted panel.

6.5.1. Behaviour under Uniaxial Compression Normal to the Bed Joints

The behaviour of solid clay masonry panels of 580x 610x130 [mm^3] retrofitted with bands parallel to bed joints and normal to the applied load is presented in this section. The study variables were the width of the reinforcement strips and the fabric type, uni and bidirectional wrap. First, masonry panels reinforced with unidirectional bands of constant length, but variable width (50 mm, 70 mm and 560 mm) were analyzed.

Maximum axial load ($P_{máx}$) and maximum axial displacement (δ_l) versus strip width are respectively are represented in Figure 20a and b.

It can be observed that, whatever the strip width is, even in the case of the panel entirely reinforced, the strength is not increased. On the other hand, the vertical deformation capacity is notably increased when the strip width increases.

Finally, the case of entire reinforcement with unidirectional and bidirectional (0º-90º) CFRP wrap was compared. The total fibre volume ratio and thickness of the resulting composite was kept constant and equal to 1 mm.

Half the fibres volume was oriented parallel to bed joints and the other half in the orthogonal direction. The evolution of axial and transverse displacements is shown in Figure 21. It can be seen that there is neither increase in the maximum load nor in the stiffness with any of the two wraps. The axial deformation capacity increases with respect to the unreinforced masonry and it is the same for both types of wrap (Rougier 2007). Furthermore, when the strength was not increased, whatever the strip width is, even in the case of the panel entirely reinforced.

On the other hand, the vertical deformation capacity is notably increased when the strip width increases (Luccioni and Rougier 2011).

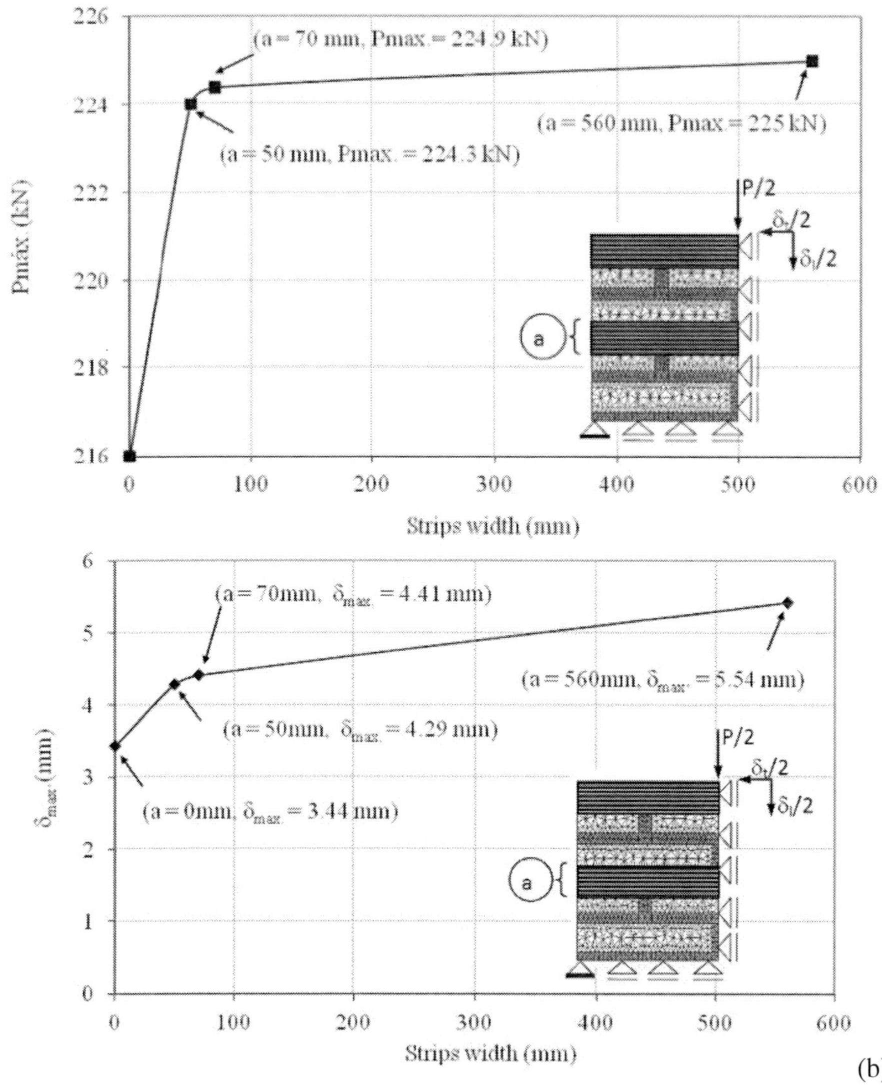

(a)

(b)

Figure 20. Uniaxial compression test normal to the bed joints of CFRP strips reinforced masonry: a) Variation of the maximum axial load with the strips width; b) Variation of the maximum axial displacement with the strips width.

6.5.2. Diagonal Compression

The behaviour of solid clay masonry panels, reinforced with unidirectional PRFC strips with different anchorage lengths is numerically analyzed in this section in order to define the best length to achieve the minimum cost-benefit relationship. Three strip lengths are analyzed: 420 mm (minimum length), 640 mm and 840 mm (total length of the tension diagonal). In all cases, the length of the other two strips is kept constant.

The load versus longitudinal and transverse displacements curves are presented in Figure 23 together with the experimental results for the longest strip. It can be concluded that, the longer the central strip, the higher the maximum load reached.

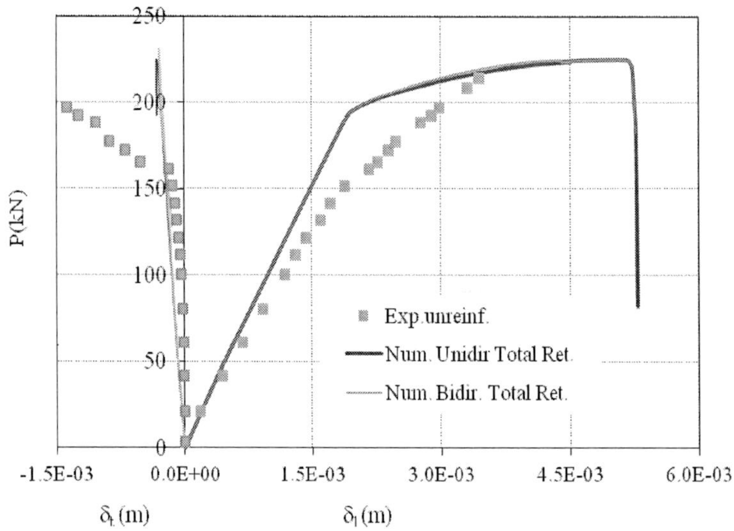

Figure 21. Load versus axial and transverse displacements for CFRP entirely retrofitted panels with unidireccional and bidireccional wrap.

However, between the 420 mm and 640 mm strips there are no significant differences in strength values. Maximum load (Pmax) achieved by the specimen and versus anchorage length of strips reinforcement is presented in Figure 22. It can be seen that the longer the central strip, the higher the maximum load reached. However, between the 420 mm and 640 mm strips there are no significant differences in strength values From Figs. 22 it can be concluded that the bond length L = 420 mm is appropriate and a minimum bond length based on bricks and mortar dimensions is proposed as follows (Luccioni and Rougier 2011),

$$L_{\min} = 2(brick\ length + mortar\ thickness)\cos 45^{\circ} \tag{14}$$

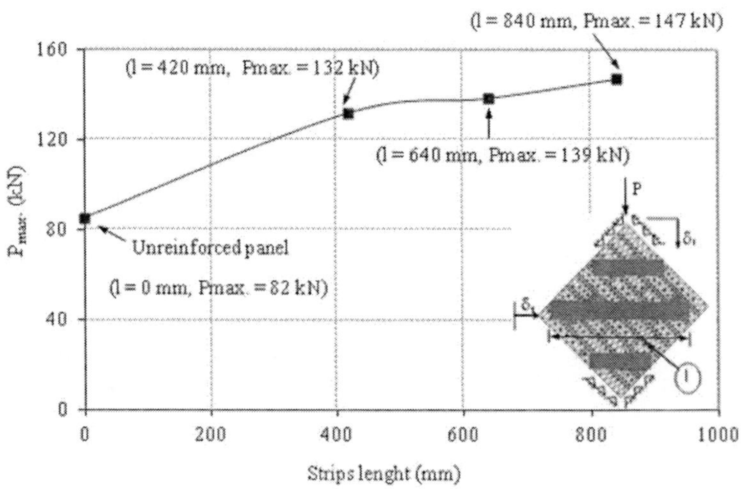

Figure 22. Diagonal compression test corresponding to CFRP strips reinforcement masonry. Variation of maximum load with central strip lenght.

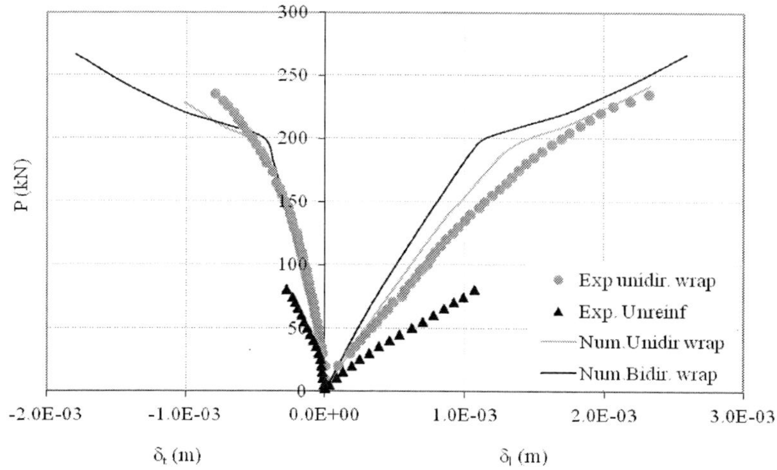

Figure 23. Load versus axial and transverse displacements curves for an entirely CFRP reinforced panel with unidireccional and bidireccional wrap under diagonal compression.

Finally, another comparative numeric analysis on panels entire reinforced with unidirectional and bidirectional carbon wrap and subjected to diagonal compression is presented. The load- axial and traverse displacement curves obtained for both types of reinforcement are presented in Figure 23. The comparison with experimental values of an unreinforced masonry panel and an entirely reinforced panel with unidirectional CFRP is also presented in Figure 23. It can be observed that a higher increase is obtained in masonry stiffness, ductility and strength with the bidirectional wrap than with the unidirectional composite (Rougier 2007). When the study variable is the anchorage length of the central strip, the longer the central strip, the higher the maximum load reached. There is no much difference in stiffness between the different retrofitting schemes (Luccioni and Rougier 2011).

6.6. Behaviour of a CFRP Reinforced Masonry Wall

The behaviour of full scale masonry walls of 1975 mm by 2000 mm x 135 mm, made of solid clay bricks, retrofitted with CFRP and subjected to quasi-static cyclic lateral load is simulated in this section. The objective of this numerical study is to quantify the improvement in the in plane shear strength of masonry walls of actual dimensions CFRP reinforced according to different configurations and with different reinforcement quantities (Rougier 2007).

Due to the size of the wall simulated, masonry is modelled in this case using only one equivalent orthotropic material, with the overall mechanical properties obtained experimentally over small specimens tested under perpendicular and parallel compression to the bed joints and diagonal compression and described in previous sections (Luccioni and Rougier 2011). Aforementioned properties are presented in Table 9. The mechanical properties of CFRP are presented in Table 3. Loading and boundary conditions, retrofitting schemes and FE meshes are shown in Figure 24.

Table 9. Masonry mechanical properties

Properties	
Elasticity Modulus, E (MPa)	2447
Poisson's ratio, □	0.15
Tension ultimate strength, $\square_{ut}$ (MPa)	0.385
Compression ultimate strength, $\square_{uc}$ (MPa)	3.85
Uniaxial compression elastic threshold, $\square_{fc}$ (MPa)	2.90
INITIAL COMPRESSION/TENSION STRENGTH RATIO, R^p_0	10
R_{bc}	1.16
γ	3.
Plastic damage variable for the peak stress, κ^p_{comp}	0.20
Fracture energy, $G_f^{\ p}$ (MPa.m)	3.0E-5
Crushing energy, $G_c^{\ p}$ (MPa.m)	3.0E-3
YIELD CRITERION	Mohr Coulomb

Plane stress triangular finite elements were used and the complete wall was modelled. Firstly, a constant vertical load of 98 kN was applied and then cyclic quasi static lateral load, with three load-discharge cycles, was applied. The vertical load value applied approximately corresponds to the load on the first floor of a three storey building with concrete slabs in two floors and light covering.

The specimens were reinforced on both sides with CFRP according to three schemes: 100 and 150 mm horizontal strips, 200 and 300 mm wide diagonal strips and entire wall reinforcement. The total length of strips was kept constant.

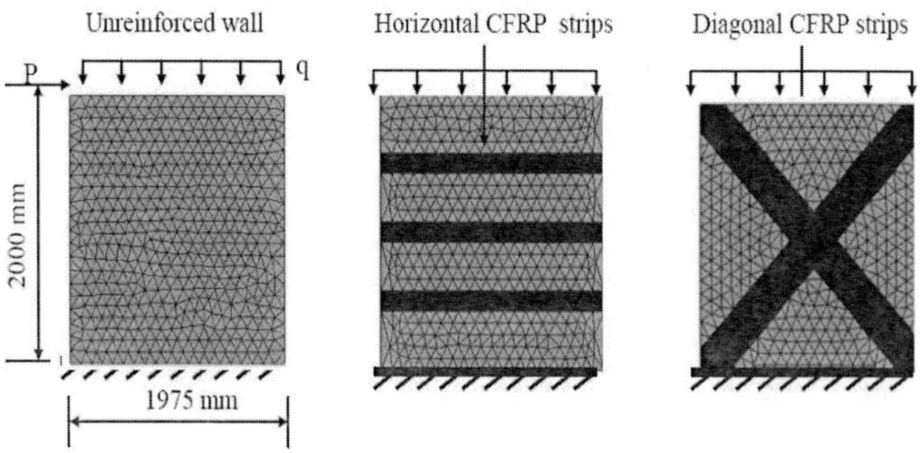

Figure 24. CFRP reinforced walls: Loading conditions, retrofitting schemes and finite element mesh.

For the horizontal strips and total reinforcement, composite material was applied with fibres in horizontal direction, while for diagonal strips; the fibres were applied in diagonal direction.

Lateral load versus total tip lateral displacement curves for the reinforced walls, and the comparison with the curve of an unreinforced wall for increasing lateral loading until failure and solid clay masonry, are represented in Figure 25.

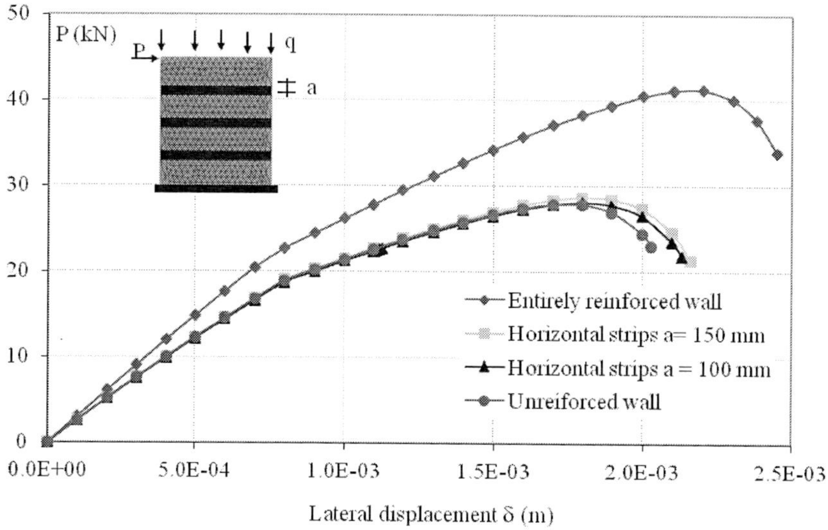

Figure 25. Load versus lateral displacements for masonry walls retrofitted with CFRP horizontal strips with constant length and variable width.

It can be seen in the Figure that, for the horizontal strips, reinforcement slightly improves the shear strength (2%) for a width of 150 mm while a significant strength increase (43%) is achieved when the wall is entirely reinforced. The 100 mm strip practically does not represent any improvement in strength. Regarding ductility, for the three retrofitting schemes an increment of the order of 4%, 6% and 20% for 100, 150 mm and entire reinforcement respectively is observed (Rougier 2007).

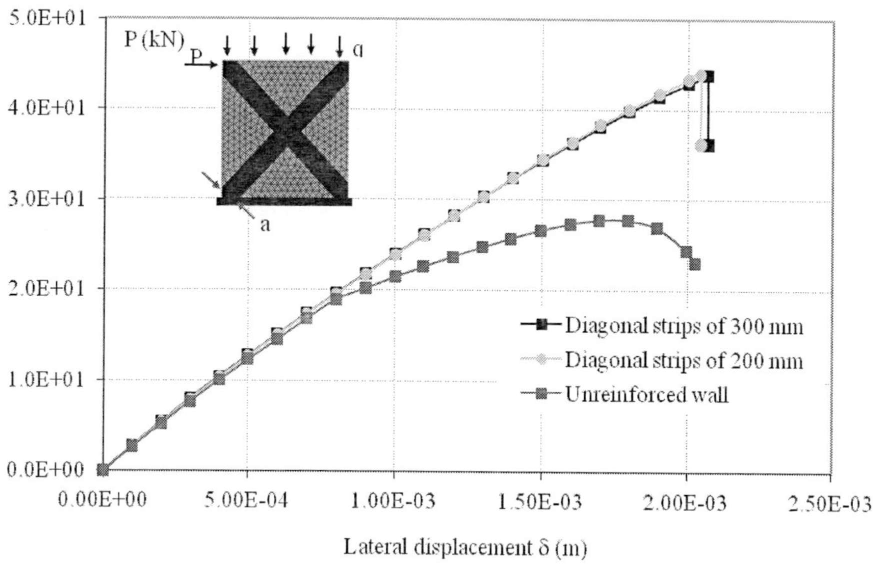

Figure 26. Load versus lateral displacements for masonry walls retrofitted with CFRP diagonal bands with constant length and variable width.

As shown in Figure 26, for 200 and 300 mm diagonal strips, considerable increase in strength, of the order of 57% for both strip widths is achieved. However, no ductility enhancement can be appreciated with respect to the unreinforced wall. The failure turns out to be less ductile than in reinforcement with horizontal bands and entire reinforcement (Rougier 2007). From the comparison of the three reinforcement schemes, it can be seen that, with a considerable saving of material, diagonal strips reinforcement allows reaching a greater ultimate resistance than the total reinforcement (7% more), although deformation capacity is smaller (20% less) (Rougier 2007). Regarding ductility improvement, the 150 mm horizontal reinforcement seems to be the most suitable one because, although the resisting capacity does not increase substantially, it allows obtaining a deformation capacity almost similar to the entire reinforcement (Rougier 2007).

These results are qualitatively coincident with those obtained in the tests by Santa María et al. (Santa María et al., 2006, Alcaino and Santa María 2008), which were carried out over walls of the same size as those presented in this study, but with masonry and reinforcement material properties slightly different. The walls were reinforced with horizontal and diagonal CFRP strips and subjected to in plane cyclic load. Unreinforced wall had a brittle failure with one wide diagonal crack. Retrofitting specimens shown several spread cracks with small thickness. The diagonal reinforced masonry walls had an increase in strength of 63% and 84% and brittle failure with sudden loss of strength. Horizontal reinforced wall had 57% and 61% increase of strength and a less brittle failure. In both cases, the increase was larger in the walls with a larger amount of reinforcement. It can be conclude that, for this type of load, numerical model allows to reproduce with good agreement failure mode of unreinforced and FRP retrofitting full scale masonry walls.

Conclusion

The behaviour and strength of unreinforced masonry made of solid clay bricks present a high variability depending on the mechanical and geometrical properties of units and mortar, the characteristics of their interfaces, the geometrical arrangement of masonry, the boundary conditions and the quality of workmanship. Generally, masonry capacity to support uniaxial compression loading normal to bed joints is good presenting also a ductile failure mode.

The failure of unreinforced specimens under shear loads without normal compression is produced by sliding of the mortar joints and it is both brittle and sudden. In this case, the adherence between units and mortar joints is very important and depend on (en gran medida) the quality of workmanship. If the correct retrofitting scheme is used, CFRP improves the behaviour of masonry, increasing ductility, ultimate strength and stiffness in some cases.

Although retrofitting with CFRP does not significantly increase strength under compression normal to bed joints, it improves ductility and failure mode. The increase in deformation capacity can reach 240% of the original value for total retrofitting. For this type of load, an optimum band width can be calculated. This width can be obtained from tests on specimens retrofitted with CFRP band of different widths and depends on the materials used and on the bricks and joints dimensions. The band should cover at least the joint width and part of the upper and lower bricks.

Depending on the dimensions and orientation, retrofitting with CFRP increases ductility under in-plane shear, prevents the joints sliding and increases the ultimate strength. Total reinforcement significantly increases the stiffness and the strength of the panel and it modifies the failure mode.

Regarding the increase in load capacity and ductility, retrofitting with diagonal CFRP bands is effective and relatively economic when compared with total reinforcement. This fact is important when full sized walls must be retrofitted. Moreover, an optimum length of the reinforcing bands could be obtained in order to balance ductility and strength increase with the amount of CFRP.

Even though more tests are needed, it can be concluded that the specimens repaired behaved satisfactorily. This repairing technique presents the same benefits of the CFRP retrofitting. Most of the specimens repaired with CFRP bands presented an important improvement in ultimate load capacity and a considerable stiffness increase.

From the numerical study performed, it can be concluded that under compression normal to the bed joints, the reinforcement with FRP strips increases neither the masonry strength nor the stiffness. However, with a wide strip, an increase in the deformation capacity can be obtained. For specimens under diagonal compression, a greater anchorage length of the central of the central reinforcement strip improves the resistance capacity of the element.

Due to the cost, extension and complexity that experimental programs may have, it is important to have a numerical tool with which the FRP retrofitted masonry behaviour under different in plane stress states can be satisfactorily reproduced. Once the model has been adjusted, a numerical study allows the analysis of different loading conditions and repair or reinforcement assemblages, which is translated into smaller number of laboratory tests.

ACKNOWLEDGMENTS

The financial support from CONICET, National University of Tucumán and Technological University is gratefully acknowledged. The authors also wish to thank Ings. Jorge Rendón and Paulino Maldonado, from Sika Colombia and Sika Argentina, respectively, for composite material donations, to carry out experimental tests and the collaboration of Ms Amelia Campos in the English revision.

REFERENCES

Betten J. Application of tensor functions to the formulation of yield criteria for anisotropic materials. *International Journal of Plasticity*, 1988; 4: 29-46.

Ehsani M, Saadatmanesh H, Al-Saidy A. Shear behavior of URM retrofitted with FRP overlays. *J. Comp. Constr.*, 1997; 1 (1):17-25.

Cecchi A, Milani G, y Tralli A. In- plane loaded CFRP reinforced masonry walls: mechanical characteristics by homogenization procedures., *Composites Science and Technology* 2005; 65:1480-1500.

Chuang S, Zhuge Y, Wong T, Peters L. Seismic retrofitting of unreinforced masonry walls by FRP strips. In: Pacific conference on earthquake engineering; 2003.

ElGwady M. Seismic retrofit of URM walls with fiber composites. Ph.D thesis, IS-IMAC, EPFL, Switzerland. 2004.

ElGawady M, Lestuzzi P, Badoux M. In-plane seismic response of URM walls upgraded with FRP. *J. Comp. Constr.*, 2005; 9(6):524-535.

ElGawady M, Lestuzzi P, Badoux M. Shear strength of URM walls retrofitted using FRP. *Engineering Structures*, 2006; 28:1658-1670.

Gabor A, Ferrier E, Jacquelin E, Hamelin P. Analysis and modelling of the in-plane shear behaviour of hollow brick masonry panels. *Constr. Build Mat.*, 2006; 20:308–321.

Gabor A, Benani A, Jacquelin E, Lebon F. Modelling approaches of the in-plane shear behaviour of of unreinforced and FRP strengthened masonry panels. *Compos. Struct.*, 2006; 74:277–288.

Gallegos H, Albañilería Estructural. Diseño y Cálculo de Muros", Pontificia Universidad Católica del Perú, 1993.

Giambanco G, Rizzo S, Spallino R. Numerical analysis or masonry structures via interface models. *Computer Methods in Applied Mechanics and Engineering,* 2001; 190: 6493–6511.

Grande E, Milani G, Sacco E. Modelling and analysis of FRP-strengthened masonry panels. *Engineering Structures*, 2008; 30: 1842-1860.

Grande E, Imbimbo M, Sacco E. Simple model for bond behavior of masonry elements strengthened with FRP. *Journal of Composites for Construction*, 2011; 15(3):354-363.

Hollaway L. A review of the present and future utilisation of FRP composites in the civil.

infrastructure with reference to their important in-service properties. *Construction and Building Materials*, 2010; 24:2419–2445.

INPRES-CIRSOC 103. Argentinian design rules for earthquake resistant buildings; 1991.

Instituto Argentino de Racionalización de Materiales. IRAM 12586. Ladrillos y bloques cerámicos para la construcción de muros, Método de ensayo de resistencia a la compresión; 2004.

Instituto Argentino de Racionalización de Materiales. IRAM 1622. Determinación de la resistencia a la compresión y a la flexión del cemento Portland; 1962.

Luciano R., Sacco E. Damage of masonry panels reinforced by FRP sheets, *Int. J. Solids Struct.*, 1998; 35:1723-1741.

Luccioni B., Oller S, Danesi R. Plastic Damaged Model for Anisotropic Materials, *Applied Mechanics in the Americas*, 1995; I:124-129.

Luccioni B, Oller S, Danesi R. Coupled plastic-damaged model. *Computer Methods in Applied Mechanics and Engineering*, 1996; 129:81-89.

Luccioni B, Rougier V. A plastic damage approach for confined concrete", *Computer and Strutures*, 2005; 83:2238–2256.

Luccioni B. Constitutive model for fiber reinforced composite laminates, *Journal of Applied Mechanics*, 2006; 73 (6):901-910.

Rougier V, Luccioni B. Numerical assessment of retrofitting systems for reinforced concrete elements. *Engineering Structures*, 2007; 29:1664-1675.

Luccioni B, Rougier V. In-plane retrofitting of masonry panels with fibre reinforced composite materials. *Construction and Building Materials*, 2011; 25:1772–1788.

Marfia S, Sacco E. Modeling of reinforced masonry elements. International *journal of Solids and Structures*, 2001; 38: 4177- 4198.

Milani G. Kinematic FE limit analysis homogenization model for masonry walls reinforced with continuous FRP grids. *International Journal of Solids and Structures*, 2011; 48: 326-345.

Page A. Finite element model for masonry. *Journal of the Structural Division, ASCE*, 1978; 104 (8): 1267-1285.

Prakash S, Alagusundaramoorthy P. Load resistance of masonry wallettes and shear triplets retrofitted with GFRP composites. *Cement and Concrete Compos*, 2008;30:745–61.

Papanicolaou C, Triantafillou T, Lekka M. Externally bonded grids as strengthening and seismic retrofitting materials of masonry panels. *Constr. Build Mater.*, 2011;25(2):504–14.

Rougier V. Refuerzo de muros de mampostería con materiales compuestos. PhD Thesis, Structures Institute, Nacional University of Tucumán, Argentina, 2007.

Saadatmanesh H. Extending service life of concrete and masonry structures with fibre composites. *Construction and Building Materials*, 1997;11:327–35.

Grande E, Milani G, Sacco E. Modelling and analysis of FRP-strengthened masonry panels. *Engineering Structures*, 2008; 30:1842-1860.

Santa María H, Alcaino P, Luders C. Experimental response of masonry walls externally reinforced with carbon fiber fabrics. Proceedings of the 8th U.S. National Conference on Earthquake Engineering, San Francisco, California, USA, 2006.

Alcaino P, Santa María H. Experimental response of externally retrofitted masonry walls subjected to shear loading. *Journal of Composites for Contruction*, 2008; 12(5):489-498.

Shrive NG. The use of fibre reinforced polymers to improve seismic resistance of masonry. *Construction and Building Materials*, 2006; 20:269–77.

Su Y, Wu Ch, Griffth MC. Modelling of the bond–slip behavior in FRP reinforced masonry. *Construction and Building Materials*, 2011; 25(1):328–34.

Toledo M, Nallim L, Luccioni B. A micro-macromechanical approach for composite laminates. *Mechanics of Materials*, 2008; 40:885-906.

Valluzzi MR, Tinazzi D, Modena C. Shear behavior of masonry panels strengthened by FRP laminates. *Construction and Building Materials*, 2002; 16:409-416.

Zhuge Y. FRP-Retrofitted URM Walls under In-Plane Shear: Review and Assessment of Available Models. *Journal of Composites for Construction*, 2010; 14 (6): 743-753.

In: Brick and Mortar Research
Editors: S. Manuel Rivera and A. L. Pena Diaz

ISBN: 978-1-61942-927-7

Chapter 4

HIGH TEMPERATURE EFFECTS ON MASONRY MATERIALS

Salvatore Russo and Francesca Sciarretta
IUAV University of Venice, Italy

ABSTRACT

Research on masonry structures is very complex and manifold. A quite novel branch of research for masonry materials is here addressed, i.e. the residual behavior of brick, cement mortar and masonry after exposure to high temperatures. Masonry buildings – especially old and historic ones – are often very vulnerable to the attack of fire, and the need for fire protection may be in conflict with preservation of architectural heritage.

The whole matter of high temperature exposure is rich in physical, mechanical and chemical issues, which are mutually connected. Moreover, masonry material is composite and involves a variety of combinations of materials, geometry and textures which are identifiable in masonry buildings through different ages and countries; this leads to possible expensiveness in testing and difficulties in theoretical and experimental modeling. The point of view of the mechanical characterization of materials after high temperature exposure is here taken into consideration; about this peculiar subject, very few theoretical as well as experimental studies are currently available.

First, the state-of-the-art of such research is briefly outlined. Real events as well as standard fire tests often demonstrate that masonry walls and structures can excellently withstand the high temperatures that can be reached during a fire event; on the other hand, the residual mechanical performance of a structure after exposure may need to be evaluated, if high levels of fire safety are required.

Then, the results of recent investigations are here reported, which have given a first contribution to the experimental knowledge of the residual behavior of masonry and separate components (solid clay bricks and cement mortar) after high temperature exposure. The theoretical elaborations of such outcomes have been useful to set a first basis for the establishment and calibration of analytical tools for the prediction of the residual mechanical performance of masonry, brick and mortar.

1. INTRODUCTION

Masonry buildings are often very vulnerable to fire, and – in the frequent case of historic masonry buildings – the need for fire protection may be in conflict with preservation issues; moreover, in the case of fire events in historic-cultural heritage buildings – which are unique and of public interest – the safeguard of the building and the safety of people should be effectively given the same importance [1]. On the other hand, the excellent behavior of masonry walls and structures in fire and high temperature conditions is often highlighted by real events [2] as well as fire testing [3]. However, in some cases, code prescriptions (for both new and existing buildings) require also the structural safety after a fire event to be evaluated, whenever high levels of fire safety are required by the client [4]. Historic-cultural heritage buildings should need such evaluation, in order to ensure the preservation of a public property after fire events; as well, new buildings made of traditional masonry materials need careful design, expensive construction and expert workmanship, resulting in precious and representative works. In Italy, which is a most peculiar case because of the richness in historic architectural heritage, fire events have damaged a number of historic and monumental masonry buildings in the recent years, e.g. the Palazzina of Stupinigi (1989), the Royal Palace and Dome with the Holy Syndone's Chapel (1997) in Turin, the theatre La Fenice (1996) and the 'Molino Stucky' (2003, Figure 1, [2]) in Venice. The vulnerability of buildings of historic-artistic interest is basically due to the following causes [5]: 1) historic buildings' typical features of construction and occupancy (e.g. presence of wooden members, abundance of combustible materials, lack of fire compartments); 2) difficulties in complying with fire prevention measures, because of preservation, landmark and aesthetics requirements; 3) frequent maintenance works (fires in historic buildings very often start in repair works yards); 4) entire historic town centers are frequently landmarked; 5) seriousness of the possible consequences of a fire that threatens unique cultural assets of public interest. In Europe, no specific code regulates the fire protection of cultural heritage; Recommendation of the European Council n°9 of 23 November 1993 deals with the safety of cultural assets in a very general way. Single Countries' codes, like Italian laws for the fire protection of libraries and museums, are generally based on a prescriptive approach. The USA code NFPA 914, updated in 2007, is the first fire protection code that features a performance-based approach to the fire safety of buildings of cultural heritage; in fact, it is aimed to the safeguard of all the historically and culturally relevant elements of the buildings together with fire protection and people's safety [6]. NFPA 914 also contains guidelines to evaluate the fire performance of ancient materials, assuming Harmathy's rules as a model of fire resistance [7] such principles were elaborated in the '60s to set up an approach to the qualitative evaluation of fire resistance as an alternative to direct testing. The problem of evaluating ancient materials' fire performance was also addressed by the U.S. Department of Housing and Urban Development, that has been collecting data and possible methods [8]. Concerning the field of civil engineering, the applicability of a performance-based approach to historic structures involves, first of all, the uncertainty of the structural response due to the lack of information about the building and reliable data of materials' properties at high temperatures and after exposure.

Figure 1. Consequences of a fire on a historic building: Molino Stucky in Venice.

The post-fire reliability of masonry structures is thus a challenging field of research; the main difficulties lie in the expensiveness of testing, the complexity of physical and numerical modeling of a composite material, and the extensibility of experimental and numerical results. A few information is available as a basis to evaluate the effects of a fire – meaning an accidental exposure to high temperatures – on the residual safety of a masonry building. Research in this field would be useful to structural design in a general sense, including new constructions and repair of existing buildings, and would deal with theoretical and experimental study of masonry, testing-based design and evaluation of reliability with respect to accidental situations. Being a composite material, masonry features a great variety of block-joint combinations; this can limit the extent of experimental results, depending on the chosen materials, wall thickness and texture; cracking and damage phenomena must be investigated with accurate displacement control, in order to clarify the post peak behavior in compression and in shear. The theoretical or theoretical-experimental research on constitutive laws for masonry in compression is nowadays a lively branch of study. For the above mentioned reasons, the mechanical behavior after exposure to high temperatures is much less investigated than fire endurance and performance of structural members under fire conditions.

2. Effects of Fire on Masonry Structures

The main branches of research concerning masonry and fire under the point of view of civil engineering are the following:

Experimental and/or theoretical research on masonry members' behavior in fire conditions

Masonry walls are traditionally and very commonly used as fire compartment members; this kind of studies investigate, from a scientific point of view, the causes of the proven reliability of masonry constructions in fire, and often focus on the peculiar masonry types of a Country or region. The concept of fire resistance, expressed in time units and established by one or more failure criteria under fire conditions - mainly: loss of load capacity, integrity and insulation, i.e. R, E and I criteria - is the basis of this branch of research. The testing programs aimed to orientate or update fire building codes or prescriptions also belong to this branch of studies.

Experimental physical and mechanical characterization of materials during fire exposure (masonry, bricks, blocks, mortars, concretes etc...)

Fire resistance testing methods do not encompass investigation at material scale. Since the 1960s, experimental research on the temperature-dependency of thermal (conductivity, specific heat, thermal expansion coefficient) and mechanical parameters of materials (strength, elastic modulus, peak strain, ultimate strain) has fulfilled the need for tools and methods of predicting the fire behavior of structures from the material point of view. Moreover, the growing diffusion of finite element calculation tools in professionals' practice has arisen the necessity of reliable material data to be implemented in numeric models.

Experimental physical and mechanical characterization of materials after fire exposure

Most of available information about this topic refers to cement-based materials, because of their frequent use in strategic buildings, which need high fire performance requirements to be fulfilled; in particular, structural integrity after the fire event is always required for such buildings.

2.1. Structural Behavior of Masonry Exposed to Fire Conditions

In its first stages (about 1920-60), particularly in USA, UK and Germany, research on the fire resistance of building members provided the preliminary experimental background of today's standard testing methods of Europe (ISO 834), USA (ASTM E 119) and Australia (AS 1530); such experimental data are still a large part of the available empirical knowledge today. Standard fire testing is still the basic tool to validate calculation methods and update tabular data that engineers frequently have recourse to. Generally, available studies focus on the problem of fire-separating walls; standard testing protocols don't account for the non-separating fire wall, i.e. exposed to fire on both sides, but this case may be treated as the situation of one-dimensional vertical elements (columns) in fire. There are also analytical tools to calculate the temperature distribution along the thickness of a member exposed to temperature increments on both sides [9]. In Italy, an important testing campaign was done in the 1990s [10] on masonry walls made of various types of clay units, which differed from each other in density and void percentage. These tests were aimed to support a proposal of update for code tabular data of fire resistance. During this testing program, all the masonry specimens, regardless of the type and thickness, reached failure by I (insulation) limit state criterion, following three basic types of behavior depending on the thermal inertia expressed by Equation 1 and illustrated in Figure 2:

$$\alpha = \lambda \times \rho \times c_p \tag{1}$$

where α is the thermal inertia, λ the thermal conductivity of the wall, ρ the density of the material and c_p is the specific heat at steady pressure.

1) walls of low thermal inertia (thickness: 6 to 8 cm, void percentage > 45%): the temperature on the unexposed side starts to grow after about 10 minutes from the beginning of test (i.e. the effect of thermal inertia is cleared), proportionally to the increase in temperature inside the oven, up to fire resistance values REI=30÷60 min;
2) walls of medium thermal inertia (thickness: 12 to 16 cm, void percentage > 45%): the temperature on the unexposed side starts to grow after about 20 minutes from the beginning of test; just before 100°C, evaporation of water content forces the temperature to remain almost steady on the unexposed side for about 15-30 minutes, until the specimen is completely dried, up to fire resistance REI=120 min;
3) walls of high thermal inertia (thickness: ≥ 16 cm, all void percentages): the temperature on the unexposed side starts to grow after about 60 minutes from the beginning of test; subsequently, the combined effect of thermal inertia and water migration keeps the temperature of the unexposed side in a steady state. The fire resistance then corresponds to the maximum test duration (REI=180 min).

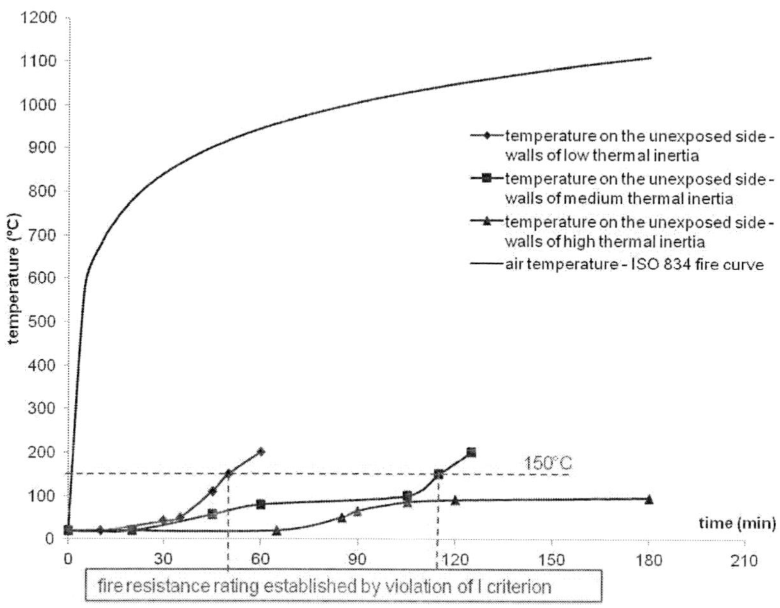

Figure 2. Experimental behavior of walls of different thermal inertia (data from [10]).

Fire testing standards explicitly state that fire resistance classification cannot be construed as having determined a member's suitability for use after exposure. From the point of view of the residual behavior, it can be said that the occurrence of integrity or insulation failure is certainly a consequence of fire-induced physical damage, especially development of micro-cracks, whose effects on the residual stiffness, stress and strain distribution of a member need to be evaluated.

From the end of 1980s, Finite Elements Method analysis was applied to the study of fire behavior of structures. In the case of masonry, few of these researches (developed mostly in UK and Australia) developed numeric models for data extrapolation and parametric studies on the structural fire behavior. Generally, the evaluation of fire resistance, maximum temperature, and strain evolution under fire condition is performed on a two-dimensional model of the wall section. The situation of the wall exposed to fire on one side is thus referred to the problem of the laterally-loaded column with axial load (weight + possible permanent load), where geometry, boundary conditions and load eccentricity are the parameters that lead to structure's displacements in a given loading condition.

The out-of-plane bowing towards the heat source due to thermal expansion on the heated surface, that happens to walls exposed to high temperatures on one side, is referred to in literature as 'thermal bowing'. Data about this phenomenon are not very abundant, since standard testing procedures do not prescribe to measure samples' displacements during the test; anyhow, Eurocode 6 [4] allows designers to assume additional fire resistance criteria based on strain under fire conditions, by client's request. Experimental research on thermal bowing has been done by Cooke [11] on brick-mortar masonry walls of various thickness, following standard testing protocol BS 476 (analogous to European ISO 834). Several tests performed in Australia in the 1980s have allowed to investigate the fire behavior of masonry walls made of cement or clay units, loadbearing or non-loadbearing, and single- or double-wythe. The following qualitative results were pointed out by Gnanakrishnan and Lawther [12]:

- the insulation capacity of single-wythe masonry walls increases with water content and equivalent thickness of the units (i.e. thickness x solid volume fraction); anyhow, the presence of voids brings in a discontinuous heat conduction; walls made of units with high insulation capacity can reach insulation failure due to the lesser properties of the mortar joints.
- integrity failure (E criterion) affects most of all cement unit walls, because of the shrinking phenomenon that starts beyond 150°C; hollow clay blocks are also sensitive to integrity loss and can show the onset of spalling. On the other hand, boundary conditions prescribed by standard procedures have a retaining effect on the sample, and it is objectively difficult that integrity limit state is reached before insulation failure.
- generally, during high temperature conditions, loadbearing walls do not undergo crushing failure; in fact, instability failure is likely, due to the horizontal displacement induced by thermal bowing. The collapse is generally reached when the displacement at midspan is about 0.8 times the thickness of the wall.

Experimental studies on loadbearing members have also enlightened the beneficial effect of the imposed load, that can mitigate the effect of thermal expansion and make the increase in displacement slower with increasing temperature; the mechanical decay induced by high temperature on the exposed side forces the compressive stress to migrate towards the unexposed side [13].

Other numeric-experimental studies address thermal-mechanical modeling of masonry. In the case of non-loadbearing walls of hollow clay units and mortar, simple models based on heat conduction, convection and radiation have proven to be reliable without accurate modeling of water migration inside the materials' porous structure; this phenomenon can be effectively accounted for by including temperature-dependent specific heat in the numeric analyses [14].

Numeric models, generally based on a two-dimensional simulation of the wall section, were also developed at the same time with the experimental studies cited above.

The main parameter involved in such numeric simulations is the maximum horizontal displacement as a function of time or temperature. Cooke's model [11] represents a column of a homogeneous material with a parabolic compressive and a bilinear tensile constitutive law, which is similar to the typical material model for concrete at high temperatures adopted in Eurocode 2, and assumes mechanical properties and thermal expansion coefficient independent from temperature.

Within this assumption, the correspondence between numeric and experimental results is worse with increasing wall thickness. Gnanakrishnan and Lawther [12] proposed a finite element model where units and joints are represented separately with plane strain elements. The model accounts for temperature-dependent thermal expansion coefficients, but it does not account for material nonlinearity; mechanical properties' temperature dependency is assumed on the grounds of the available constitutive models of concrete at high temperatures. Dhanasekar et al. [15] developed a thermal-structural FE model aimed to analyze the effects of thermal bowing in masonry walls, on the grounds of a simplified calculation of heat transfer across the wall thickness. Materials' nonlinearity is represented by the bilinear failure criterion commonly used for concretes; mechanical properties' temperature dependency is accounted for on experimental grounds.

Most of the recent FE models of brick-mortar masonry in fire conditions apply material models and empirical relationships of thermal-induced decay previously elaborated for concretes. The plane stress model by Nadjai et al. [16] also accounts for geometric nonlinearity due to great displacements of the axially loaded masonry column, and for the beneficial effects of loading in retaining displacements. This model enlightens the prevalent influence of geometric slenderness on displacements due to thermal strain.

Besides the updating of fire resistance class tables, empirical data of fire resistance of masonry walls were also useful in defining relationships to calculate fire resistance, based on the above mentioned Harmathy's rules. Such relationships can be useful in avoiding or reducing direct tests for evaluating the fire resistance of multi-wythe and/or cavity walls, whose variety can't often be encompassed by tabular data.

Moreover, it must be noticed that, until relatively recent times, computer-aided calculation methods were hardly accessible by the majority of professionals for reasons of expensiveness and difficulty in use [17].

As an example, US regulations allow engineers to use a calculation method derived from standard tests' results, with the following basic formulation [3]

$$R = (cV)^n \quad (2)$$

where R is the fire resistance period (hours), V the effective volume (solid volume) per surface unit, c a coefficient depending on the material, wall texture and units of measurement, and n depends on the heating rate on the exposed side. In the case of multi-wythe walls, Equation 2 becomes

$$R = (c_1V_1 + c_2V_2 + \ldots + c_iV_i)^n = (R_1^{1/n} + R_2^{1/n} + \ldots + R_i^{1/n})^n \quad (3)$$

where R_n are the known values of the fire resistance of every single wythe. Whenever a cavity is present, its favorable contribution to fire resistance is represented by the A_i factor, that means an increase of 20 minutes.

$$R = (R_1^{1/n} + R_2^{1/n} + \ldots + R_i^{1/n} + A_1 + A_2 + \ldots + A_i)^n \quad (4)$$

Moreover, US regulations allow to account for external plaster by turning its thickness into an equivalent wall thickness to be added, with a multiplication coefficient depending on the type of plaster (unexposed side), or by explicitly adding a contribution to fire resistance, expressed in minutes (plaster on the exposed side).

2.2. Temperature-Dependent Material Properties

The effects of high temperatures and fire on construction materials have been investigated since a long time, to improve the comprehension of structures' behavior in fire with specific and detailed information that can't usually be acquired from standard fire tests. Moreover, the switch from structure's to material's scale is useful to assess the inevitable differences between testing and real working conditions of a structural member [18]. From the point of view of fire reaction (i.e. the extent of participation of a combustible material to fire), masonry materials, being clay-based and generally having a prevailing content of metallic oxides or inorganic compounds, are incombustible. The physical and chemical phenomena that arise in clay-based materials under high temperature exposure (Table 1) and originate the mechanical temperature-dependent decay, have been investigated mainly in concretes and cement-based materials, due to the need for mix-design solutions to prevent spalling. Phenomena indicated in Table 1 refer to Portland-type concrete [19], but can be of reference also for common types of mortars and also for bricks with a high content in silicates.

Reliable data have thus been available, since the 1970s, on the temperature-dependent properties of a lot of types of concretes and cement-based materials (normal, lightweight and high performance concretes, cement units and mortars); empirical formulations of properties' decay during high temperature conditions were also derived for concretes only [20]. Recently, attention was specifically paid to the codification of standard procedures of mechanical testing of concrete samples under high temperature conditions, that mean: 1) during exposure without imposed load, 2) during exposure with imposed load, 3) after exposure without imposed load, or residual condition [21]. As far as our subject is concerned, few data are available on temperature-dependent thermal and mechanical properties of cement mortars and thermal properties of clay units; useful considerations could also be made on the results of research on types of normal strength concrete and cement masonry units. It must be clarified that generally, and mainly from the point of view of the properties in the residual condition, the temperature of reference is the maximum external (air) temperature that the material sample is exposed to.

Table 1.

Temperature (°C)	Phenomena
100	Start of dehydration - Increment of thermal inertia
300-400	Dissociation of calcium hydroxide Triple point of water Dehydration of some types of aggregates Start of mechanical decay Possible explosive spalling
573	Expansive inversion of α- to β-quartz Strong increase in creep strain Start of materials' ineffectiveness
700	Dissociation of calcium carbonate
800	Complete dehydration of the material
1200-1400	Material fusion

2.2.1. Temperature-Dependent Material Properties under Fire Exposure

- Elastic modulus

Nguyen et al. [14] pointed out an increase in the elastic modulus of bricks, that is greater between 400 and 750°C and is followed by a rapid decrease; the parameter can be considered as null at 1000°C.

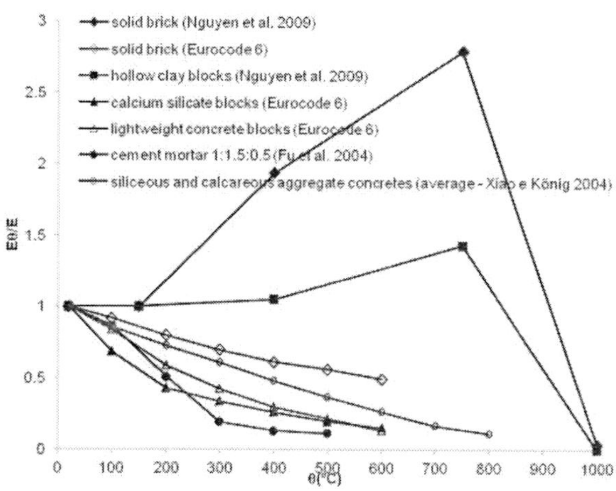

Figure 3. Experimental data of elastic modulus under high temperatures.

Exponential or linear decrease was observed in concretes, regardless of the type of aggregates [22]. Decay with increasing temperature of exposure is even more rapid in cement mortar [23]. The data are presented in Figure 3, where θ indicates temperature and Eθ/E the ratio between the elastic modulus at 20°C and at temperature θ.

- Compressive strength

The study by Nguyen et al. [14] indicates an increase in compressive strength with increasing temperature, followed by a decrease at around 750°C; at such temperature, where elastic modulus is about three times the original value, compressive strength is a little greater than twice.

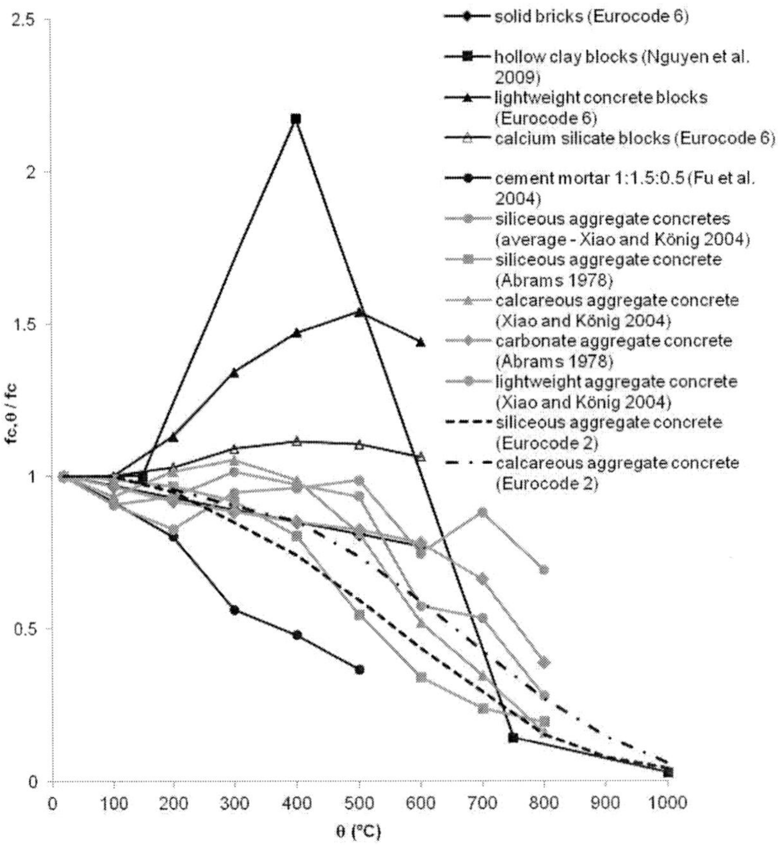

Figure 4. Experimental data of compressive strength under high temperatures.

Eurocode 6 assumes a steady decay of the parameter with increasing temperature. Cement mortar shows a clear decrease with increasing temperature [23]. Concretes have shown very variable results and a marked influence of the aggregate type [18, 22]. The favorable action of the load level in slowing the process of decay was investigated only in concretes. Figure 4 reports the experimental data of temperature - compressive strength relationships for different materials (in unloaded conditions), with analogous symbols as in the previous graph.

- Tensile Strength

All the available data are about concretes only, and point out a fast decay of the tensile strength of cylindric samples with increasing temperature [24]. Nadjai et al. [13] assumed a tri-linear law of decay derived from Thelandersson's data to account for temperature dependent tensile strength of masonry to the purpose of numeric modeling.

- Stress-Strain Compressive Law and Strain Behaviour

Concerning masonry materials, the data by Nguyen et al. [14] point out that bricks maintain the original elastic-brittle behavior even under high temperatures.

The graph in Figure 5 collects such data, with the temperature-dependent bilinear constitutive law of bricks of Eurocode 6 and Harmathy and Bernd's data about lightweight concrete masonry blocks [25].

The stress is related to the compressive strength at the temperature accounted for. It can be noticed that Eurocode 6 assumes that, at increasing temperatures, the elastic modulus and strength of brick decrease, while the peak and ultimate strain increase and this is in conflict with the experimental data reported in the same graph.

Concerning cement mortar, experimental evidence points out that the peak strain increases at increasing temperatures, while compressive strength and elastic modulus, as above said, decrease; the stress-strain law holds its nonlinear path ([23], Figure 5).

Theoretical-experimental models (i.e. expressions of the stress-strain law) of concrete under fire conditions have been elaborated as well, in order to express the stress-strain behavior as a function of temperature.

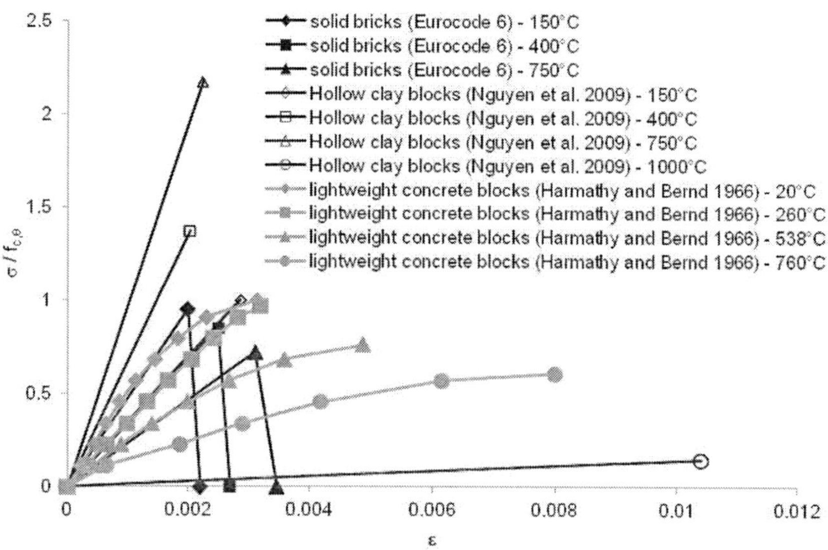

Figure 5. Experimental data of stress-strain behavior under high temperatures.

Usually, the law of the undamaged material is expressed as a function of temperature and may be modified by empirical parameters which account for other important factors, e.g. type of material, loading level or water content. Generally, the stress-strain relationship of a material as a function of temperature can be expressed in the following way [26]

$$F(\sigma, \tilde{\sigma}, \varepsilon, \tilde{\varepsilon}, \theta, \tilde{\theta}) = 0 \tag{5}$$

where σ is the instantaneous stress state, the stress history, the instantaneous strain state, the instantaneous temperature, the temperature history and *t* the considered instant. In FE thermal analyses, where nodal displacements are function of the total strain, the relationship can be simplified by taking the strain state as a function of the total strain, that can be decomposed in the following way [26]

$$\varepsilon_{tot} = \varepsilon_{\sigma}(\sigma, \theta, \tilde{\sigma}) + \varepsilon_{th}(\theta) + \varepsilon_{cr}(\sigma, \theta, t) + \varepsilon_{tr}(\sigma, \theta) \tag{6}$$

where is the instantaneous load-induced strain (function of stress, instantaneous temperature and stress history); is the thermal strain (function of instantaneous temperature); is the basic creep (function of stress, instantaneous temperature and instant); is the transient creep function of stress and instantaneous temperature). On the grounds of this theory, theoretical-experimental constitutive laws can be elaborated and calibrated for materials in high temperature conditions; currently, only experimental-based models for concrete are available (e.g. [20, 27]).

These material models differ from each other mainly in accounting for nonlinearity and in the choice and relevance of empirically-derived parameters.

Anderberg-Thelandersson's model [20] was frequently assumed and adapted in mechanical and thermal-mechanical finite elements analyses, to model masonry as a homogeneous material as well as its separate components, e.g. by Gnanakrishnan e Lawther [12] and Nadjai et al. [13]. Currently, a specific theoretical-experimental elaboration for brick-mortar masonry, analogous to the models above cited but focused on the residual behavior of masonry, brick and mortar after high temperature exposure, is in progress after the experimental results presented in the following Section 3.

Concerning Poisson's ratio, this parameter is essential in numerical modelling, but it is of minor importance in experimental assessment of structures and is generally not accounted for as an outcome of standard tests on masonry. The temperature-dependency of Poisson's ratio is neglected in numerical models of masonry in fire conditions, since it is generally assumed [e.g. 12, 13] as independent from temperature .Finally, data are available on the temperature dependency of linear thermal expansion coefficient and thermal strain of bricks and cement mortar under exposure to high temperatures. Westman's detailed research, done in the 1920s [28], demonstrates that bricks with a high content in silicium oxide show a quick increase in the linear expansion coefficient around the point of expansive inversion of α- to β-quartz (i.e. 573°C).

The empirical data reported in Eurocode 6 [4] don't account for this phenomenon, since it depends on the particular composition of the bricks. Portland cement-based mortar shows an increase in linear thermal expansion coefficient until 500°C and then a progressive decrease [12, 29]; the influence of the level of loading under high temperatures was also investigated [29] and was found to induce a strong reduction of the thermal strain of cement mortars.

2.2.2. Temperature-Dependent Material Properties after Fire Exposure (Residual)

Research on the residual mechanical properties after fire exposure has been producing a great amount of data concerning concretes and other cement materials [18, 22, 30-34]. Recommendations about test parameters and procedures in post-exposure conditions are also available for concrete [21], but not for masonry materials. Recently, research done by the Authors on the residual behavior of masonry damaged by exposition to high temperatures [36, 37] has brought in new information on this topic, in the field of civil engineering. The outcomes are reported in the following section, while the most relevant information from previous research is outlined here below.

- Concretes

It can be said that cement masonry materials (e.g. mortars, concrete or calcium silicate blocks) generally undergo a decrease in mechanical performances after high temperature exposure [18, 31-34]. Comparisons were made between 'under' and 'after' fire conditions of normal- and high-strength concretes, which reveal that residual mechanical properties after exposure to a certain maximum temperature are generally lower than under exposure to the same temperature [18]. Generally, research on concretes after fire exposure underlines the major role of the maximum temperature of exposure in the residual mechanical properties and point out the additional importance of other factors (e.g. hydration rate and cooling regime) [18, 32]. Besides, these researches have assessed the relevance of the component materials in the residual mechanical performance, e.g. concretes with different types of aggregate often show different performances under the same conditions.

- Cement Mortars

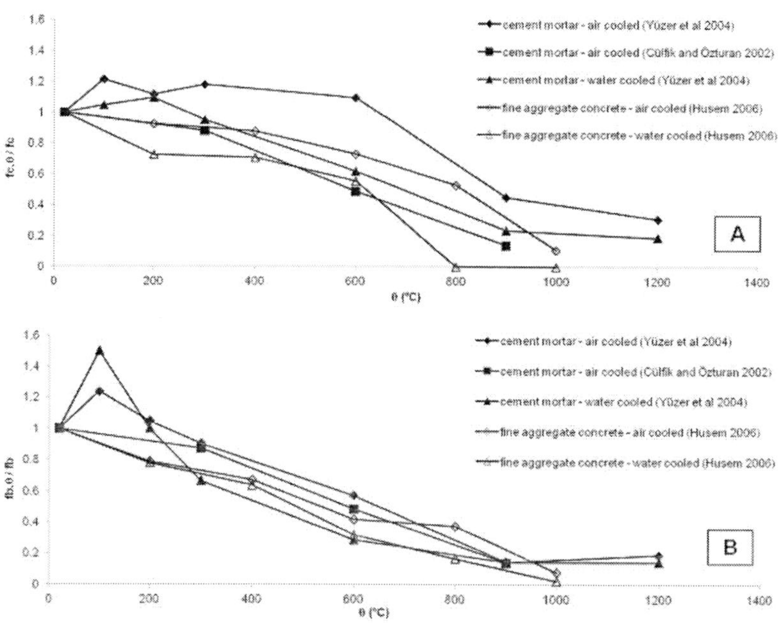

Figure 6. Experimental data of residual compressive (A) and flexural strength (B) of cement mortars and fine aggregate concrete, following different authors.

Generally, cement mortars in residual conditions behave like normal strength concretes.

The graphs in Figure 6, where f_c indicates compressive strength, f_b flexural strength and subscript θ the property after exposure to temperature θ, collect data on the residual compressive and flexural strength of mortar and fine aggregate concrete, according to [30], [31] and [35].

- Bricks and Masonry

The tests performed by Russo in the 1990s [2] reported data on clay bricks of a historic building hit by a severe fire, whose duration was about 3 hours and maximum temperature about 1000°C.

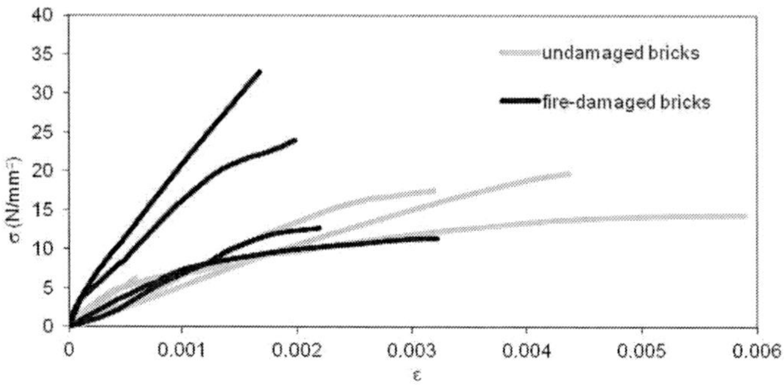

Figure 7. Experimental data of residual properties of bricks (from [2]).

The compressive tests on samples of bricks indicated that the action of a real fire and direct exposure to flames carried an increase in stiffness and strength and a decrease in ultimate strain; tests on masonry prisms assembled with exposed and unexposed bricks demonstrated a slightly lower compressive strength of the prisms made with damaged bricks. Figure 7 shows the experimental stress-strain graphs of the tests.

3. Research on Residual Temperature-Dependent Properties of Masonry

The results reported in this section refer to a specific physical model of masonry, that is a fire-separating wall 25 cm thick (i.e. exposed to fire on one side), made of solid clay bricks and cement mortar [36, 37]. The influence of maximum temperature on the residual mechanical performance in compression and in shear can be evaluated on the ground of the results. The choice of the mechanical parameters to be investigated was driven by two reasons. First, a quick and simple mechanical characterization of masonry and its components was needed, resulting in the knowledge of the basic parameters of masonry design (compressive strength and pure shear strength) for each case of exposure, with the usual standard mechanical tests; then, additional experimental information (elastic modulus and strain properties both in compression and in shear), was yielded – by means of displacement control – for evaluation of linear and nonlinear behavior of masonry. Concerning the parameters of exposure, it was decided to focus only on maximum temperature, and to fix the heating rate and duration at maximum temperature in order to simulate an accidental exposure with fast heating and short-term exposure at temperatures in the range 20-750°C at least, according to the RILEM recommendations for concrete [21]; two values of maximum temperature were thus taken into account, i.e. 300 and 600°C; such values belong to a low-medium range of high temperatures, since real fires in civil buildings can often reach higher temperatures (1000-1200°C). The two exposure conditions do not mean to represent all possible severe situations, but to provide a basis for the first investigation at a low-medium level of exposure. Based on the experimental results here presented, functions of parameters' mechanical decay have been set up for masonry materials after high temperature exposure.

3.1. Testing Program

The physical model refers to masonry 25 cm thick, made of hand-made type clay bricks (sized 250 x 120 x 55 mm) and cement mortar, with joints 1 cm thick, assembled in the so-called 'gothic bond' pattern (in each row, alternate 'heads' and the 'sides' of the bricks are visible); square specimens of 51 x 51 x 25 cm were built (Figure 8). Attention was paid to the selection of materials, masonry thickness and pattern, in order to represent a type of load-bearing masonry wall which could be common both to traditional and new buildings. These bricks are suitable for load-bearing walls, as they are similar to ancient ones in composition, dimensions and crafting process. A cement mortar classified as M10 following the standard UNI EN 998-2 was chosen because of its wide diffusion in existing load-bearing masonry structures in Italy; it is made of cement, hydraulic lime and sand in the proportions of 1 : 0.5 : 4.

Usual and simple tests following UNI standard procedures for masonry, bricks and mortar were planned, to get a quick mechanical characterization. Compressive and diagonal compressive tests were thus performed on masonry samples, flexural and compressive tests on mortar samples and compressive tests on brick samples. The initial elastic moduli in compression were also measured on samples of brick and mortar. Each type of test was performed on exposed and unexposed materials.

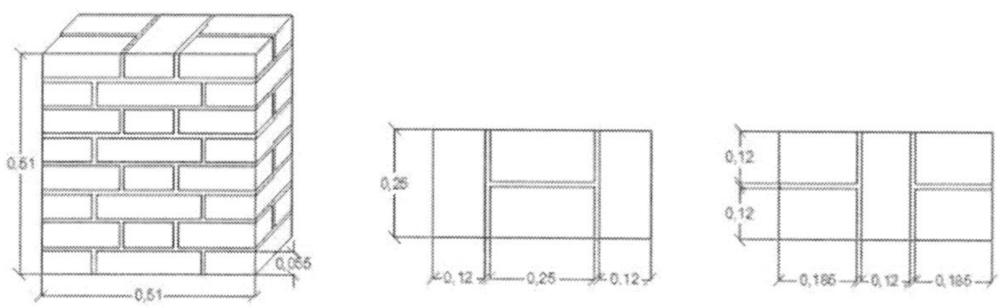

Figure 8. 25 cm thick masonry specimens.

Two time-temperature curves were set up, each one expressing an exposure condition represented by the maximum temperature (300 and 600°C), with the same heating rate (about 19°C/min) and duration at maximum temperature (1 hour). The heating rate was decided to be as high as possible, in order to simulate a fast rise of temperature; as well, in order to reflect an accident condition such as fire, a very short exposure was established. The time-temperature curves representing the two conditions are shown in Figure 9.

Ten samples were exposed to high temperatures following each of the two curves. In order to simulate the case of a wall exposed to fire on one side, i.e. load-bearing separating wall, as it is contemplated in Eurocode 6 1:2 [4], the samples were covered in fire-resistive rock wool claddings which left just one of the largest faces free to exposure, as it is shown in Figure 10. At the same time, samples of bricks and mortar were also exposed to the two thermal cycles; the largest faces of the bricks were insulated too, so to simulate the heating of an element within a masonry assembly. Since the material samples had been stored indoor within a dry environment, they were cured in water – masonry samples for 12 hours, bricks and mortar prisms for 2 hours – before the thermal cycle, so to attain a uniform level of water content. The high-temperature exposure cycle was performed by means of a brick furnace, normally used for baking hand-made clay products. This oven ensured a quick heating process, a duration at a well approximately constant maximum temperature, and a uniform internal temperature; it allowed for all ten samples in each thermal cycle to be exposed. The temperature of the specimens during the exposure was recorded by means of nine thermocouples applied to three masonry specimens – one on a brick, one on the central bed joint and the last on the unexposed side, each inserted in a 7 mm deep hole (Figure 11); another thermocouple was used to record the air temperature inside the oven. It can be noticed that the real time-temperature curve fits with good approximation the prescribed one, until cooling begins; the temperature on the insulated side remains about 400°C lower than the air temperature, so that an appropriate simulation of the load-bearing separating wall is attained (Figure 12).

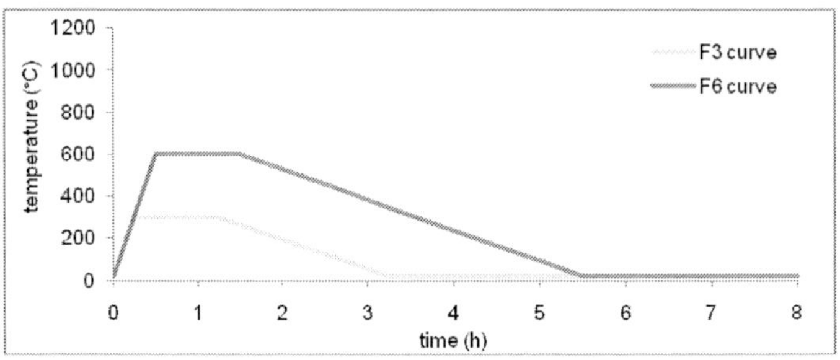

Figure 9. Time-temperature curves of the two exposure conditions.

Figure 10. Masonry specimens being put inside the oven.

Figure 11. Masonry specimen with applied thermocouples.

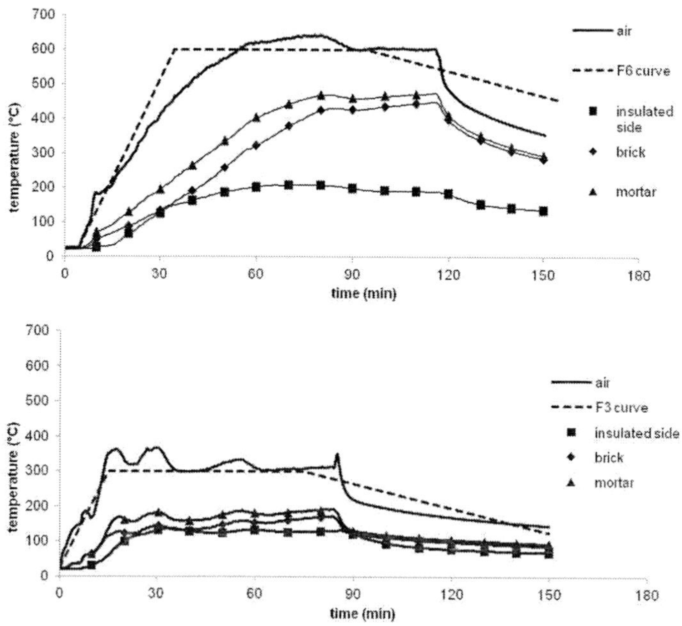

Figure 12. Time-temperature graphs of the cycles compared to the prescribed curves.

Finally, a slow air-cooling process simulated a fire self-extinguishment, by turning off the furnace just after the duration at maximum temperature was completed; the door of the oven was opened twelve hours later. The time-temperature curves of air and Just after the 300°C cycle, the exposed faces of the masonry specimens showed very little damage, i.e. some detachments at vertical brick-mortar interfaces; after the 600°C exposure, vertical interface cracking and vertical micro-cracking in all materials (including bricks and mortar specimens) were clearly visible (Figure 13). After removal of rock wool claddings, in both cases of exposition, the insulated faces of the masonry samples looked undamaged.

Figure 13. Cracking and interface detachment on 600°C (F6) exposed masonry specimens.

3.2. Experimental Results

The mechanical testing program was carried out at the Laboratory of Strength of Materials (LabSCo) at the IUAV University of Venice. In the following subsections, unexposed masonry, bricks and mortar are denoted with NF, while materials exposed to 300°C and 600°C are referred to as F3 and F6 respectively.

3.2.1. Brick and Mortar Specimens

Brick specimens and 4 x 4 x 16 cm prismatic mortar samples were employed for testing exposed and unexposed materials. All these tests were performed with a 200 kN maximum load testing machine (Galdabini SUN/20).

Table 2. Experimental data of compressive tests on brick samples

Compressive tests - BRICKS		
sample	dimensions (mm)	$f_{bc,\theta}$ (N/mm^2)
B-NF-1	48 x 48 x 49	19.69
B-NF-2	45 x 45 x 45	18.58
B-NF-3	47 x 47 x47	19.25
average NF		19.17
standard deviation		0.456
relative standard deviation		0.024
B-F3-1	53 x 52.5 x 52.5	16.73
B-F3-2	54 x 53 x 53	18.32
B-F3-3	54 x 54 x 53	18.44
B-F3-4	54 x 55 x 53	16.84
B-F3-5	54 x 55 x 52.5	16.64
average F3		17.39
standard deviation		0.80
relative standard deviation		0.046
B-F6-1	54 x 54 x 54.5	13.76
B-F6-2	54 x 55 x 55	12.48
B-F6-3	55 x 55 x 54.5	12.02
B-F6-4	56 x 54 x 56	11.87
B-F6-5	53 x 54 x 55	9.67
average F6		11.96
standard deviation		1.324
relative standard deviation		0.1107

Compressive tests on bricks were carried out following UNI EN 772-1 standard procedure, to obtain the mean compressive strength $f_{bc,\theta}$; cubic samples were cut from exposed and unexposed bricks and dried before testing. Table 2 reports the compressive strength data. The elastic modulus of bricks $E_{b,\theta}$ was evaluated on 2 x 2 x 5 cm specimens, according to UNI 9724 prescriptions; the vertical strain of each specimen was recorded by two electric strain gages.

Table 3 reports the results of those tests, and Figure 14 shows the corresponding stress-strain diagrams up to specimens' failure; the strain is the average of the two series of values recorded.

The tests show a decrease in strength (-9% for F3 and -38% for F6 bricks) and a slight increase in stiffness (+15% for F3 and +1% for F6 bricks).

The progressive decay in compressive strength of bricks is similar to the trend observed in concretes, and especially in siliceous aggregate concrete; this can be due to the bricks' high content in silicates; concerning elastic modulus, an increase after exposure to a real fire – whose temperature history was not precisely known – was already observed in the bricks of a historic structure, although accompanied by an increase also in compressive strength [2].

The tests on mortar were carried out following UNI-EN 1015-11 standard; first, the prismatic samples were subjected to bending test, then the two halves of each one were tested in compression. The denominations and data about the mortar samples are recorded in Table 4, where $f_{mf,\theta}$ and $f_{mc,\theta}$ are respectively the flexural and the compressive strength of mortar.

Table 3. Experimental data of elastic modulus testing on brick samples

Elastic modulus - BRICKS	
Specimen	$E_{b,\theta}$ (N/mm^2)
B-NF-1E	6606
B-NF-2E	5082
B-NF-3E	5162
B-NF-4E	5991
average NF	5710
standard deviation	646.004
relative standard deviation	0.113
B-F3-1E	7934
B-F3-2E	5754
B-F3-3E	5986
average F3	6558
standard deviation	977.578
relative standard deviation	0.149
B-F6-1E	5986
B-F6-2E	5405
average F6	5765
standard deviation	4079.465
relative standard deviation	0.708

Elastic moduli $E_{m,\theta}$ of exposed and unexposed mortar were tested on the same specimens before the destructive tests; the vertical strain of each specimen was recorded by two 60 mm electric strain gages and the load was applied until half of the expected compressive strength was reached, so not to damage specimens before flexural and compressive tests. The stress-strain diagram of these tests is depicted in Figure 15 and the corresponding data are listed in Table 5. The tests point out a clear increase in compressive strength (+29%) of F3 samples and an equal decrease of F6 samples; while the flexural strength decreases in F3 (-23%) and especially in F6 samples (-61%) which is greater in bending than in compression; the elastic modulus undergoes a considerable decrease as well (-10% in F3 and -51% in F6 samples).

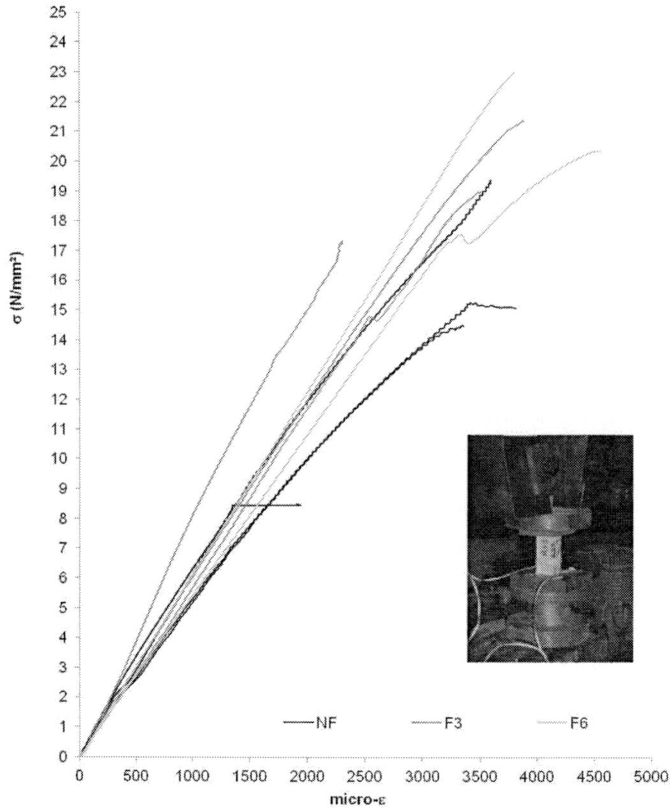

Figure 14. Stress-strain diagram of elastic modulus tests on bricks.

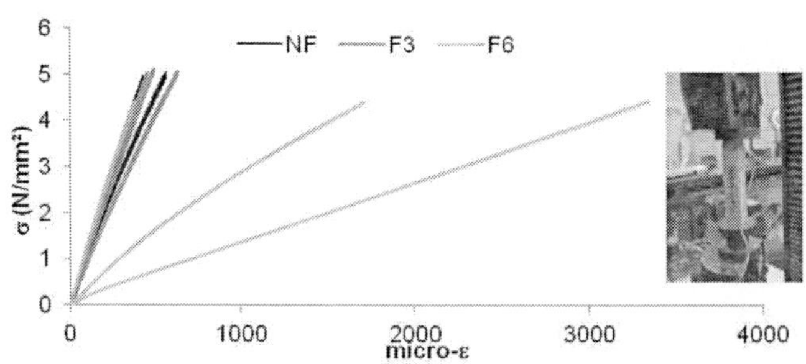

Figure 15. Stress-strain diagram of elastic modulus tests on mortar samples.

Table 4. Experimental data of flexural and compressive tests on mortar samples

Flexural and compressive tests - MORTAR		
Sample	$f_{mf,\theta}$ (N/mm^2)	$f_{mc,\theta}$ (N/mm^2)
M-NF-1	5.60	12.38; 12.91
M-NF-2	5.97	14.96; 13.44
M-NF-3	4.58	10.61; 13.72
average NF	5.38	13.00
standard deviation	0.587	1.333
relative standard deviation	0.109	0.103
M-F3-1	3.87	15.03; 14.29
M-F3-2	4.66	18.65; 18.36
M-F3-3	3.91	17.75; 16.73
average F3	4.14	16.80
standard deviation	0.363	1.643
relative standard deviation	0.088	0.098
M-F6-1	2.03	9.53; 8.60
M-F6-2	2.16	9.29; 10.03
M-F6-3	2.09	9.12; 9.01
average F6	2.09	9.26
standard deviation	0.053	0.444
relative standard deviation	0.026	0.048

Table 5. Experimental data of elastic modulus testing on mortar samples

Elastic modulus - MORTAR	
Sample	E_m (N/mm^2)
M-NF-1E	11784
M-NF-2E	12190
M-NF-3E	9408
average NF	11127
standard deviation	1227
relative standard deviation	0.110
M-F3-1E	8257
M-F3-2E	11191
M-F3-3E	10541
average F3	9996
standard deviation	1258.2
relative standard deviation	0.126
M-F6-1E	11731
M-F6-2E	2822
M-F6-3E	1331
average F6	5295
standard deviation	4591.7
relative standard deviation	0.867

The increase in compressive strength at 300°C here observed (and subsequent decrease at 600°C) agrees with what is reported by Yüzer et al.[30] on the residual behavior of a cement mortar. As well, the decrease in elastic modulus with increasing temperature could in some way be expected, although only based on the current information about the residual parameters of concretes.

3.2.2. Masonry Specimens

Uniaxial compressive tests under displacement control, following UNI EN 1052-1 standard, and diagonal compressive tests following ASTM E 519-81 standard were carried out on the samples for the characterization of masonry in compression and shear. Five exposed and three unexposed samples were employed for each kind of test. All these tests were performed by means of a 6000 kN maximum load press by Metrocom Engineering, with data control system.Before the compressive tests, the upper and lower faces of the samples were rectified by a cement covering to improve the load transfer from the machine to the sample; an uniform load distribution was ensured as well by interposing a 40 mm thick steel plate between the sample and the upper platen. The vertical displacement of the platen was controlled by means of a LVDT transducer, while four extensometers were placed on the two largest faces of each tested sample to record the vertical displacements; the loading velocity was taken as 0.05 mm/s, and each test was terminated when half the peak load value was reached within the post-peak branch. For each test, the values of compressive strength $f_{c,\theta}$, initial elastic modulus E_θ , peak strain $\varepsilon_{c1,\theta}$ and ultimate strain $\varepsilon_{c2,\theta}$ are listed in Table 6. $\varepsilon_{c2,\theta}$ is the strain value at which, in the post-peak branch, half the value of peak stress is reached; since at this value the tests were conventionally considered as ended, it represents the ultimate value of strain. Figure 16 shows all the stress-strain diagrams, where the values of increasing strain are calculated from the average of the displacements recorded by the four extensometers; from these diagrams, the value of initial elastic modulus was calculated as the average of the stress/strain ratio at one and two third of the maximum stress. Concerning unexposed specimens, the increase in strain in the softening phase is generally small. Immediately before the peak load, vertical cracks occurred in the bricks and at brick-mortar interfaces following the vertical alignment of joints. During the performing of tests on F6 samples, cracks occurred first on the exposed side, immediately followed by superficial detachment of bricks; at the end of the tests, the damage of the exposed face was fairly greater. In a less evident way, this was also the behavior of F3 samples. Moreover, at the peak stress, a vertical crack appeared along the middle of both lateral faces; after this was noticed, two additional extensometers were applied horizontally on the lateral faces of the remaining samples. The strength and stiffness of F3 masonry samples are slightly higher than NF (respectively +4% and +10%), while F6 samples show a decrease of -13% in strength and -7% in stiffness with respect to unexposed ones. The peak strain decreases (-12% in F3 and -10% in F6 cases), while the ultimate strain considerably increases with increasing temperature (+51 in F3 and +158% in F6 tests). Taking into consideration all the investigated properties except ultimate strain, their variations between NF and F3 specimens can be considered small; on the opposite side, the great increase in ultimate strain of both F3 and F6 samples points out that the softening phase is markedly influenced by exposure effects. The properties of F6 samples seem to indicate that at 600°C a not negligible influence of thermal damage on the mechanical strength begins to appear.Figures 17a and 17b show an exposed and an unexposed specimen after testing, with cracking pattern highlighted by black lines. It can be seen that the vertical cracks along the central vertical joints are present on both sides of the unexposed specimen (Fig. 17a), as expected; such cracks can be seen also on the exposed face of the F6 specimen, but the opening of the cracks following the parallel plane to the exposed face is far greater, as it is emphasized by the different thickness of lines.The pre-peak and peak behavior was analyzed in detail by evaluating the distribution of vertical strain at $0.33f_{c,\theta}$, $0.66f_{c,\theta}$ and $f_{c,\theta}$ following the recordings of each vertical extensometer; the distribution of horizontal strain, where it was measured, was evaluated too. A representation of this analysis is given in Figures 18-20, on the cross-section of each sample, where abbreviations are: v, h: vertical, horizontal exstensometer; ex , un: exposed, unexposed side; sx, dx: left-, right-positioned. The positioning on the samples' faces is also schematically represented in each figure. The data of vertical strain at maximum stress $f_{c,\theta}$, which refer to the last rows of schemes in Figures 18-20, are also collected in Table 7. This analysis was carried out in order to clarify, with the help of recorded strain data, the concentration of mechanical damage at the exposed side which was evident in some cases.

Table 6. Experimental data of compressive tests on masonry samples

Compressive tests - MASONRY				
Sample	$f_{c,\theta}$ (N/mm²)	E_θ(N/mm²)	$\varepsilon_{c1,\theta}$	$\varepsilon_{c2,\theta}$
2T-NF-1	9.64	2723	0.0037	0.0038
2T-NF-2	9.97	3085	0.0040	0.0039
2T-NF-3	9.13	2360	0.0044	0.0051
average NF	9.58	2723	0.0040	0.0043
standard deviation	0.346	295.98	0.00029	0.00059
relative standard deviation	0.036	0.109	0.072	0.138
2T-F3-1S-1	8.94	3293	0.0027	0.0056
2T-F3-1S-2	10.46	3246	0.0037	0.0054
2T-F3-1S-3	9.79	2874	0.0033	0.0069
2T-F3-1S-5	10.21	2438	0.0046	0.01
2T-F3-1S-6	10.31	3128	0.0030	0.0047
average F3	9.94	2996	0.0035	0.0065
standard deviation	0.548	314.45	0.00066	0.01358
relative standard deviation	0.055	0.105	0.189	2.089
2T-F6-1S-3	5.73	2687	0.0045	0.0265
2T-F6-1S-4	9.13	2246	0.0042	0.0085
2T-F6-1S-5	9.84	2736	0.0034	0.0083
2T-F6-1S-8	8.14	3131	0.0022	0.0064
2T-F6-1S-10	8.75	1771	0.0038	0.0058
average F6	8.32	2515	0.0036	0.0111
standard deviation	1.407	465.57	0.00080	0.00700
relative standard deviation	0.169	0.185	0.223	0.529

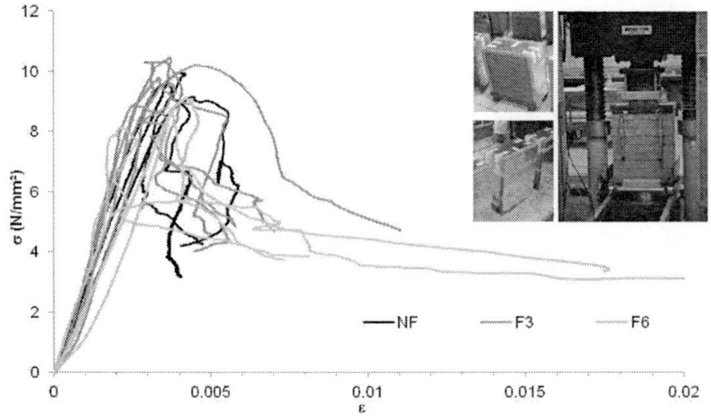

Figure 16. Stress-strain diagram of compressive tests on masonry samples.

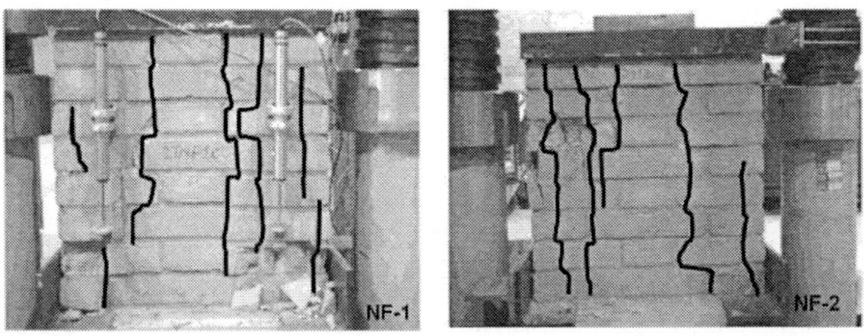

Figure 17a. Both sides of an unexposed (NF) specimen after the compressive test.

Figure 17b. Exposed (1) and unexposed (2) sides of an exposed specimen after the compressive test.

Provided that the uniformity in the distribution of compressive load at the top of the sample had been assured, the uniformity of strain distribution, which is good in the unexposed samples (Figure 18), is related to the temperature of exposure. The results confirm that exposed specimens have generally larger vertical deformations at the exposed side. This asymmetrical distribution (Figures 19-20) can be seen as a consequence of thermal damage that affected the exposed side. The samples with the most uneven strain distribution (e.g. 2T-F3-3, 2T-F6-8) have most notably shown the concentration of mechanical damage on the exposed side at failure; at the end of the test, the detachment of the thermally damaged thickness of the specimen was visible on the exposed side (see also Figure 17 above). Looking at the last two columns of Table 7, reporting the mean values of strain on each of the two faces of the samples, the occurrence of greatest compressive deformation at the exposed side is frequent; in such cases, a decrease in elastic modulus in the thermally damaged thickness can be inferred, likely due to thermally induced cracking which mostly occurred at brick-joint interfaces and, in 600°C exposed samples, also in bricks. The detachment of the thermally damaged thickness after exposure was already observed in ceramic materials by Gei et al. [38].

The presence of this high-temperature affected thickness can thus noticeably influence the strain behavior of exposed masonry – the difference between the mean strain values of the two sides are greater for exposed masonry – even if the failure mode is the same as for unexposed (vertical cracks first occurring along the joint alignment, as above said).

The compressive failure is also accompanied by asymmetrical horizontal strain on the shortest sides, which accounts for cracking in the planes parallel to the exposed side.

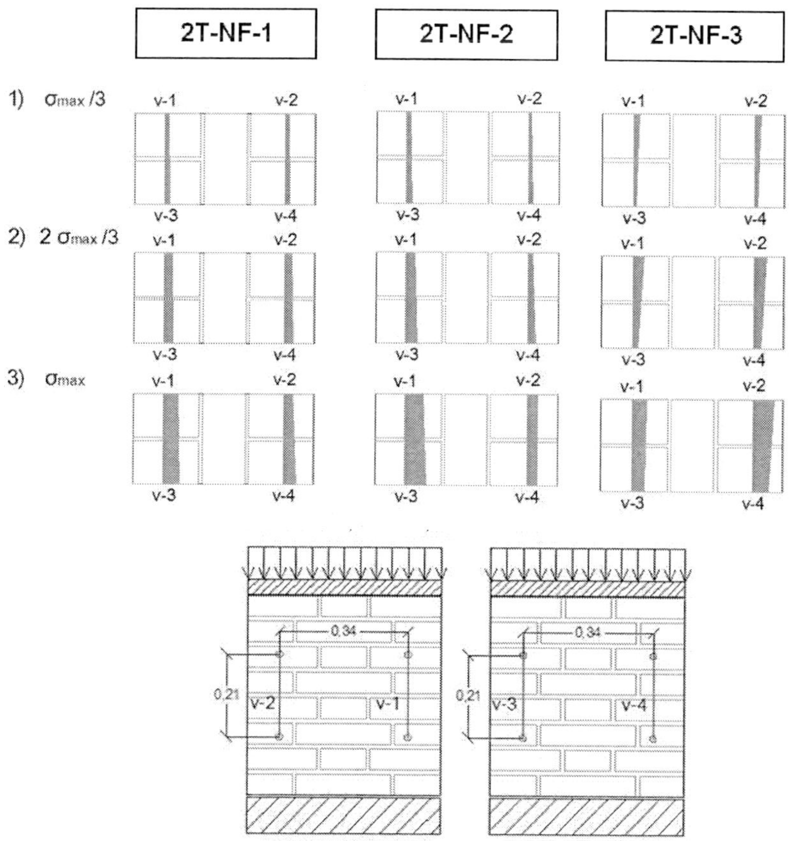

Figure 18. Distribution of strain – NF specimens.

Diagonal compressive tests on unexposed samples were performed at a velocity of 0.008 mm/s; the vertical and horizontal displacements were recorded by cross-placed extensometers along the diagonals of the two faces. The shear stress τ and shear strain γ were calculated from the recorded data following the reference standard ASTM E 519-81.

Figure 21 shows the corresponding stress-strain diagrams, while the results of diagonal compressive tests, i.e. first peak stress $\tau_{1,\theta}$, shear strength $f_{v0,\theta}$ (corresponding to the maximum stress value), shear modulus G_θ and peak strain $\gamma_{1,\theta}$ are listed in Table 8. The shear modulus was calculated as the average of the stress-strain ratio at one and two third of the first maximum stress value.

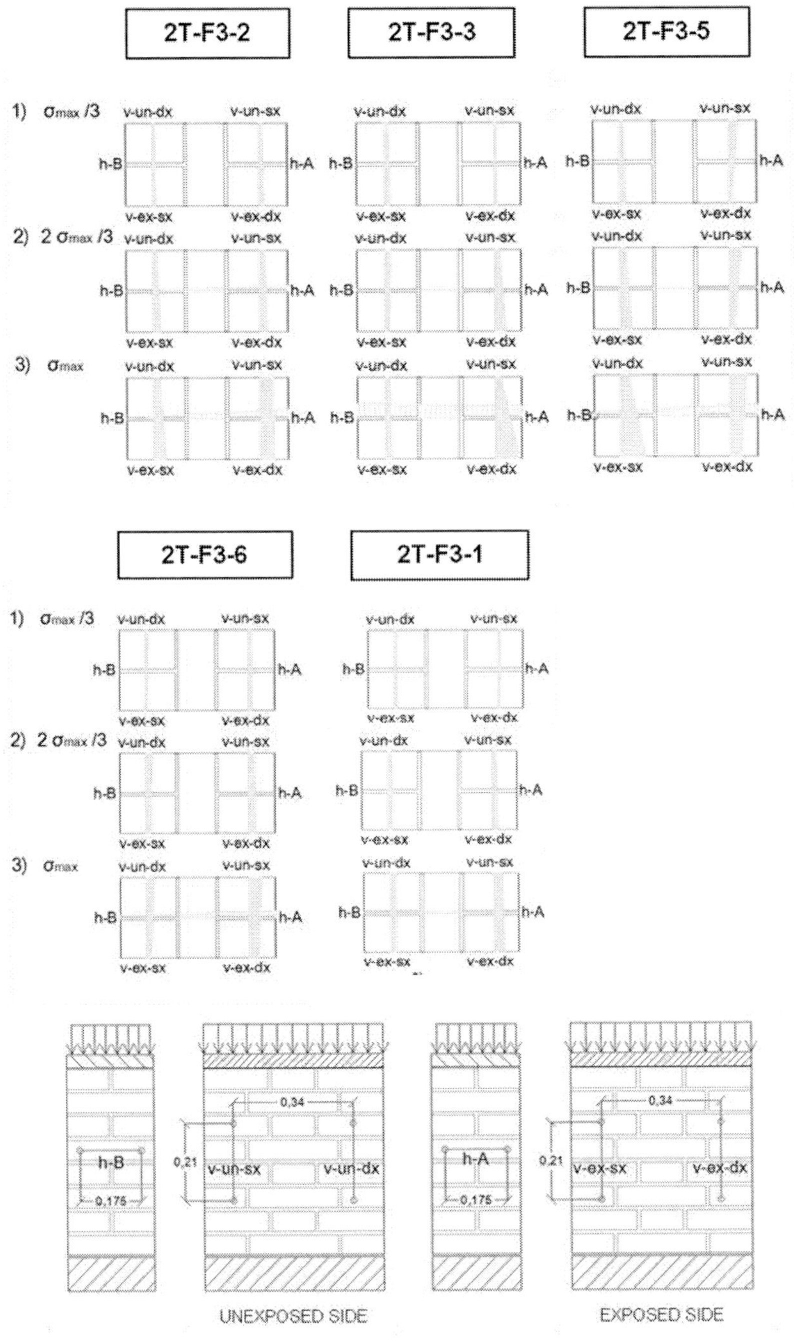

Figure 19. Distribution of strain – F3 specimens.

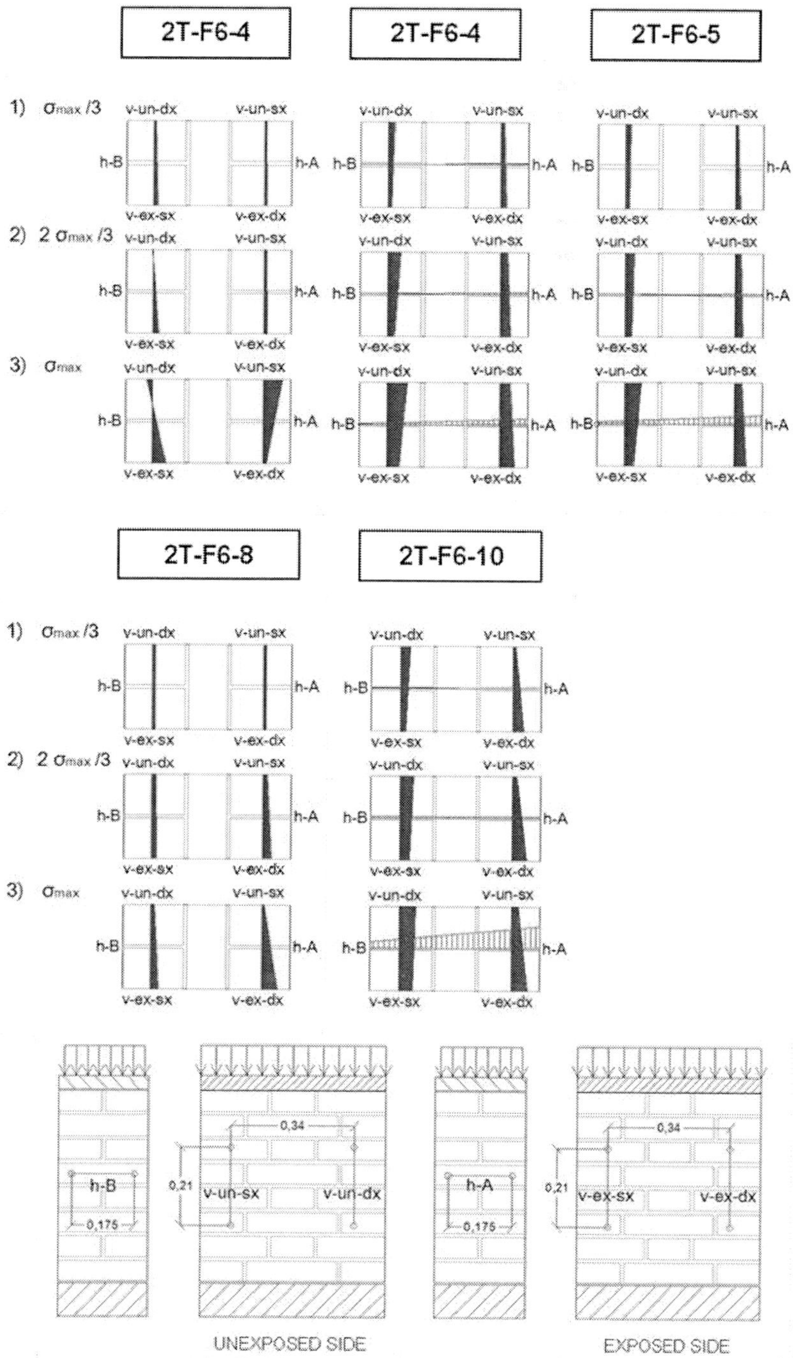

Figure 20. Distribution of strain – F6 specimens.

Like it happened in compressive tests, the shear strength slightly increases in F3 (+12%) and decreases in F6 samples (-26%). The shear modulus suffers a significant decay with increasing temperature of exposition (F3: -65% and F6: -82%). Generally, concerning the shear stress-strain behavior, both exposed and unexposed masonry samples underwent a stress drop after a first peak, and then slowly reached the maximum shear stress; at the first peak, cracks appeared both in bricks and at brick-vertical joint interfaces in unexposed samples, while interface cracking prevailed in the exposed samples, especially in F6 cases.

Table 7. Compressive strain distribution

Vertical strain distribution at collapse - MASONRY						
Sample	v1	v2	v3	v4	(v1+v2)/2	(v3+v4)/2
	unexposed sides					
2T-NF-1	0.0041	0.0025	0.0046	0.0036	0.0033	0.0041
2T-NF-2	0.0049	0.0027	0.0057	0.0029	0.0038	0.0043
2T-NF-3	0.0039	0.0061	0.0034	0.0046	0.0050	0.0040
	unexposed side		exposed side		(v1+v2)/2 unexposed	(v3+v4)/2 exposed
2T-F3-1S-1	0.0031	0.0020	0.0021	0.0036	0.0026	0.0029
2T-F3-1S-2	0.0040	0.0022	0.0042	0.0044	0.0031	0.0043
2T-F3-1S-3	0.0018	0.0014	0.0022	0.0080	0.0016	0.0051
2T-F3-1S-5	0.0052	0.0022	0.0075	0.0035	0.0037	0.0055
2T-F3-1S-6	0.0042	0.0029	0.0015	0.0033	0.0036	0.0024
2T-F6-1S-3	0.0060	-0.0018	0.0043	0.0013	0.0021	0.0028
2T-F6-1S-4	0.0031	0.0059	0.0035	0.0044	0.0045	0.0040
2T-F6-1S-5	0.0021	0.0050	0.0026	0.0037	0.0036	0.0032
2T-F6-1S-8	0.0006	0.0011	0.0026	0.0046	0.0009	0.0036
2T-F6-1S-10	0.0021	0.0050	0.0042	0.0049	0.0036	0.0046

Moreover, this occurred at remarkably lower shear stress values for exposed samples. Then, the subsequent slow load increase was accompanied by slipping along the bed joints, until the maximum shear stress was reached and the specimen showed wide gaps between bricks and vertical joints as well as indentation due to sliding along the middle horizontal joint. The tests were interrupted on the subsequent descending branch; in the case of diagonal compression, a common test end criterion could not be observed for instruments' safety reasons. Figure 22 shows a comparison between an exposed and an unexposed sample after testing.

The concentration of cracking at interfaces, together with lower first peak stress values, indicates a loss of brick-mortar cohesion in the exposed samples; this phenomenon can be linked to the interface cracking, as an effect of fire exposure, observed in the samples' exposed surfaces just after the thermal cycle. In this case again, the fairest difference resorting from the diagrams lies in the longer post peak branches of exposed masonry samples. Moreover, F3 and F6 samples reached the maximum stress at much larger strain values than unexposed ones.

The repeated increase-decrease in stress, which originate a saw-shaped diagram, is a well-known phenomenon in diagonal compressive tests. It can be ascribed to the combined strength contributions of brick-mortar cohesion (decreasing with the growth of interface cracks) and friction along the bed joints.

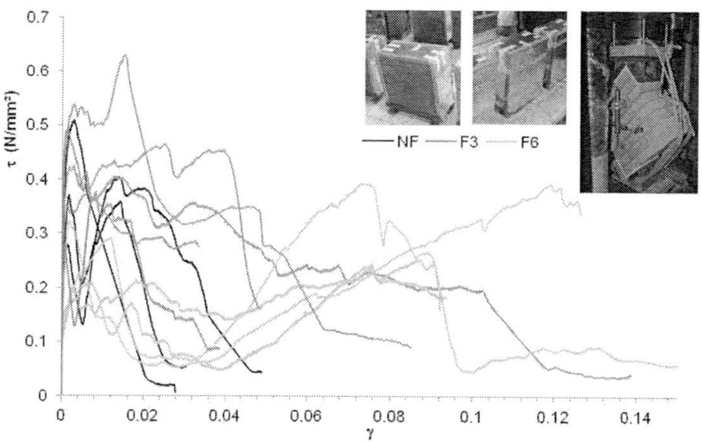

Figure 21. Shear stress-strain diagram of diagonal compressive tests on masonry samples.

Table 8. Experimental data of diagonal compressive tests on masonry samples

Diagonal compressive tests - MASONRY				
Sample	$\tau_{1,\theta}$ (N/mm²)	$f_{v0,\theta}$ (N/mm²)	G_θ (N/mm²)	$\gamma_{1,\theta}$
2T-NF-4	0.51	0.51	3295	0.0029
2T-NF-5	0.37	0.39	2052	0.0016
2T-NF-6	0.28	0.40	2273	0.0140
average NF	0.39	0.43	2540	0.0062
standard deviation	0.0946	0.060	541.44	0.00556
relative standard deviation	0.243	0.140	0.2131	0.8975
2T-F3-1S-4	0.49	0.49	1235	0.001
2T-F3-1S-7	0.35	0.47	882	0.025
2T-F3-1S-8	0.42	0.43	852	0.0029
2T-F3-1S-9	0.28	0.38	606	0.0082
2T-F3-1S-10	0.54	0.63	833	0.015
average F3	0.42	0.48	882	0.0104
standard deviation	0.094	0.083	202.03	0.0088
relative standard deviation	0.225	0.175	0.22	0.84
2T-F6-1S-1	0.18	0.25	516	0.0867
2T-F6-1S-2	0.32	0.39	503	0.1193
2T-F6-1S-6	0.29	0.39	142	0.0735
2T-F6-1S-9	0.20	0.24	666	0.0744
average F6	0.25	0.32	457	0.0885
standard deviation	0.0590	0.073	192.68	0.0185
relative standard deviation	0.236	0.227	0.421	0.2096

This could be seen in all the tested samples NF, F3 and F6, but the phenomenon appears quite magnified with increasing temperature of exposure; in fact, in F6 samples (Figure 21) noticeable increases in stress begin at strain values around 0.03. In those cases, since the first dropping occurs at lower values of stress than NF and F3 samples, evidence can be found that the high temperature exposure has most affected the brick-mortar cohesion, while friction still appears to give an appreciable contribution to the strength of the sample. The high values of strain generally reached by the exposed samples at the highest stress values can be interpreted as a reinforcement of the shear softening instability of the specimen due to the effect of thermal damage. Finally, the displacement data revealed no appreciable difference between exposed and unexposed faces.

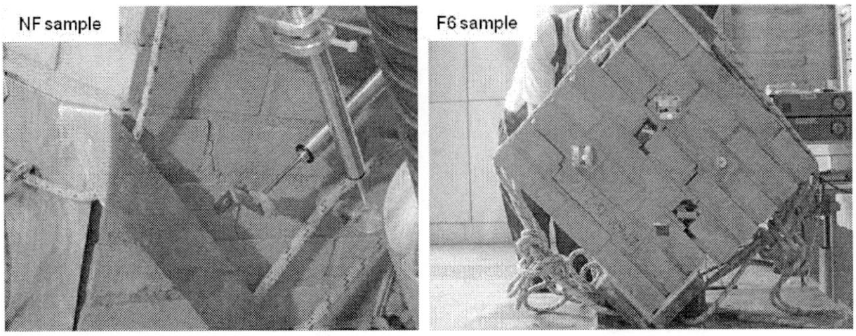

Figure 22. Exposed and unexposed specimens after diagonal compressive tests.

3.2.3. Functions of Decay

Table 9. presents the ratios between original and residual values of all the mechanical properties

FACTORS OF ORIGINAL PROPERTY			
PROPERTY		F3	F6
BRICK	$f_{bc,\theta}/f_{bc}$	0.91	0.62
	$E_{b,\theta}/E_b$	1.15	1.01
MORTAR	$f_{mf,\theta}/f_{mf}$	0.77	0.39
	$f_{mc,\theta}/f_{mc}$	1.29	0.71
	$E_{m,\theta}/E_m$	0.90	0.48
MASONRY	$f_{c,\theta}/f_c$	1.04	0.87
	E_θ/E	1.10	0.93
	$\varepsilon_{c1,\theta}/\varepsilon_{c1}$	0.88	0.90
	$\varepsilon_{c2,\theta}/\varepsilon_{c2}$	1.51	2.58
	$f_{v0,\theta}/f_{v0}$	1.12	0.74
	G_θ/G	0.35	0.18
	$\gamma_{1,\theta}/\gamma_1$	1.68	14.27

Table 9. Ratios of original/residual material properties.

The factor of original property k_θ indicates the ratios between each residual mechanical parameter and the original one, for brick, mortar and masonry. For each property, a function of mechanical decay with increasing temperature of exposure (k_θ -θ functions) was defined based on the above reported data on 300°C and 600°C exposure. These functions express a possible evolution of the respective k_θ ratio on the basis of the known values; parabolic and exponential functions were chosen according to the identified trend of each parameter, as follows (subscripts $_{300}$ and $_{600}$ denote residual values at 300°C and at 600°C):

1) $k_{300} > 1$, $k_{300} > k_{600}$ ($f_{c,\theta}$, E_θ , $f_{v,\theta}$, $f_{mc,\theta}$, $E_{b,\theta}$)

2) $k_{300} < 1$, $k_{600} > 1$ (ε_{c1})

3) $k_{300} < 1$, $k_{600} < 1$ (G_θ, $f_{bc,\theta}$, $f_{mf,\theta}$, $E_{m,\theta}$)

4) $k_{300} > 1$, $k_{600} > 1$ ($\varepsilon_{c2,\theta}$, $\gamma_{1,\theta}$)

Thus the following k_θ - θ relations express the temperature-dependent decay for each mechanical parameter:

- Compressive Strength of Masonry ($f_{c,\theta}$)

$$k_\theta = \frac{f_{c,\theta}}{f_c} = -(1.0 \cdot 10^{-6})\theta^2 + (5.0 \cdot 10^{-4})\theta + 0.9898 \qquad (7)$$

- initial Elastic Modulus of Masonry (E_θ)

$$k_\theta = \frac{E_\theta}{E} = -(2.0 \cdot 10^{-6})\theta^2 + (9.0 \cdot 10^{-4})\theta + 0.9833 \qquad (8)$$

- peak compressive strain of masonry ($\varepsilon_{c1,\theta}$)

$$k_\theta = \frac{\varepsilon_{c1,\theta}}{\varepsilon_{c1}} = (9.0 \cdot 10^{-7})\theta^2 + (7.0 \cdot 10^{-4})\theta + 1.0144 \qquad (9)$$

- ultimate compressive strain of masonry ($\varepsilon_{c2,\theta}$)

$$k_\theta = \frac{\varepsilon_{c1,\theta}}{\varepsilon_{c1}} = (9.0 \cdot 10^{-7})\theta^2 + (7.0 \cdot 10^{-4})\theta + 1.0144 \qquad (10)$$

- shear strength of masonry ($f_{v0,\theta}$)

$$k_\theta = \frac{f_{v0,\theta}}{f_{v0}} = -(0.3 \cdot 10^{-5})\theta^2 + (1.4 \cdot 10^{-2})\theta + 0.9739 \qquad (11)$$

- tensile strength of masonry ($f_{t,\theta}$)

$$k_\theta = \frac{f_{t,\theta}}{f_t} = -(0.3 \cdot 10^{-5})\theta^2 + (1.3 \cdot 10^{-3})\theta + 0.9762 \qquad (12)$$

- shear modulus of masonry (G_θ)

$$k_\theta = \frac{G_\theta}{G} = 0.9826 \cdot e^{-0.003\theta} \qquad (13)$$

- peak shear strain ($\gamma_{1,\theta}$)

$$k_\theta = \frac{\gamma_{1,\theta}}{\gamma_1} = 0.7006 \cdot e^{-0.0046\theta} \tag{14}$$

- compressive strength of brick ($f_{bc,\theta}$)

$$k_\theta = \frac{f_{bc,\theta}}{f_{bc}} = -(1.0 \cdot 10^{-6})\theta^2 + (3.0 \cdot 10^{-5})\theta + 0.9998 \tag{15}$$

- initial elastic modulus of brick ($E_{b,\theta}$)

$$k_\theta = \frac{E_{b,\theta}}{E_b} = -(2.0 \cdot 10^{-6})\theta^2 + (1.1 \cdot 10^{-3})\theta + 0.9789 \tag{16}$$

- compressive strength of mortar ($f_{mc,\theta}$)

$$k_\theta = \frac{f_{mc,\theta}}{f_{mc}} = -(5.0 \cdot 10^{-6})\theta^2 + (2.7 \cdot 10^{-3})\theta + 0.9486 \tag{17}$$

- flexural strength of mortar ($f_{mf,\theta}$)

$$k_\theta = \frac{f_{mf,\theta}}{f_{mf}} = -(8.0 \cdot 10^{-7})\theta^2 - (6.0 \cdot 10^{-4})\theta + 1.0118 \tag{18}$$

- initial elastic modulus of mortar ($E_{m,\theta}$)

$$k_\theta = \frac{E_{m,\theta}}{E_m} = -(2.0 \cdot 10^{-6})\theta^2 + (2.0 \cdot 10^{-4})\theta + 0.9964 \tag{19}$$

These functions are graphically represented in the diagrams of Figures 23-24, in the range 20-600°C.

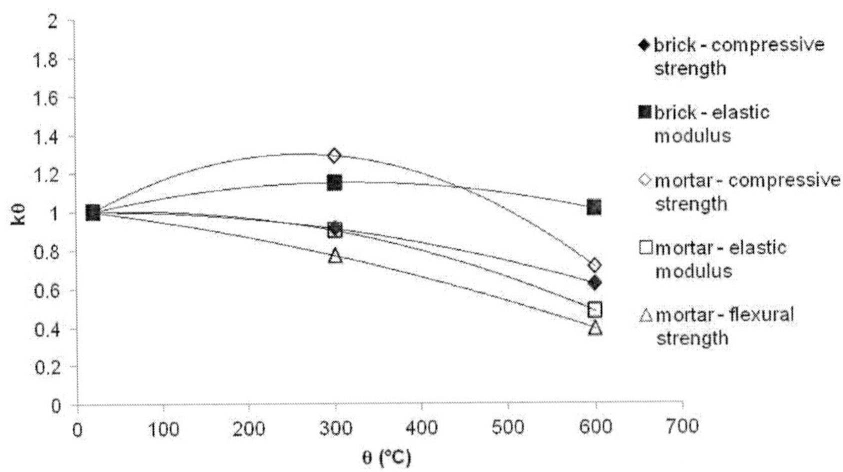

Figure 23. k_θ - θ diagrams for the mechanical properties of brick and mortar.

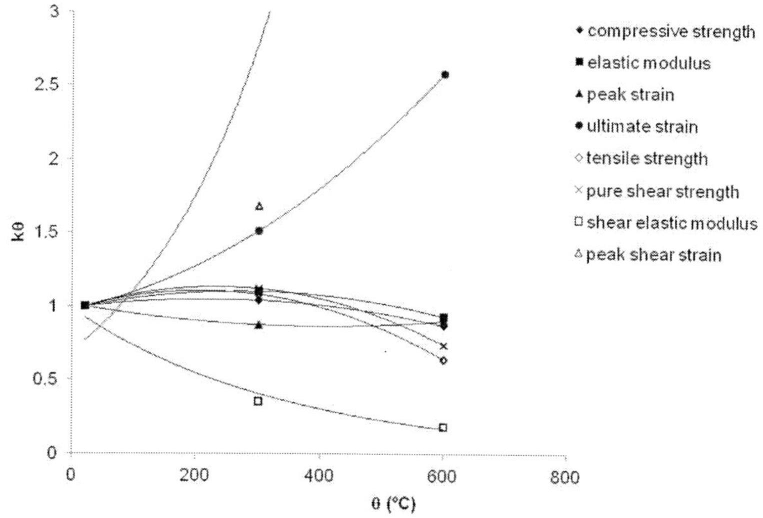

Figure 24. k_θ - θ diagrams for the mechanical properties of masonry.

As it can be seen from the graphs, $f_{c,\theta}$ and $f_{v0,\theta}$ increase at 300°C and decrease at 600°C similarly to the strength of the mortar $f_{mc,\theta}$, while the elastic modulus $E_{c,\theta}$ follows the same trend as $E_{b,\theta}$. In both cases, the variations in the properties of masonry are smaller than those of components. As well, the tensile and shear strength of masonry show a trend similar to the compressive strength of mortar; the shear modulus progressively decays similarly to the elastic modulus of mortar. The great increase in peak strain both in compression and in shear ($\varepsilon_{c1,\theta}$, $\varepsilon_{c2,\theta}$, $\gamma_{1,\theta}$) is of such entity that can't totally be ascribed to the variations in stiffness and strength of the two components; more likely, a different mechanical stress redistribution in the exposed masonry assemblies could have been induced by thermal micro-cracking. As above said, local losses of brick-mortar cohesion were actually detected in the masonry samples just after thermal cycles as the prevailing phenomenon of thermal damage. Moreover, this may also be the reason of the great decrease in shear modulus with increasing temperature; diagonal compressive tests showed that, since the early stage of collapse, interface detachment becomes more frequent than brick cracking as the temperature of exposition increases.

The observed collapse behavior of exposed masonry samples under uni-axial compression – with the asymmetrical distribution of vertical strain at the peak – put into evidence that the thickness of weakened material at the exposed side is markedly increased after 600°C exposure.

The mechanical decay is quantified by the k_θ factor, meaning the ratio between residual and original property. The values of k_θ at 300°C and 600°C for each property provided the basis for the formulation of the respective equations of mechanical decay in function of the external temperature θ. These functions are a first attempt to define the development of temperature-dependent decay, since they are based on the experimental values at 20, 300 and 600°C; their reliability may be improved by further experimental research taking into account intermediate and higher temperatures.

CONCLUSION

Currently, there is a lack of experimental information about the residual mechanical performances of masonry and clay bricks after high temperature exposition. The experimental program here described, which is the first part of a wider research, provides new experimental data on the residual mechanical properties of one type of masonry, made of traditional bricks and cement mortar and having a thickness of 25 cm, considered as a load-bearing wall having the function of fire compartment, i.e. exposed to high temperature on one side. Two maximum temperature conditions, i.e. 300 and 600°C, have been accounted for, in order to represent low-medium levels of exposition; to the purpose of setting up the functions of properties' decay with increasing temperature (k_θ - θ functions), it was preferable not to investigate severe situations, in order to enlighten the initial evolution of residual parameters in function of temperature. The heating rate (19°C/min) and duration at maximum temperature (1 hour) were chosen in order to simulate 'short-fire' conditions as features of an accidental exposure. The thermal cycles adequately reproduced, as it could be seen from temperature recordings, the situation of a load-bearing separating masonry wall.

After the performing of thermal cycles, the masonry samples showed clear signs of thermal damage on the exposed side, i.e. interface cracking and, in 600°C-exposed samples, micro-cracks in bricks and mortar joints. During the subsequent mechanical tests, the mechanical behavior up to failure could be observed in the different cases:

- *Uniaxial compressive tests*: In all unexposed and exposed cases, vertical cracks along the alignment of central mortar joints appeared at incipient failure; this feature of behavior should be ascribed to the peculiar masonry pattern. Subsequently, superficial detachments of bricks at the exposed surfaces of F3 and particularly F6 samples were observed, due to the entity of the superficial thickness damaged by high temperature exposure. The decay in compressive strength can be considered of relevance in 600°C-exposed samples. The stress-strain diagrams showed longer descending branches in both cases of exposed masonry, leading to high values of ultimate strain.
- *Diagonal compressive tests:* As the first signs of failure, exposed samples showed a clear prevalence of brick-joint interface cracking; vertical cracks in bricks were almost absent in F6 samples. The decay of mechanical strength after exposure to 600°C is considerable; the shear stiffness was found to decrease dramatically even after 300°C-exposure. A huge increase in strain at maximum stress values could be observed in all the exposed samples.

The following Tables 10-12 recollect the per cent differences in the mean values of all the investigated properties of exposed materials with respect to the unexposed. Concerning bricks, the increase in stiffness after 300°C-exposure (which is almost completely lost after 600°C-exposure) can be related to the parallel decrease in compressive strength; the bricks' strain capacity in the elastic field is thus certainly reduced by exposure to low-medium levels of high temperature.

The experimental results of mortar samples have shown to agree with data from other similar researches; a progressive decay was found to affect the flexural strength and the compressive stiffness with increasing temperature of exposure. After 600°C-exposure, the decrease in all the properties is clearly appreciable. The tendency of compressive strength - being higher than the original in F3 and lower in F6 samples - was found not to be uniform at low-medium levels of high temperature, as it was already demonstrated by other researches on cement materials. The differences in compressive strength and stiffness of masonry remain relatively small if compared to those of components. The general mechanical decay after 600°C-exposure was clearly ascertained. The remarkable increase in ultimate compressive strain and peak shear strain ($\varepsilon_{c2,\theta}$ and $\gamma_{1,\theta}$ parameters) at growing temperatures can be related to the observed features of thermally induced damage on the exposed surfaces of masonry samples. Pre-existing cracking and micro-cracking can well originate a loss of stiffness affecting the whole stress-strain behavior in diagonal compression and the post-elastic field in compression (where brick-mortar cohesion gives a less important contribution to strength); this can lead to the abnormal augmentation of $\varepsilon_{c2,\theta}$ and $\gamma_{1,\theta}$ and thus to a fictitious increase in strain capacity.

Indeed, as above said, a refinement of the functions can be attained by tests on brick, mortar and masonry after exposure at other values of maximum temperature; the initial evolution of the residual mechanical parameters (especially the compressive and shear strength of masonry and compressive strength of mortar) could thus be clarified in a better way, as well as the entity of decay beyond 600°C. Moreover, a parametric study on the exposure conditions could add information on the residual behavior of masonry after high temperature exposure; reasonably, increases in duration at maximum temperature will result in decreases in mechanical properties, but the effects of a longer duration can be expected to be less important than those of a higher maximum temperature; investigations on different cooling regimes could be very useful to understand the effects of the extinguishment of a fire, since the thermal shock due to rapid water cooling can likely worsen the mechanical decay with respect to slow air cooling.

Table 10. Per cent differences in residual mechanical properties - bricks

BRICKS		
Property	% F3	%F6
compressive strength	-9	-38
elastic modulus	+15	+1

Table 11. Per cent differences in residual mechanical properties - mortar

MORTAR		
Property	% F3	%F6
flexural strength	-23	-61
compressive strength	+29	-29
elastic modulus	-10	-51

Table 12. Per cent differences in residual mechanical properties - masonry

MASONRY		
Property	% F3	%F6
compressive strength	+4	-13
elastic modulus	+10	-7
peak strain	-12	-10
ultimate strain	+51	+158
tensile strength	+8	-36
shear strength	+12	-26
shear elastic modulus	-65	-82
peak shear strain	+68	+1327

References

[1] Nassi L, Marsella S (2008) *Sicurezza antincendio per i beni culturali*, Hoepli, Milano [in Italian].

[2] Russo S, Boscato G, Sciarretta F (2008) Behaviour of a historical masonry structure subjected to fire. Masonry International 21(1): 1-14.

[3] BIA (2008) Fire Resistance of brick masonry. Technical Notes 16, Brick Industry Association, Reston, VA, March 2008.

[4] CEN/TC 250, (2005b), *Eurocode 6: Design of masonry structures – Part 1-2: General rules – Structural fire design*, UNI EN 1996-1-2.

[5] Rossi V (2003) La sicurezza del patrimonio culturale in caso di incendio. In: Built heritage and its protection – 43th International Fire-fighters' Workshop, Moreton in Marsh, UK, 30th September–2nd October 2003.

[6] Solomon R (2008) NFPA 914 code for fire protection of historic structures. In: International Congress Innovations in Building a Safer World, Cairo, Egypt, november 2008.

[7] Harmathy T Z (1965) Ten rules of fire endurance rating. *Fire technology*, 1: 93-102.

[8] HUD (2000) *Fire ratings of archaic materials*, U.S. Department of Housing and Urban Development – Office of Policy Development and Research, Washington, D.C.

[9] Kreith F (1975) *Principi di trasmissione del calore*, Liguori, Napoli [Italian ed.].

[10] CSE-ANDIL (1995) *Ricerca sperimentale per la determinazione della resistenza al fuoco di varie tipologie di solai e pannelli murari con elementi di laterizio, svolta in collaborazione con ANDIL – AssoLaterizi*. Centro Studi ed Esperienze Antincendi – Roma (in Italian).

[11] Cooke G M E, Virdi K S, Jeyarupalingam N (1996) The thermal bowing of brick walls exposed to fire on one side. In: Interflam '96 International Conference, Cambridge, 26-28 March: 915-919.

[12] Gnanakrishnan N, Lawther R (1990) Performance of masonry walls exposed to fire. In: Proceedings of the Fifth North American Masonry Conference, University of Illinois at Urbana-Champaign USA, June 3-6, 901-914.

[13] Nadjai A, O'Garra M, Ali F A, Laverty D (2003) A numerical model for the behaviour of masonry under elevated temperatures. *Fire and Materials*, 27: 163-182.

[14] Nguyen Th-D, Meftah F, Chammas R, Mebarki A (2009) The behaviour of masonry walls subjected to fire: modelling and parametric studies in the case of hollow burnt-clay bricks. *Fire Safety Journal*, 44: 629-641.

[15] Dhanasekar M, Chandrasekaran V, Grubits S J (1994) A numerical model for the thermal bowing of masonry walls. In: 10th International Brick Masonry Conference, Calgary, Canada, 5-7th July: 1093-1102.

[16] Nadjai A, O'Garra M, Ali F A, Jurgen R (2006) Compartment masonry walls in fire situations. *Fire Technology*, 42: 211-231.

[17] Hosser D, Dorn Th, Richter E (1994) Evaluation of simplified calculation methods for structural fire design. *Fire Safety Journal,* 22: 249-304.

[18] Abrams M S (1979) Behavior of Inorganic Materials in Fire, Design of Buildings for Fire Safety. ASTM Special Publication 685, *American Society for Testing and Materials*, Baltimore: 14-75.

[19] Khoury G A (2000) Effect of fire on concrete and concrete structures. *Progress in Structural Engineering Materials*, 2: 429-447.

[20] Li L-y, Purkiss J (2005) Stress-strain constitutive equations of concrete material at elevated temperatures. *Fire Safety Journal*, 40: 669-686.

[21] RILEM TC 200-HTC (2007) Recommendation of RILEM TC 200-HTC: mechanical concrete properties at high temperatures – modelling and applications. *Materials and Structures*, 40: 841-864.

[22] Xiao J-Zh, König G (2004) Study on concrete at high temperature in China – an overview. *Fire Safety Journal*, 39: 89-103.

[23] Fu Y-F, Wong Y-L, Poon Ch-S, Tang Ch-A, Lin P (2004) Experimental study of micro/macro crack development and stress-strain relations of cement-based composite materials at elevated temperatures. *Cement and Concrete Research*, 34: 789-797.

[24] Thelandersson S (1972) Effect of high temperature on tensile strength of concrete. Nordisk Betong.

[25] Harmathy T Z, Bernd J E (1966), Hydrated Portland cement and lightweight concrete at elevated temperatures. *ACI Journal*, 63(1): 93-112.

[26] Purkiss J A (1986) High temperature effects. In: Design of structures against fire, Aston University of Birmingham, England, Elsevier Applied Science Publishers, New York, 41-51.

[27] Gawin D, Pesavento F, Schrefler B (2004) Modelling of deformations of high strength concrete at elevated temperatures. Materials and Structures / Concrete Science and Engineering, 37: 218-236.

[28] Westman A E R (1928), The thermal expansion of fireclay bricks, Engineering Experiment Station Bulletin 181, University of Illinois, Urbana.

[29] Černý R, Maděra J, Poděbradská J, Toman J, Drchalova J, Klečka T, Jurek K, Rovnaníkova P (2000), The effect of compressive stress on thermal and hygric properties of Portland cement mortar in wide temperature and moisture ranges. *Cement and Concrete Research*, 30: 1267-1276.

[30] Yüzer N, Aköz F, Öztürk L D (2004), Compressive strength-colour change relation in mortars at high temperature. *Cement and Concrete Research*, 34: 1803-1807.

[31] Husem M (2006) The effects of high temperature on compressive and flexural strengths of ordinary and high-performance concrete. *Fire Safety Journal*, 41: 155-163.

[32] Bingöl A F, Gül R (2008) Effect of elevated temperatures and cooling regimes on normal strength concrete. *Fire and Materials*, 33: 79-88.

[33] Cülfik M S, Özturan T (2002) Effect of elevated temperatures on the residual mechanical properties of high-performance mortar. *Cement and Concrete Research*, 32: 809-816.

[34] Arioz O (2007) Effect of elevated temperatures on properties of concrete. *Fire Safety Journal*, 42: 516-522.

[35] Baker G (1996) The effect of exposure to elevated temperatures on the fracture energy of plain concrete. *Materials and Structures*, 29: 383-388.

[36] Russo S, Sciarretta F (2009) Residual strength of traditional brick masonry subjected to high temperatures. In Mazzolani F (ed) Protection of Historical Buildings – ProHiTech 09, Rome, 2009, Balkema: 1417-1422.

[37] Russo S, Sciarretta F (2011) Experimental and theoretical investigation on masonry after high temperature exposure. *Experimental Mechanics.* doi: 10.1007/s11340-011-9493-0.

[38] Gei M, Bigoni D, Guicciardi S (2004) Failure of silicon nitride under uniaxial compression at high temperature. *Mechanics of Materials,* 36: 335-345. doi:10.1016/S0167-6636(03)00063-2.

Reviewed by:
prof. Mario Como, University of Tor Vergata, Rome, Italy
prof. Enzo Siviero, IUAV University of Venice, Italy

In: Brick and Mortar Research
Editors: S. Manuel Rivera and A. L. Pena Diaz
ISBN: 978-1-61942-927-7

Chapter 5

DATING BRICKS AND MORTARS OF ANCIENT AND HISTORICAL BUILDINGS

Jorge Sanjurjo-Sánchez[1]
University Institute of Geology "Isidro Parga Pondal".
University of A Coruña, Spain)

ABSTRACT

Traditional building materials provide valuable information on the past of ancient and historical buildings. Studies on the characterisation of such materials provide information on the origin of raw materials, manufacture and building technologies, or decay of materials. Also, they provide information on different building phases and periods (chronology). Among traditional materials, bricks and mortars are particularly interesting.

One of the most useful information provided by bricks and mortars is the possibility of date them (relative and absolute dating). This is very useful as information from written historical documents is unusual. Different procedures have been developed for the study of building chronologies. Relative dating methods are the study of building stratigraphy, chronotypology, mensiochronology and chemical analysis of building materials. Other methods provide absolute ages. Luminescence dating is the most used method for dating bricks but it has been applied on mortar dating in the last decade. Archaeomagnetism has been tested to date bricks in some specific cases. Other methods, such as radiocarbon, have been tested on lime mortars with relative success. Different approaches have provided useful information on the history of ancient buildings due to brick dating. Dating mortars include some methodological problems partially overcome in the last years. This paper reviews the advances occurred on the use of dating methods on dating bricks and mortars, and specialy focuses on luminescence methods.

---

[1] Address: Edificio de Servizos Centrais de Investigación, Campus de Elviña, 15071 A Coruña (Spain).E-mail: *jsanjurjo@udc.es*.

1. INTRODUCTION

1.1. Building Materials

Building materials are natural occurring substances or man-made products used for construction purposes. They exist from early prehistoric time (even Paleolithic) when stone were used on different types of early human-made building structures (e.g. housing, tombs). Mud, wood or rocks were typically natural occurring materials used on ancient (and even modern) buildings, some of them can be used to manufacture man-made (composite) materials.

Every natural geological formations or sediment of the solid Earth's crust can be used as (or to produce) stony building materials. Rocks have been commonly used depending on their local presence in a geographic area, but their use has also depended on their properties. Most of them are polymineral rocks (e.g. granite, shale) but also monomineral rocks exist (e.g. gypsum, limestone). However, the classification of rocks depends on their origin, and they show very variable chemical and physical properties.

1.2. Mortars

Binders or mortars are man-made stony materials. They have been used from early human civilizations (Elert 2002) to fill the gaps between stony blocks. Mortars are workable pastes resulting in rigid aggregate structures typically composed of mixture of an aggregate (typically sand or gravel) and a binder agent. Depending on the mortar type binders can be lime, clay, gypsum, cement or others. Natural raw materials used for produce binders are very variable, including calcium manganese, carbonate rocks, anhydrite, and muddy rocks (e.g. clay, shale, loess). Sometimes, organic agents can be used to provide specific features (water repellent, greater resistance, etc.).

The use and manufacture of binders and mortars have changed depending on the geographic area, since prehistoric and historic times until the 20th century. The manufacture of mud mortars is the most simple. However, some other mortars include burning of the raw materials during manufacture. Burning cause dehydration of minerals of raw materials (e.g. gypsum) or dissociation, as occurs in air-hardening lime. For gypsum mortars, gypsum rock is crushed and burned above 280ºC, and the product is mixed with water to get the binder. When dissociation, the formation of new mineral compounds is the goal. In the case of lime mortars, crushed limestone is burned in a kiln at 900-1000 ºC to form quicklime (calcium oxide). The quicklime is slaked thoroughly mixed with water to give lime putty (calcium hydroxide) or with less water to produce hydrated lime. Before use, hydrated non-hydraulic lime is usually left in the absence of carbon dioxide (usually under water) to mature. Putty can be matured in air for anything from a few hours to many years, an increased maturation time improving the quality of the putty. Hydraulic lime mortars set under water. It is produced in the same manner but the raw limestone contains clay and other impurities. During the burning process, calcium reacts with the clay minerals to produce silicates that enable the lime to set without exposure to air. Any no reacted calcium is slaked to calcium hydroxide. Hydraulic lime is used for providing a faster initial set than ordinary lime in some extreme conditions (Vitruvius, 2011).

1.3. Bricks

Ceramic materials have been produced from Neolithic (about 12,000 years). They are artificial burned stone materials manufactured from clays. Thus, they can be defined as inorganic (non-metallic) solids produced by cooking and subsequent cooling of inorganic fine grain minerals. They have crystalline structure or may be amorphous.

The most common ceramic material in building is fired brick. The ancient methods to produce bricks started with the raw clay, often mixed with 25-30% sand to reduce shrinkage. The clay is ground, mixed with water and pressed into moulds. The shaped material is then fired (burned) at 900-1000 °C to achieve strength. The burning of clay is accompanied by essential changes in the constituents of the raw materials, such as a change in colour, density, strength, a decrease in volume (firing shrinkage), sintering at the partly molten-to-liquid phase transition boundary (Shestoperov, 1988). Other building materials, such as tiles, are also ceramic materials. Before developing such procedure, mudbrick was used in ancient and historic times (and currently in some parts of the World). Mudbricks are unfired bricks made of a mixture of mud, sand, and water mixed with a binding material such as rice husks or straw. They are let dry under the sun for about 30 days before use.

2. Dating Ancient Buildings: Methods and Problems

2.1. Historical Data and Archaeological Methods

Finding information on written historical documents is the most suitable dating technique for ancient buildings. Historical data on construction, reconstruction or modification of buildings provide the more precise and accurate age data. However, it is usual to find fragmentary or indirect data on construction (but not reconstruction or further modifications) instead of complete and detailed information. Also, historical data are available for important historical buildings, but not for small secondary buildings. In other cases, if building remains are very ancient, no written data exist. Thus, other procedures have been developed for the study of building chronologies. Such procedures are the study of building stratigraphy, chronolypology, mensiochronology, chemical analysis of some architectural elements or absolute dating methods applied on some building materials.

Stratigraphic techniques have been developed for archaeological studies on ancient buildings in the last two decades. They are applied in a similar way than stratigraphic techniques in classical archaeology. Stratigraphic techniques used on building archaeology are used to reconstruct the history of existing buildings by direct observation (including constructive and destructive chronological sequences). It allows obtaining relative chronological sequences of building phases. The construction of sequences should be performed considering the stratigraphy of façades (including horizontal and vertical stratigraphy), surface and subsurface elements. In this technique, stratigraphic units (S.U.)

and architectural elements (A.E.) are defined and used to denote constructive and destructive phases, identifying the points of contact between units (Bortolotto et al. 2005). The technique provides relative dating of construction phases, and dating methods usually provide absolute dates for the obtained chronological sequence. Integrating stratigraphic techniques and absolute dates usually provide a complete chronological and spatial sketch of the history of a building.

Performing building archaeological analysis requires an interdisciplinary study. Geometric surveys and topographic approaches are necessary for processing of stratigraphic data. Destructive techniques are avoided, although sometimes necessary. The stratigraphic techniques combined with analysis of materials, building techniques, observation of demolition tracks and even available (despite fragmentary) historical data can allow reconstructing the whole chronological sequence of a building. Analysis of materials usually includes physicochemical characterisation of the building materials, typically bricks, mortars and stones.

Chronotypology is dating method based on the observation of building features typically used in different historical periods. It consists on the reading of the masonry techniques. The technique applies to element types, not on the whole building. Architectural elements can be dated due to stylistic features, when compared of similar know features in a known geographic area. From data of multiple buildings of different historical periods local chronotypological catalogues or curves can be built as a reference database. Distinctive features include materials (mortars, bricks, stones), construction techniques, dimensions, marks and shapes (different types of doors, windows).

Chronotypolgy is based on boundaries in the use of some materials, building features or techniques. A boundary between two periods is marked by the oldest and most recent occurrences of an element or feature in a homogeneous territory. Such technique requires the use of databases including information on the definition of a homogeneous geographical context for the use of an element or feature, the indexing of a large number of elements, the identification and classification of distinctive features, the identification of a group of instances that share the same features and the identification of a period of time for each group of features (Ferrando et al, 1989; Boato and Pittaluga, 2000). The use of a large number of samples and features minimize the deviation and provide more accurate results.

Mensiochronlogy is a kind of chronotypology applicable when the main dating features taken into account are the dimensional characters of the elements. It is based on the statement of irreproducible building materials. Building materials can be investigated as an individual identity regarding the perspective of irreproducible complexity. Thus, the characteristics of brick or stonewalls provide information on the period when a wall was constructed based on the morphology and type of bricks or stones. In fact, brick mensiochronology typically uses brick dimensions to obtain chronological dates for bricks and therefore for walls of buildings constructed with them, as the bricks have been manufactured with different dimensions at different historical periods in some geographical areas. The method is based on databases, as occurs with chronotypology and the precision of the data is strongly dependent on the geographical area and the quality of the database. It has been speculated that deviation ranges from 5 to 20 years can be obtained in some areas (Boato and Pittaluga, 2000). Stone mensiochronology (Figure 1) is more complex and the age range can be of one or two centuries, although the mensiochronological analysis should be compared with other chronotypology data to get a reliable result.

Figure 1. Example of mensiochronlogy applied on a church façade. Two types of rocks have been used in the constrution and two building phases can be differentiated.

2.2. Absolute Dating Methods

Absolute dating methods provide absolute ages for different kind of geological and archaeological objects. When historical documents are not available or they do not provide historical dates, and archaeological of building stratigraphy is not possible (due to the absence of databases or reference data) or do not solve the different chronological hypothesis proposed, absolute dating methods provide a reliable, and even definitive tool. Even, it is common that some parts of a building cannot be surely dated, and absolute dating methods provide an extremely useful way to solve the problem. Such techniques provide an absolute age for an architectural material, but not for a building. Thus, some problems can be found when these techniques are used. One of them is the reuse of materials. When materials are reused, usually provide the age of the first use, and thus provide overestimated ages. Also, deliberated of accidental man-made modifications on some materials can distort the properties of the dated materials, resulting in underestimated or overestimated ages.

The main absolute dating techniques used for building materials are dendrochronology, radiocarbon and luminescence dating. In some cases, other techniques such as archaeomagnetic dating can be used for some materials. Dendrochronology is based on tree-ring dating of wood used as building material. The pattern of tree-rings provides the time at which rings were formed. Radiocarbon is commonly applicable on organic materials. Thus, it is useful to date wood, but also mortars containing organic matter (e.g. charcoal, bones, vegetal fragments). It has also successfully applied to date lime mortars in some cases. Luminescence dating has extensively been applied to date ancient bricks. Thermoluminescence dating provides the age of the last heating of a brick, commonly due to the cooking of the brick in the manufacture process. Optically Stimulated Luminescence (OSL) has been developed in the 90's and it has been succesfuly applied on fired bricks and tested to date lime mortars with variable results. Archaeomagnetic dating should also be applicable for brick dating in some cases, if they are in situ magnetized during an archaeological time. The characteristics and applicability of such methods are explained in the next points of this chapter.

3. Brick Dating

Fired brick is today one of the most extended building materials. Its use has been very frequent in past times, from the Roman Empire. From that period it has been an ubiquitous building material in many areas of Europe and the Mediterranean Basin. Thus, the study of bricks is useful to assess the historical construction and evolution of buildings. Bricks allow dating by variable approaches, by means of methods related above (see point 2) that provide relative chronologies to absolute dating methods, namely luminescence dating.

Brick dating is possible due to existing absolute dating techniques applicable to ceramic materials. Among absolute dating methods, luminescence dating is the most extended. The thermoluminescence (TL) as a dating tool was reported in the 1950s (Daniels et al. 1953), and developed in the 1970s by a research group of the Oxford University, headed by Martin Aitken, to date ceramics, other archaeological objects and even sediments, the latter with uneven results (Aitken, 1985).

3.1. Luminescence Dating

Luminescence is the emission of light from crystalline materials (minerals). Naturally occurring radioactivity causes the excitation of atoms within a mineral crystal lattice. As a consequence electrons are activated at higher energy states and some of them are captured at levels called 'electron traps'. With increased time, more and more electrons will be captured at the traps and so the luminescence signal will increase at a constant rate (Aitken, 1985). Since the number of traps is limited the luminescence will reach a saturation value (saturation dose). The release of trapped electrons occurs in the form of light (luminescence) and requires a stimulus.

The intensity of light emitted by a mineral is proportional to the amount of electrons trapped, and therefore it is proportional to the energy received by the mineral due to radioactive exposure since the signal was zeroed for last time. Such correlation allows calculating the time period elapsed from the last zeroing (Aitken, 1985). The trapped electrons stored within minerals can be released in the laboratory producing a luminescence signal. Heating the sample releases the trapped electrons and the resulting signal emission is called thermoluminescence (TL). The TL signal of a mineral typically comprises several peaks, due to single or composite traps. It is considered that TL peaks obtained at higher temperatures are originated from deeper traps.

When a mineral is stimulated by light, traps are emptied in a few seconds. The emited (and measured) signal is termed optically stimulated luminescence (OSL). The OSL signal decreases very fast, but faster for quartz than feldspars. The light used for stimulation depends on the use of one or other mineral: blue diodes (LEDs) are commonly used for quartz while infrared stimulation (IRSL) is used for feldspars.

Although different minerals can be used for luminescence dating, this method is usually limited to quartz and feldspars. The use of these minerals is due to their ubiquity in geological and archaeological objects (including bricks) and as they are resistant to weathering.

Quartz is considered as the most adequate mineral for luminescence dating. The luminescence properties of quart are relatively well-known, above all regarding the fact that it is not affected by the phenomenon of anomalous fading, as occurs with some feldspar. There are a few circumstances when quartz may not be a suitable mineral for dating: it is absent, it contain feldspar inclusions that complicate the detection of the luminescence, it shows low luminescence signal of sensitivity changes, it displays anomalous fading (volcanic quartz) (Bonde et al. 2001), or the dose is saturated. This last fact does not typically occur in archaeological dating, as usually saturated doses exist on ancient (Pleistocene) sediments.

The use of feldspars shows disadvantages when compared with quartz, as the luminescence signal usually exhibit anomalous fading. This consists on the loss of part of the luminescence signal with time, and ages must be corrected by different possible approaches. However, potassium-rich feldspars are commonly used as they have bright signals and a higher saturation dose than quartz (they allow dating older Pleistocene sediments). Also, K-feldspars (potassium-rich feldspars) contain substantial proportions of ^{40}K, which lowers the uncertainty from external dose rate.

It is common in archaeological ceramics, including bricks, the use of polymineral fine grains for luminescence dating. Fine grains are a mixture of unknown minerals containing variable amounts of quartz and feldspars. This technique is advantageous when quartz inclusions are not common in the samples. Moreover, the alpha contribution to the dose rate is more important in fine grains, diminishing the importance of the gamma dose in heterogeneous environments (as some buildings). However, the unknown nature of the minerals can result in problematic signals and anomalous fading is frequent when this mineral fraction is used (Aitken, 1985).

3.1.1. Luminescence Age Equation

The luminescence age equation is the ratio between the total dose absorbed by the minerals (estimated as equivalent dose) and the dose-rate of ionizing radiation in the environment surrounding the dated material. The calculated age of the dated material is the time elapsed since the last exposure to sunlight or heat before burial (Aitken, 1985). Thus:

$$\text{Age} = \text{ED} / \text{DR}$$

where ED is the Equivalent Dose and DR the Dose Rate. The equivalent dose is an estimation of the total energy accumulated by the measured minerals since the last exposure to light or heat. It usually measured in Gray (Gy).

The dose rate is the energy delivered each unit of time (year) from ionizing radiation from the surrounding environment. Naturally ionising radiation occurs in the form of alpha and beta particles, gamma radiation and cosmic rays. Such ionizing radiation occurs due to the natural radioactivity of the dated materials and the surrounding environment. The radioactivity can be assessed in terms of dose rate as radioactivity received per unit of time (Aitken, 1985).

TL dating of brick is possible as bricks are ceramic materials, made by cooking clays but they are also composed of a number of crystalline inclusions (mainly quartz and feldspars) in the ceramic matrix. Such inclusions can be used as dosimeters as their atoms are excited due to natural radioactivity mainly from the ^{238}U and ^{232}Th decay series and ^{40}K content within the ceramic material and surrounding building materials (e.g. stones, mortars). The absorbed dose is the dose accumulated since the cooking occurred during the brick manufacture.

The development of OSL protocols from the 1980s (Huntley et al. 1985; Hütt et al. 1988) has resulted in an exponential growing of luminescence laboratories and dates. OSL shows important advantages over TL requiring smaller sample size, a more variable number of protocols exist, and laboratory techniques allow recording weak luminescence emitted from minerals. Also, single aliquots of small number of grains (Duller 1991; Murray and Wintle 2000) and even single-grains can be used to date (Murray and Roberts 1997). The irruption of the SAR (Single Aliquot Regeneration) protocol (Murray and Wintle 2000) has provided an increasing number of dates and more precise results for luminescence dating.

The use of OSL for dating bricks has been revealed as a very useful technique (Bailiff, 2008). Among advantages of the OSL for dating bricks several can be highlighted: it requires a few sample amounts (few grains) as OSL is more sensitive, the SAR protocol allows test the reliability of each measured aliquot, it provides equivalent doses by interpolation and not extrapolation, providing lower deviation.

3.1.2. Dose Rate Estimation

The dose rate is the result of four types of ionizing radiation: alpha, beta, gamma and cosmic. Alpha and beta radiation are particles while gamma and cosmic are rays. The three first are produced by the radioactive content of the natural radioactivity of the dated and surrounding materials. Such content is due to uranium (U), thorium (Th) and potassium (K). ^{40}K decays to stable isotopes by the emission of beta particles and gamma rays while U and Th decay are more complex. As an example, they decay of ^{238}U to unstable isotopes (with radioactive emission) forming a chain (decay series, see table 1) until reach a final chain stable isotope (^{210}Pb). Cosmic rays are produced by ionising cosmic radiation and can be estimated from geographic position, altitude and burial depth of the sampled material (Prescott and Hutton, 1994). The effect of the different types of radiation on the dose rates is not the same. Thus, alpha particles typically travel a few microns through geological and archaeological materials (such as ceramics), while beta particles reach a few millimetres. Gamma rays reach about 30 cm. Thus, gamma and cosmic rays constitute the external dose while beta and alpha (and partially gamma) the internal dose.

Table 1. Main radioactive isotopes of uranium and thorium (decay series)

Uranium series			Thorium Series		
Isotope	Half-life	Decay	Isotope	Half-life	Decay
^{238}U	4.468x10^{9} years	alpha	^{232}Th	1.405x10^{10} years	alpha
^{234}Th	24.1 days	beta	^{228}Ra	5.75 years	beta
^{234}Pa	1.17 min	beta	^{228}Ac	6.25 hours	beta
^{234}U	2.48x10^{5} years	alpha	^{228}Th	1.9116 years	alpha
^{230}Th	7.7x10^{4} years	alpha	^{224}Ra	3.6319 days	alpha
^{226}Ra	1600 years	alpha	^{220}Rn	55.6 days	alpha
^{222}Rn	3.82 days	alpha	^{216}Po	0.145 sec	alpha
^{218}Po	3.05 min	alpha	^{212}Pb	10.64 hours	beta
^{214}Pb	26.8 min	beta	^{212}Bi	60.55 min	alpha+beta
^{214}Bi	19.8 min	beta	^{212}Po	299 nsec	alpha
^{214}Po	162 μsec	alpha	^{208}Tl	3.053 min	beta
^{210}Pb	22.3 years	beta	^{208}Pb	stable	.
^{210}Bi	5.01 days	beta			
^{210}Po	138.4 days	alpha			
^{206}Pb	stable				

There are three methods to determine the dose rate: direct in situ measurement using dosimeters, measuring the dose rate using counting devices or by determination using conversion factors after chemical analysis of U, Th and U concentrations in the materials.

In any case, some assumptions are made to get reliable final ages (Aitken, 1985): the radionuclide concentrations are constant in time, the material is homogeneous and the system is an infinite matrix.

In situ measurements provide valuable data on the radiation of the materials surrounding the dated object, even in heterogeneous environments. Two methods provide in situ data. Portable gamma spectrometry provide a spectrum of lower quality than high resolution gamma spectrometry but sufficient to assess the activity of U and Th series, and radioactive K. Another possibility is the use of TLDs. They are artificial phosphors sensitive to ionizing radiation. TLDs are situated in the sampling point for a limited time (from few weeks or months to ~~a~~ few years) and the accumulated luminescence signal is estimated by TL.

Counting devices are another method to assess the dose rate. Alpha particles can be counted by using a thick source alpha counter (TSAC). This method allows measuring the total number of alpha particles (coming from U and Th decay series) with simple sample preparation. Also, beta counting allows assessing the dose rate. However, alpha and gamma spectroscopy are the best methods to measure the dose rate of the U and Th decay series if disequilibrium exist. Alpha spectroscopy requires complex and time consuming sample preparation but provides measures of each isotope undergoing alpha decay. High resolution gamma spectrometry provides results of activity of the whole decay series of U and Th, after simple sample preparation.

Different chemical methods allow knowing the concentration of K, U and Th on the dated materials to asses the dose rate. For this purpose, the material is usually crushed and analysed by neutron activation analysis (NAA) or inductively coupled plasma mass spectrometry (ICP-MS). Such methods allow measuring the concentration of trace elements such as U and Th, while K is a minor elements (typically concentrations oscillate between 0.5 and 5% in ceramic materials) and can also be measured by other methods such as atomic absorption spectrophotometry (AAS), flame photometry or X-Ray fluorescence (XRF). From the element concentration, the dose rate can be calculated by using conversion factors (Adamiec and Aitken 1998). Some problems can arise from the use of chemical methods. The concentration of U and Th can be close to the detection limits of the used method (and thus they can be overestimated). Moreover, the measurements are made on subsamples, not representative of the whole sample if some heterogeneity exists. Also, some methods (such as ICP-MS) include etching steps with strong acids, not reliable when highly resistant minerals. The use of conversion factors also arises some problems as they assume that the U and Th decay changes are in equilibrium (the concentration of parent and daughter isotopes is the same) (Preusser et al, 2008). However, equilibrium not always exists and some daughter isotopes of ^{238}U are more soluble than others and they can be leached, while other are gaseous and can be lost. Despite this problem, ceramic materials are a relatively closed system. Regarding U, two decay U series exist (^{238}U and ^{235}U) and chemical methods do not allow separate them.

To calculate the dose rate from counting devices and chemical methods (but not from in situ measurements) it is necessary to estimate the water content of the dated samples, as water absorbs ionizing radiation. It is very difficult to estimate the water content, but it is possible to measure the saturation water content on the dated samples and estimate a range of reasonable water content oscillation.

3.1.3. Limitations of Dating Bricks by Luminescence

Dating bricks by luminescence is a very advantageous technique due to the availability of large quantities of material (large brick walls provide abundant samples), the homogeneity of the environmental radioactivity (in most historical buildings bricks were purposely manufactured) and the low influence of environmental water fluctuations (open-air objects such as bricks at certain height are scarcely affected by water oscillations from the foundations and depends on the diurnal and seasonal oscillations or air humidity) (Martini and Sibilia, 2006). However, different questions regarding the history of buildings can affect to the reliability of their age assessment by luminescence dating of bricks.

In some buildings fire occurred in the past, and building materials (such as bricks) could be affected by high temperature. Such exposition to high temperatures causes the 'zeroing' of the luminescence signal and thus, the luminescence age of such bricks will be underestimated.

Architectural modifications due to destructions, modifications and reconstructions can distort the dosimetry and subsequent age calculated for bricks. If changes occurred in the surrounding of a dated bricks, changes in the surrounding radioactivity (in the gamma dose rate) can be introduced, causing inaccurate age estimations. It is also known that bricks have been typically reused in the past. The reuse of bricks has been found in different studies of brick dating on buildings of Europe (Bailiff, 2008; Blain, 2010; Blain et al., 2010; Martini and Sibilia, 2006). This reuse has specially been evident during the middle ages. If bricks are reused, important changes in the gamma dose rate have been introduced in the past, and the zeroing of the luminescence signal corresponds to the brick manufacture before its first position. Thus, the age of a wall where a reused brick is dated will be overestimated. Despite this problems and limitations TL/OSL dating of bricks is considered as a very resolving technique in most cases, as different studies have demonstrated (Bailiff, 2008; Guibert et al., 2009; Blain et al., 2010).

3.2. Other Methods: Archaeomagnetism

Other absolute dating techniques than luminescence have not been developed for bricks or generally for ceramic materials. However, archaeomagnetic dating is sometimes possible. Archaeomagnetic dating is a field of the paleomagnetism developed to date archaeological or geological materials magnetized during an archaeological time. This is possible because both the direction and the intensity of the Earth's magnetic field vary spatially across the Earth surface over time. Thus, archaeomagnetism requires the study of the Earth's magnetic field during archaeological time (Thellier and Thellier, 1959).

Archaeomagnetic dating is possible on ceramic samples that remained in position on cooling down from firing by means of thermoremanent magnetism (Aitken, 1974), which is induced by the Earth's magnetic field (at temperatures near to 700ºC). Thus, this technique allows dating bricks, but it is limited to bricks used in kilns and furnaces (where in situ heating has occurred) or fires that affected to buildings. In the first case, the archaeomagnetic age provides the age of the last (or lasts) heating of the kiln. In the second case, the result provides the age of the fire. Dating the cooking process of the brick is not possible (as occurs with luminescence dating).

Archaeomagnetic dating is possible on certain ferromagnetic minerals: in rocks or materials able to acquire permanent magnetization. Such minerals are commonly hematite and magnetite. Therefore, archaeomagnetic dating is possible when some clays are baked in hearths, ceramic artefacts are fired or walls or floors are burned. After firing, the materials cool down acquiring a thermoremanent magnetisation (Aitken, 1974). Secular variation, combined with the recording mechanism of thermoremanet magnetisation, provides the basis of both archaeodirectional and archaeointensity dating techniques. However, archaeomagnetism can also been referred not as an absolute dating method but a "derivative" dating method o a correlation method (Aitken, 1974; Sternberg, 1997). This is because dating is performed by matching the archaeomagnetic measurement to the relevant phenomenon established before.

Temporal variations of the geomagnetic field are due to different phenomena. The direction of the magnetic field undergoes secular variations. Such direction at a surface point changes at irregular rates but over the last 150 years the average change in direction has been around 1º every 1 decade. The intensity of the field also shows changes of several percent every century (it has varied by about 0.5% per decade). Secular variations involve changes on an age range from decades to millennia. Moreover, the polarity of the field occasionally exchanges north and south poles, randomly (at an average of once every 250 ka) (Clark et al., 1988; Sternberg, 1997).

Archaeomagnetic dating requires information on declination, inclination and intensity, but studies using all the three parameters are not common. However, in most cases only the directional information supplied by declination and inclination of the remanence vector is available. To determine the intensity is more difficult than estimations of the direction, and detailed studies on the intensity are scarce. Thus, theoretical models of the evolution of the Earth's magnetic field must be built up from direct observations (Aitken, 1974; Sternberg,

1997). Archaeomagnetic dating requires master curves of the Earth's magnetic field extended over the past. Master curves have been built up indirectly using magnetic measurements made on independently dated archaeological and geological materials. In some parts of the World direct observations of the magnetic field have also been made, extended over a few centuries (Casas et al., 2007; Morales et al., 2011).

4. Mortar Dating

Mortars are the most used building material from ancient times. They have been used as join mortars (to join other building materials such as bricks, stones or wood), plasters (to cover or coat mansonries) or renders (as protecting covers of plasters). They are artificial mixtures of different materials or modified raw materials. Thus, they have been very interesting objects for archaeometric and architectural studies, as the characterisation of ancient mortars has been an interesting source of data from past cultures and technologies. Studies on the characterisation of ancient mortars are variable and they are not standardized. However, they have provided information on the provenance of raw materials used to manufacture them (and them information on commercial exchanges and transport in the past) and on the technology (crushing, sieving, mixtures, cooking) of manufacture (Casadio et al., 2005; Middendorf et al., 2005a, 2005b; Barba et al., 2009; Elsen, 2006; Sanjurjo-Sánchez et al. 2010).

Different kinds of mortars have been commonly used in historical buildings from mud to gypsum or more generally lime. Most mud mortars are made of mud or mixtures of mud and sand. They are considered as not suitable for dating purposes, as usually do not contain organic materials (radiocarbon dating is not possible). However, a recent insight on the OSL dating of ancient mortars has provided valuable information on the possibility of dating mud mortars (Feathers et al., 2008). Gypsum mortars are more common and widely used, although their use have been reported in early human civilizations in Mesopotamia (Sanjurjo-Sánchez and Montero Fenollós, 2012). However, gypsum mortars are considered today as not suitable for absolute dating.

Other more modern mortars are Portland cement. Such kind of mortars has been used from the nineteenth century and at present it is the most used mortar. This mortar consists on a mixture clinker (more than 90%) with calcium sulfate and other minor constituents. The clinker is a hydraulic material made of calcium silicates (about 66%) and other aluminium- and iron- containing phases (the ratio of CaO to SiO_2 is ≥ 2). It is not possible to date Portland cement today, but it has not been used in buildings older than 19th-20th centuries. Portland cement is very similar to the Roman cement or concrete. Roman concrete (also called *Opus caementicium*) is hydraulic cement made using pozzolana (volcanic sands). Vitruvius explained its manufacture mode by mixing lime and pozzolana. Some attempts to date such kind of mortars has been partially successful by OSL (Goedicke, 2011).

Lime mortars have been the most historically used worldwide. They are commonly used as joint mortars, plasters, renders or even whitewash. They are composed of lime binder and aggregate sand, although sometimes gravel, crushed brick or rock are used as aggregate. Some attempts to date them have been performed in the last 40 years. Attempts have been partially successful with several techniques, although none of them is commonly performed for routine dating. The most intense research on the absolute dating of mortars has been based in three techniques: geochemical, radiocarbon and OSL dating.

4.1. Geochemical Procedures for Dating

Geochemical dating is based on the comparison of the geochemistry and mineralogy of the mortars of a building, when the age of a part of the building phases is known due to independent data (such as written documents). This method is based on the fact that the Ca cation of the calcite structure may be partially substituted by other elements including Mg, Fe, Sr, Rb and Ba (Deer et al, 1992), depending on the palaeoenvironment. Clay minerals contained in limestones used for binder in lime mortars react with the CaO during the calcination process to produce silicates such as gehlenite ($Ca_2Al_2SiO_7$), providing hydraulic properties to the mortar. The Ca of calcites may be partially substituted by the elements referred above. As a result, the relative concentration of such elements in the original limestone is reflected in the hardening products of the mortar. Therefore, when two mortars have been manufactured from two different limestones (or limestones with different geochemistry) it should be possible to distinguish them considering the concentration of these minor elements (Vendrell-Saz et al., 1996).

It can be noticed that this method is not very different of geochemical methods used for relative dating. Indeed, this is a relative dating method, but provides absolute ages when we know that a building has been built in different periods and we have ~~written~~ documentation on the ages correspond to the different phases, but we do not know the correspondence among phases and ages.

The use of the method is very limited as it allows obtaining the absolute ages of building phases when historical documentation about the age of a building is available, and the chronology obtained is usually relative. Other similar procedures have been proposed with variable success (Casadio et al., 2005; Barba et al., 2009; Sanjurjo-Sánchez et al. 2010) with variable results but no further discussion has been performed. However, such methods have been proven as useful to assess the origin of raw materials and the recognition of different building phases in a same building. Absolute dating is not possible today with geochemical analyses.

4.2. Radiocarbon Dating

Radiocarbon dating has been used from 1950 to date biological and some inorganic materials. The age range of this technique is ~ 50 ka BP, so it is very useful for archaeological studies (and its use has been extended in archaeology). ^{14}C is a radioactive isotope of carbon produced in the upper atmosphere. At this height cosmic-rays cause secondary thermal neutrons that bombard ^{14}N nucleus to produce ^{14}C nucleus (by transmutation). Such carbon is quickly oxidized to form $^{14}CO_2$ (in a few hour or days). The atmospheric circulation on the Earth's surface causes the distribution and mixing of such $^{14}CO_2$ (mainly by wind). The molecule is mostly deposited in the sea (or water) surface and about 1% is incorporated to the biosphere by photosynthetic processes (Figure 2). Living organisms keep the same ^{14}C

content in biological systems (in approximate equilibrium with atmospheric ^{14}C) by metabolic processes until death, when the amount of ^{14}C begins to decrease by radioactive decay.

In the case of inorganic materials, carbonates can be used for radiocarbon dating. $^{14}CO_2$ is deposited and dissolved in water surfaces.

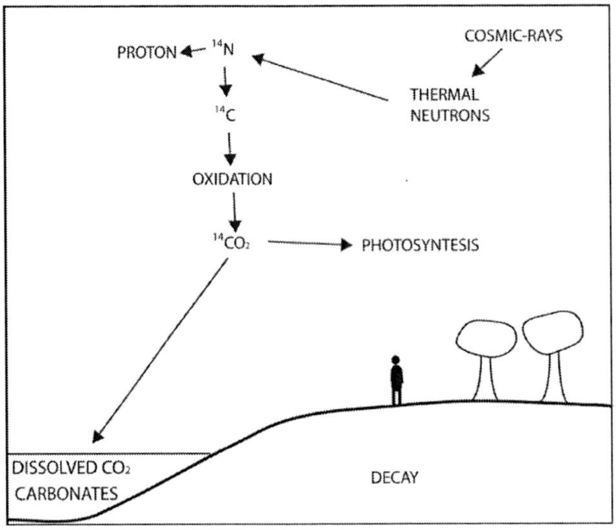

Figure 2. Model for radiocarbon dating (modified from Taylor, 1997).

Underwater, $^{14}CO_2$ reacts with some cations and usually with Ca to from carbonates. Although they can be removed by acid washing (above all in soils) they can be used for dating after precipitation. The carbonates in water (seawater or water bodies) are formed by exchange reaction, so that CO_2 leaves as well as enters the ocean or water body again except for a small portion locked up for long periods. Thus atmosphere, biosphere and water bodies form exchange reservoirs throughout the carbon circulate quickly and the residence time in any of them (except in the deep ocean) is much shorter than the mean lifetime of a ^{14}C atom.

The half-life for used radiocarbon dating is 5568 (±30) years (8033 yr mean life) although it is considered as closer to 5730 (±30). The age range comprised for radiocarbon is from 300 yr to about 40 or 60 ka BP. The statistical constraints associated with the measurement of ^{14}C concentrations are usually the dominant component of the analytical uncertainty (statistical error). Such error is usually expressed as ± one standard deviation ($\pm 1\sigma$). For radiocarbon dating of a sample some basic assumptions must be considered (Taylor, 1997):

a) the concentration of ^{14}C in each carbon reservoir has remained constant over the ^{14}C timescale;
b) quick mixing of ^{14}C occurs throughout the various carbon reservoirs,
c) carbon isotope ratios in a sample have not been changed except ^{14}C (by decay);
d) the natural ^{14}C of a sample can be measured to appropriate levels of accuracy and precision.

Samples form freshwater, such as carbonates are often affected by the known reservoir effect. In such cases, the reservoir corrected radiocarbon age can be calculated measuring the apparent age in control samples and correcting for the observed deviations.

Another problem of radiocarbon is that the ^{14}C production rates have not remained constant over the ^{14}C timescale. Thus, radiocarbon ages do not correspond to solar ages, due to the variability of the ^{14}C cosmic-ray production in the upper atmosphere. Therefore, radiocarbon ages must be "calibrated" considering international calibration curves (Stuiver and Polach, 1977).

The possibility of dating the lime binder of mortars by ^{14}C has been extensively studied since 1960s (Delibrias and Labeyrie, 1964; Folk and Valastro, 1976). Mortars can be dated by radiocarbon when they contain some organic components (such as wood, charcoal of bones) although the obtained age corresponds to the dated object, and not always to the mortar age. Theoretically, radiocarbon dating in lime mortars can be performed on the $CaCO_3$ of the lime binder. Because of this, the ^{14}C activity of the binder can be measured and it can be dated and converted to calendar years using normal calibration procedures. The binder of lime mortars is formed due to the slaking with water of the quicklime (calcium oxide) and further reaction of atmospheric CO_2 as the mortar hardens (Heinemeier et al., 1997). The quicklime is obtained after crushing and burning limestone or marble fragments in a kiln to 900°C. At such temperature CO_2 is completely released leaving the quicklime. Atmospheric CO_2 is fixed on the lime during the mortar hardening (figure 3). In such moment the ^{14}C clock begins ticking, but the practice has been proven as more difficult. Such difficulties has been found and studies during different experiences dating lime mortars.

There are some well-known problems due to the mortar content in old limestone. Such old limestone appears as lumps within the binder. They are formed by incompletely burned limestone of marble fragments. As such fragments are derived from fossil carbonate deposits, they lead to apparent radiocarbon ages that are too old (Heinemeier et al., 1997; Hale et al., 2003). An additional similar problem can be found due to the aggregate composition of the lime mortars. It is very common to use sand, gravel of crushed ceramic materials as aggregate. If limestone sand is used more fossil carbonate is added to the mortar hindering the radiocarbon dating.

Further problems dating mortars are due to the hardening process. Mortar hardening depends on the porosity, air circulation and water content within a mortar. Moreover, hardening of mortars takes from some days to some years, even decades (Elert et al., 2002) yielding radiocarbon ages younger than the building age. It must be considered that the hardening rate of a mortar is not the same on the mortar surface that in inner parts, hindering the measuring of an accurate age.

Finally, due to the exposure of building materials to variable environmental conditions a different chemical processes (e.g. dissolution, weathering) can cause changes on the lime ^{14}C content. Also, mortars are more susceptible to dissolution in urban areas due to acid rain. Thus, dissolution and re-crystallization of $CaCO_3$ in mortars is more or less common, and younger carbon can be incorporated yielding dates too young (Amoroso and Fassina, 1983; Sanjurjo-Sánchez and Alves, 2011; Heinemeier et al., 1997).

Different methodological approaches have been used to date the lime of mortars. Methods to date lime mortars by ^{14}C dating require previous analysis of the samples to study the carbonaceous character of the binder. It is necessary to remove limestone components of the binder. This step has been performed by different procedures. A mechanical procedure was tested (by Nawrocka et al., 2005) based on the fact that limestone in aggregate is stronger than the more porous mortar carbonate. Thus, the mortar is carefully crushed and repeatedly frozen and thawed to separate the binder from other carbonates. After that, fragments are treated by microwaves and collected under binolular observation. Results indicated successful ages for some samples but not for others.

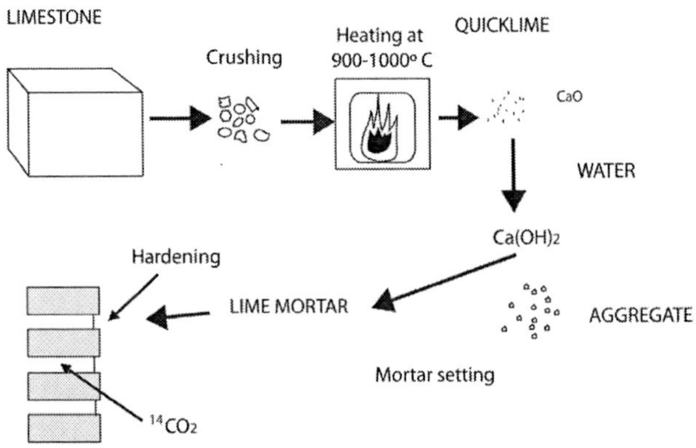

Figure 3. Process for production of lime mortars. The binder absorbs $^{14}CO_2$ from the atmosphere during mortar hardening. Such $^{14}CO_2$ is potentially suitable for radiocarbon dating.

Lindroos et al. (2007) proposed previous cathodoluminescence (CL) and mass spectrometry (MS) studies to identify and characterize carbonates. The dissolution rate, and $^{13}C/^{12}C$ and $^{18}O/^{16}O$ ratios were measured from different experiments with mortars. Such mortars were dissolved in phosphoric acid collecting successive CO_2 increments for analysis. With this method a CO_2 evolution pressure curve and a ^{14}C age and stable isotopes profiles of successive dissolution increments were represented to interpret the ^{14}C profiles and identify contaminated carbonates. The method is based on the assumption of limestone dissolution rates and their stable isotopes signatures (Lindroos et al., 2007). Considering such procedure different fractions are extracted and the radiocarbon age is calculated for them. ^{14}C profiles are plotted for each sample showing possible dating errors due to contamination. This method combined with mechanical separation, age control, and characterization with petrologic microscopy and chemical analysis have provided conclusive results for an important percentage of samples (80%) in a further study of Heinemeier et al. (2010). They presented results of this procedure based on the assumption that two crystal types correspond to lumps (rapidly dissolved) and binder (remaining crystals).

Marzaioli et al. (2011) proposed an alternative method to separate the binder from other carbonate sources of the lime, based on the procedure described by Nawrocka et al. (2005). They found good results form laboratory produced mortars at different calcination temperatures.

4.3. Luminescence Dating

Luminescence dating of some mortars has been tested providing promising results. For lime mortars different studies have been carried out on the mortar quartz of the aggregate sand. The use of the quartz sand requires the exposure to daylight of quartz grains during the mortar manufacture (before the mortar setting) enough time (or enough light intensity) to bleach the residual dose of ionizing radiation (Figure 4). This occurs during the extraction and transport of the sand, and the mortar manufacture. Also, the shielding of grains from daylight within the mortar is required. Such requirements have been shortly studied for dating purposes (Botter-Jensen et al, 2000; Zacharias et al., 2002; Goedicke, 2003; Jain et al., 2004).

A first study on the use of mortar in retrospective dosimetry was reported by Botter-Jensen et al. (2000). They applied OSL to joint mortar samples from a building used for storage of radioactive materials. They separated quartz grains and found that at least some grains were well bleached during mortar mixing and manufacture. First OSL dating of an ancient lime mortar (Zaccharias et al. 2002) was successfully performed from quartz grains of the aggregate sand, opening a new way to date buildings constructed using not typically datable materials such as rock ashlars. However, related to luminescence dating some problems have just been partially resolved.

In OSL dating, the ED of a mineral grain is mainly controlled by three factors: the post-depositional or burial dose (historical or archaeological ED), its intrinsic sensitivity (luminescence emission per unit dose, it depends on the mineral sensitivity) and any residual (geological) ED remaining in the mineral grains before burial in the dated material due to incomplete zeroing (poor daylight exposure). Poor bleaching causes incomplete zeroing of the residual signal and age overestimation. In optical dating procedures, the OSL signal is derived from a number of grains that form the sub-sample (aliquot). A typical sample disc holds 5 mg of material; this amounts to 1000 grains when a 150 μm grain size is used. In this sense, Wallinga (2000) suggested that the best method to check whether the paleodose of a sample might be overestimated as a consequence of poor-bleaching is to use small aliquots (~100 grains) and to look at both the spread in the paleodoses and the symmetry of the dose distribution. Symmetrical dose distributions on very small aliquots are indicative of homogeneous bleaching, but even these doses not necessarily mean that its OSL signal was completely reset during the last transport event or before the mortar hardening in our case. Thus, obtaining paleodoses from several aliquots implies the use of statistical tools. In this sense, different methods have been developed to date geological sediments (i.e. marine, aeolian, fluvial).

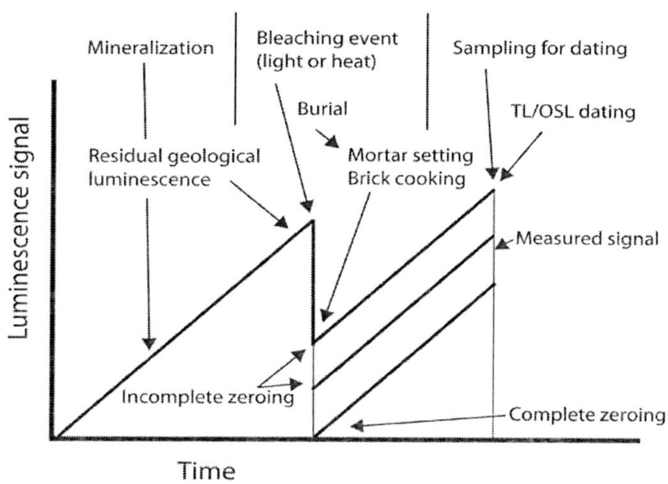

Figure 4. Model for luminescence dating of bricks and mortars. Incomplete zeroing of the residual geological signal can be responsible of age overestimation.

The existence of a residual dose in the quartz sand of mortars (due to incomplete bleaching of the geological signal because insufficient exposure to daylight) has been regarded as particularly problematic to get reliable age estimates.

In the case of the mortar sand, once used in construction, quartz grains within mortar are sufficiently protected from light impact so that an archaeological dose can accumulate (Goedicke, 2003). Also, methods of mortar preparation have remained virtually unchanged and during transport of quartz sand from the gravel pit to the mortar production site, bleaching occurs and is stopped when the grains are covered by a calcite layer. However, due to the bulk amounts of transported sand, not all grains could have been sufficiently exposed to light to bleach the luminescence acquired over geological times. Hence, the incomplete bleaching of some grains will result in a heterogeneously mixture of bleached and unbleached grains. Goedicke (2003) established a bleaching model assuming the bleaching of some grains due to the manual transport, the small tools used for work in gravel pits and the indications of Ollerhead (2001) on the study of the bleaching of not only the surface grains but grains in deeper layers due to the transparency of sand grains. Furthermore, he recommended the use of histograms and radial plots to choose the most suitable method of evaluation of the ED, as the degree of bleaching determines the shape of the distribution. Symmetrical dose distributions on very small aliquots are indicative of homogeneous bleaching. Asymmetrical distributions show bleached and unbleached grains. Also, methods of mortar preparation have remained virtually unchanged during the past.

Jain et al. (2004) investigated the OSL of different mortar samples, including render, whitewash and an inner plaster. They found that quartz sand were poorly bleached and weakly sensitive. Thus, poor precision resulted from equivalent dose calculations. However, in a previous study they found an accurate dose-depth curve in the walls of the building (Jain et al. 2002).

Goedicke (2011) investigated methods of equivalent dose estimation and to predict the datability of mortar samples. He assessed the level of bleaching of quartz for 14 mortar samples (taken from buildings of the roman city of *Mogontiacum*, the modern Mainz) with methods used to determine the degree of bleaching of sediments. He also performed single grain OSL. He found that 7 of the 14 samples were datable by standard OSL and other 5 with other procedures, while 2 samples were not datable.

Regarding other types of mortars few studies have been performed. Goksu et al. (2003) approached to retrospective dosimetry on three different types of Portland cement by fine grain polymineral extracts and quartz inclusions. They found problems on the separation and purification of quartz and feldspar. They also found insufficient zeroing of the TL signal. However, Thomsen et al. (2003) measured doses below 1 Gy on commercial concrete by using the OSL signal of quartz grains.

OSL studies have also been performed on mud mortars. Feathers et al. (2008) applied OSL dating to quartz extracted from ancient mud mortars collected from an early Andean monumental centre, in Peru. OSL signals showed that samples were only partially exposed to sunlight during mortar preparation and construction. They used a minimum age model to calculate equivalent dose distributions and obtained 14 dates around 1000 BC, in agreement with radiocarbon ages of charcoal collected on the same mortars.

Calculating the dose-rate of ionizing radiation in building materials also involves problems. Some of them are related with the heterogeneity of some mortars, particularly lime mortars. Beta microdosimetry can cause skewed dose distributions of equivalent doses, and inaccurate dose-rate measurements if few sample amounts are considered for analysis. Moreover, Goedicke (2011) recommend measuring the beta dose rate with methods that focus on the radioactive decay equilibrium (gamma spectrometry), as disequilibrium leads to non-negligible deviation from the true age.

Problems derived from the gamma dose must be considered, as building walls are heterogeneous environments. It is therefore necessary to use in situ dosimetry (by gamma spectrometry or TLDs) or to consider a model regarding the content of radioactive elements in each of the materials of the wall, depending on the volume they occupy around the sampling point (Guibert et al., 1988). Feathers et al. (2008) found good dose rate estimation form the last approach, although in situ dosimetry is also recommendable if the building environment has not been strongly modified from the building period. Sometimes the studied area of a building has been subjected to reconstructions or modifications. In such case further problems could be due to the historical variation of the gamma dose. The estimated gamma dose will result in a spread of age estimates for objects, which are actually contemporaneous, due to the unknown variation in gamma dose rates because of historical construction and reconstruction phases. The cosmic dose should be calculated depending on the situation of the samples on the architectural elements.

5. Final Considerations

Finding information on written historical documents is the most suitable dating technique for ancient buildings. However, available complete historical data is very unusual and other procedures have been developed for the study of building chronologies. Such procedures are the study of building stratigraphy, chronolypology, mensiochronology, chemical analysis of building materials and some absolute dating methods.

Stratigraphic techniques have been adapted from archaeology, providing good general approaches to the history (building phases) of historical buildings, but ~~it does~~ they do not provide absolute ages. Chronotypology is dating method based on the observation of building features typically used in different historical periods but requires the use of databases in a homogeneous geographical context (not frequently available). Mensiochronlogy is a kind of chronotypology applicable when the main dating features taken into account are the dimensional characters of the elements. It provides similar results with similar limitations than chronotypology.

Absolute dating methods provide absolute ages for different kind of geological and archaeological objects. The main absolute dating techniques used for building materials are dendrochronology, radiocarbon, archaeomagnetism and luminescence dating. Dendrochronology provides absolute ages for wood, but original woody materials are not always available in ancient buildings.

Archaeomagnetic dating should also be applicable for brick dating, if they are in situ magnetized (due to fires in buildings) during an archaeological time. Its application is very limited. However, luminescence dating has extensively been applied to date ancient bricks with success, providing the age of the last heating of a brick. Luminescence has been also applied on mortars (lime, mud and cement mortars) with variable but promising results. Alternatives exist for dating lime mortars. Radiocarbon has had relative success, although routine dating procedures are not available for archaeologists and architects studying ancient and historical buildings. The results of recent research studies indicate a promising future for the use of radiocarbon and luminescence in a mortar.

REFERENCES

Adamiec, G., Aitken, M. (1988) Dose rate conversion factors: update. *Ancient TL* 16, 37-50.

Aitken, M.J. (1974) Physics and archaeology. Clarendon Press, Oxford.

Aitken, M.J., (1985) Thermoluminescence dating, Academic Press, London.

Amoroso GG, Fassina V (1983) Stone decay and conservation: Atmospheric Pollution, Cleaning, Consolidation and Protection. Elsevier, Amsterdam.

Bailiff I. K. (2008) Methodological developments in the luminescence dating of brick from English late-medieval and post-medieval buildings », *Archaeometry* 49 (4), 827-851.

Barba, L., Blancas, J., Manzanilla, R., Ortiz, A., Barca, D., Crisci, G.M, Mirello, D., Pecci, A. (2009) Provenance of the limestone used in Teotihuacan (Mexico): a methodological approach. *Archaeometry* 51, 525-545.

Blain, S. (2010) An application of luminiscence dating to building archaeology: The study of ceramic building materials in early medieval churches in north-western France and south-eastern England. *Arqueología de la Arquitectura*, 7, 43-66.

Blain, S., Bailiff, I.K., Guibert, P., Bouvier, A., Bayle, M. (2010) An intercomparison study of luminescence dating protocols and techniques applied to medieval brick samples from Normandy (France). *Quaternary Geochronology* 5, 311–316.

Boato, A., Pittaluga, D. (2000) Building Archaeology: A Non-Destructive Archaeology. 15th World Conference on Nondestructive Testing, Roma (Italy) 15-21 October 2000.

Bonde, A., Murray, A., Friedrich, W.L. (2001): Santorini: Luminescence dating of a volcanic province using quartz? *Quaternary Science Reviews*, 20: 789-793.

Bortolotto, S., Colla, C., Mirandola, D., Sponchioni, A.: Palazzo Cittadini Stampa: role of stratigraphy and cinematic analysys in the knowledge of a mansonry building,, C. Modena, P.B. Lourenco, P. Roca (2005) Structural analysis of historical constructions. *Possibilities of numerical and experimental techniques*, London 1, 167-175.

Botter-Jensen, L., Solongo, S., Murray, A.S., Banerjee, D., Jungner, H. (2000) Using the OSL single-aliquot regenerative-dose protocol with quartz extracted from building materials in retrospective dosimetry. *Radiation Measurements* 32, 841-845.

Casas, L., Lindford, P., Shaw, J. (2007) Archaeomagnetic dating of Dogmersfield Park brick kiln (Southern England). *Journal of Archaeological Science*, 34, 205-213.

Casadio, F., Chiari, G., Simon, S., (2005) Evaluation of binder/aggregate ratios in archaeological lime mortars with carbonate aggregate: a comparative assessment of chemical, mechanical and microscopic approaches. *Archaeometry* 47 (4), 671–689.

Clark, A.J., Tarling, D.H., Noël, M. (1988) Developments in archaeomagnetic dating in Britain. *Journal of Archaeological Science*, 15, 645-667.

Daniels, F., Boyd, C.A., Saunders, D.F. (1953) Thermoluminescence as a research tool. *Science*, 117: 343-349.

Delibrias, G., Labeyrie, J. (1964) Dating of old mortars by the carbon-14 method, *Nature* pp. 742.

Deer WA, Howie RA, Zussman J (1992) An introduction to the rock forming minerals. Longman Scientific and Technical, New York.

Duller, G.A.T. (1991): Equivalent dose determination using single aliquots. *Nuclear Tracks and Radiation Measurements*, 18: 371-378.

Elert, Kerstin, Carlos Rodriguez-Navarro, Eduardo Sebastian Pardo, Eric Hansen and Olga Cazalla. Lime Mortars for the Conservation of Historic Buildings. *Studies in Conservation* 47:62-75.

Elsen, J., 2006. Microscopy of historic mortars – a review. *Cement and Concrete Research* 36, 1416-1424.

Feathers, J.K. Johnson, J.,Kembel, S.R. (2008) Luminescence Dating of Monumental Stone Architecture at Chavín De Huántar, Perú. *Journal of Archaeological Methods and Theory* 15, 266–296.

Ferrando, I., Mannoni, T., Pagella, R., (1989) Cronotipologia. *Archeologia Medievale*, 16, 647-661.

Folk RL, Valastro S Jr. 1976. Successful technique for dating of lime mortars by carbon-14. *Journal of Field Archaeology* 3:203–8.

Goedicke, C. (2003) Dating historical calcite mortar by blue OSL: results from known age samples. *Radiation Measurements* 37, 409-415.

Goedicke, C. (2011) Dating mortar by optically stimulated luminescence: A feasibility study. *Geochronometria* 38, 42-49.

Göksu, H.Y., Bailiff, I.K., Mikhailik, V.B. (2003) New approaches to retrospective dosimetry using cementitious building materials. *Radiation Measurements* 37, 323-327.

Guibert, P., Bechtel, F., Schvoerer, M., Müller, P., Balescu, S. (1998) A new method for gamma dose-rate estimation of heterogeneous media in TL dating. *Radiation Measurements* 29, 561-572.

Guibert, P., Bailiff, I.K., Blain, S., Gueli, A.M., Martini, M., Sibilia, E., Stella, G., Troja S.O. (2009) Luminescence dating of architectural ceramics from an early medieval abbey: The St Philbert Intercomparison (Loire Atlantique, France). *Radiation Measurements*, 44, 488-493.

Hale J, Heinemeier J, Lancaster L, Lindroos A, Ringbom Å. (2003) Dating ancient mortar. *American Scientist* 91, 130–7.

Heinemeier J, Jungner H, Lindroos A, Ringbom Å, von Konow T, Rud N. (1997) AMS 14C dating of lime mortar. *Nuclear Instruments and Methods in Physics Research* B 123, 487–95.

Heinemeier, J., Ringbom, A., Lindroos, A., Sveinbjörnsdóttir, A.E. (2010) Succesful AMS 14C dating of non-hydraulic lime mortars from the medieval churches of the Aland Islands, Finland. *Radiocarbon* 52, 171-204.

Huntley, D.J., Godfrey-Smith, D.I., Thewalt, M.L.W. (1985) Optical dating of sediments. *Nature*, 313: 105-107.

Hütt, G., Jaek, I., Tchonka, J. (1988) Optical dating – K-feldspars optical-response stimulation spectra. *Quaternary Science Reviews*, 7: 381-385.

Jain, M., Botter-Jensen, L., Murray, A.S., Jungner, H. (2002) Retrospective dosimetry: dose evaluation using unheated and heated quartz from a radioactive waste storage building. *Radiation protection dosimetry* 101, 525-530.

Jain, M., Thomsen, K.J., Botter-Jensen, L., Murray, A.S. (2004) Thermal transfer and apparent-dose distributions in poorly bleached mortar samples: results from sigle grains and small aliquots of quartz. *Radiation Measurements* 38, 101-109.

Lindroos, A., Heinemeier, J., Ringbom, A., Braskén, M. Sveinbjörnsdóttir, A.E. (2007) Mortar dating using AMS 14C and sequential dissolution: examples from medieval, non-hydraulic lime mortars from the Aland Islands, SW Finland. *Radiocarbon* 49, 47-67.

Martini, M., Sibilia, E. (2006) Absolute dating of historical buildings: the contribution of thermoluminescence (TL). *Journal of Neutron Research* 14, 69-74.

Marzaioli, F., Lubritto, C., Nonni, E., Passariello, I., Capano, M., Terrasi, F. (2011) Mortar radiocarbon dating: preliminary accuracy evaluation of novel methodology. *Analytical Chemistry* 83, 2038-2045.

Middendorf, B., Hughes, J.J., Callebaut, K., Baronio, G., Papayianni, I., 2005a. Investigative methods for the characterization of historic mortars – Part 1: Mineralogical characterization. *Materials and Structures* 38, 761- 769.

Middendorf, B., Hughes, J.J., Callebaut, K., Baronio, G., Papayianni, I., 2005b. Investigative methods for the characterization of historic mortars – Part 2: Chemical characterization. *Materials and Structures* 38, 771-780.

Morales, J.; Goguitchaichvili, A., Aguilar-Reyes, B., Pineda-Duran, M., Camps, P., Carcalho, C., Calvo-Rathert, M. (2011) Are ceramics and bricks reliable absolute geomagnetic intensity carriers? *Physics of the Earth and Planetary Interiors*, 187, 310-321.

Murray, A.S., Roberts, R.G., 1997. Determining the burial time of single grains of quartz using optically stimulated luminescence. *Earth and Planetary Science Letters* 152, 163-180.

Murray, A.S., Wintle, A.G., 2000. Luminescence dating of quartz using an improved single-aliquot regenerative-dose protocol. *Radiation Measurements* 32, 57–73. Prescott and Hutton, 1994.

Nawrocka, D., Michniewicz, J. Pawlyta, Pazdur, A. (2005) Application of radiocarbon method for dating of lime mortars. *Geochronometria* 24, 109-115.

Ollerhead, J. (2001) Light transmittance through dry, sieved sand. *Ancient TL* 19, 13–16.

Prescott, J.R., Hutton, J.T. (1994) Cosmic ray contributions to dose rates for luminescence and ESR dating: large depths and long term variations. *Radiation Measurements*, 23: 497-500.

Preusser, F., Degering, D., Fuchs, M., Hilgers, A., Kadereit, A., Klasen, N., Kretschek, M., Ritcher, D., Spencer, J.Q.G. (2008) Luminescence dating: basics, methods and applications. *Quaternary Science Journal* 57, 95-149.

Sanjurjo-Sánchez, J., Trindade, M. J., Blanco Rotea, R., Benavides, R.; Fernández Mosquera, D. Burbridge, C.I.; Prudêncio, M. I. Dias, M.I. (2010) Geochemistry of Rare Earth and other trace elements applied to the characterization of ancient mortars. *Journal of Archaeological Science* 37: 2346-2351.

Sanjurjo-Sánchez, J., Alves, C.A.S. (2011) Decay effects of pollutants on materials applied in the built environment. In: Environmental Chemistry for a sustainable world (Lichtfouse, E., Schwarzbauer, J., Robert, D., Editors) Springer, Berlin. (In press).

Sanjurjo-Sánchez, J., Montero Fenollós, J.L. (2012) Chronology during the Bronze Age in the archaeological site Tell Qubr Abu al-'Atiq, Syria. *Journal of Archaeological Science* 39, 163-174.

Shestoperov, S. (1988) *Road and building materials*. Vol. 1. Moscow, Mir Publishers. 273 p.

Sternberg, R.S. (1997) *Archaeomagnetic dating*. In: Chronometric dating in Archaeology (Taylor, R.E., Aitken M.J., eds). Plenum Press, New York, 323-356.

Stuiver M., Polach H.A. (1977) Discussion: reporting of ^{14}C data. *Radiocarbon* 19, 355–63.

Taylor, R. E. (1997) *Radiocarbon dating*. In: Chronometric dating in archaeology (Taylor, R.E., Aitken, M.J., editors). New york and London, Plenum Press. 65-96.

Thellier, E., Thellier, O. (1959) Sur l'intensité du champ magnétique terrestre dans le passé historique et géologique. Ann. *Géophysique*, 15, 285-376.

Thomsen, K.J., Jain, M., Boter-Jensen, L., Murray, A.S., Jungner, H. (2003) Variation with depth of dose distributions in sigle grains of quartz extracted from an irradiated concrete block. *Radiation Measurements* 37, 315-321.

Vendrell-Saz, M., Alarcon, S., Molera, J., García-Vallés, M. (1996) Dating ancient lime mortars by geochemical and mineralogical analysis. *Archaeometry* 38, 143-149.

Vitruvius M.L. (2011) The ten books of architecture. Online: *http://penelope.uchicago.edu/Thayer/E/Roman/Texts/Vitruvius/home.html*.

Wallinga, J. (2002) Detection of OSL age overestimation using single-aliquot techniques. *Geochronometria* 21, 17-26.

Zacharias, N., Mauz, B., Michael, C.T., (2002) Luminescence quartz dating of lime mortars. A first research approach, *Radiation Protection Dosimetry* 101, 379-382.

In: Brick and Mortar Research
Editors: S. Manuel Rivera and A. L. Pena Diaz
ISBN: 978-1-61942-927-7

Chapter 6

RAMAN SPECTROSCOPIC CHARACTERIZATION OF BRICK AND MORTARS: THE ADVANTAGES OF THE NON DESTRUCTIVE AND IN SITU ANALYSIS

O. Gómez-Laserna, N. Prieto-Taboada, I. Ibarrondo, I. Martínez-Arkarazo, M. A. Olazabal and J. M. Madariaga

Department of Analytical Chemistry, University of the Basque Country, (UPV/EHU), Leioa, Spain

ABSTRACT

Raman spectroscopy is becoming a popular technique to perform molecular analysis on building materials in a non destructive way, which permits the preservation of samples. Nowadays, the technical development of Raman achieved from high sensitive microscopic instruments till handheld spectrometers. This fact sets Raman spectroscopy as the technique of choice for the characterization and diagnosis works in Cultural Heritage.

In our researches different kind of building materials, including bricks and mortars belonging to several sampling locations, have been analysed. All of them were affected by different environmental stressors, such as atmospheric acid gases and particles, infiltration waters and biodecaying. Generally, calcite ($CaCO_3$) and diverse iron oxides plus quartz (sands) were identified as original compounds in mortars samples. Silicates and oxides like rutile (TiO_2) and hematite (Fe_2O_3) were the common original compounds in bricks.

Degradation products of those original compounds were identified in environments with high atmospheric pollution; gypsum ($CaSO_4 \cdot 2H_2O$) is the principal compound formed by the attack of the SOx acid gases over calcite in the mortars. Calcite, is an original compound in mortars but in the case of the bricks it comes from a decaying (hydration and subsequent carbonation of the original calcium oxide, CaO); other original compounds can react analogously with atmospheric acid gases (CO_2, SO_2, NOx) to form the respective carbonates, sulphates and/or nitrates. The formation of nitrates is also due to the impact coming from the organic matter degradation that produces ammonium nitrate, which is the precursor of the formation of other nitrate salts such as niter (KNO_3) and nitrocalcite ($Ca(NO_3)_2{\cdot}4H_2O$). The study of the biodeterioration markers, permits the identification of several pigments and other organic compounds (calcium oxalates) associated to the metabolism of the colonizing microorganisms.

In conclusion, the Raman identification of original and deterioration compounds allow us to define the conservation state and the impacts suffered by building materials like bricks and mortars. The in situ capability of the Raman technique increases its potentiality.

1. Introduction

In order to understand the benefits of Raman spectroscopy in the study of materials is necessary to start defining it as a technique based on the study of the molecular vibrations produced by light scattering [1]. Those vibrations can be seen in the Raman spectrum only if there is a change in polarizability; that is, merely a distortion of the electron cloud around the vibrating atoms is required. The Raman scattering occurs when radiation from a source is passed through a sample and some of the radiation is scattered by the molecules present. For simplicity, it is better to use a radiation of only one frequency and the sample should not absorb it. The beam of radiation is dispersed in the space. Then, three types of scattering occur. Rayleigh scattering is the most common and has the same frequency as the source of radiation. It occurs as a result of elastic collisions between the photons and the molecules in the sample although no energy is lost in collision. A slight interaction between the incident beam and the molecules can be observed and after that interaction, few photons (0.001%) are scattered. Raman-Stokes lines are produced by those photons scattered with less energy than the incident radiation and on the other hand, the Raman-antistokes lines belong to the photons scattered with more energy. Raman spectrum is composed by stokes lines as Raman shift from the Rayleigh scattering, which is placed in zero position. The Raman signals generally are represented as wavenumber (cm^{-1}) versus Raman Intensity [2, 3]. Raman instruments use a laser as light source mainly because a high intensity beam of radiation is needed, due to the scarce fraction of scattered photons that suffer a frequency change. So that, it is possible to obtain Raman scattering with high signal to noise ratio [4]. The most used lasers are UV, Visible and NIR, being most effective (in terms of sensibility) those with shorter wavelength according to the following expression: Intensity= $1/\lambda^4$. Nonetheless, the use of short wavelength lasers favors the phenomena of fluorescence (which results in a curvature of the baseline, with the subsequent difficulties on band assignment) and photodecomposition of the sample [5]. If the frequency of the laser beam is close to the frequency of an electronic transition, the resonance effect is produced and scattering enhancements of up to 10^4 units have been observed. When this resonance condition occurs, new Raman bands are visible in the spectrum. For all of this, Raman spectroscopy becomes a much more sensitive technique and, since only the chromophore molecule is the most efficient scattering, it will also be selective for this molecule of the sample.

Due to the use of dispersive systems, with a visible laser as a source, the most common detectors are photomultiplier tubes. Instead, the multichannel instruments use photodiode array (PDA), current injected detector (CID), or charge-coupled device (CCD) detectors

(Figure 1). Nowadays, the selection of the suitable detector determines the speed of the measurement, the spectral range and sensitivity of the technique [6].

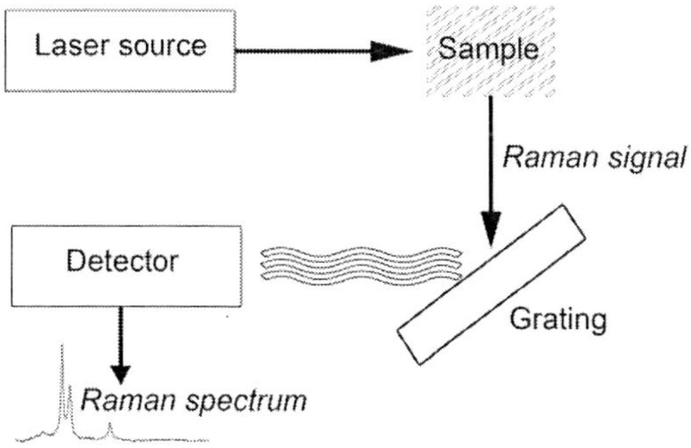

Figure 1. Raman spectrometer diagram.

Concerning applications of Raman spectroscopy, it presents several advantages respect to other analytical tools. It is a non-destructive technique that requires little or no preparation of the sample. In the field of Heritage, one of the biggest advantages is the possibility of carrying out in situ analysis [2]. It is possible to measure different type of samples like compact solids (amorphous and crystalline) with the simple placement under the laser beam or even liquid, gels, films and gases. It has a wide spectral range with a high chemical specificity, therefore the Raman peaks are easily related to the structure of compounds. It also allows the differentiation between different cations bound to the same anion and polymorphic compounds such as calcite and aragonite as can be observed in figure 2. Both have the same molecular formula, $CaCO_3$, but they belong to different crystal systems (trigonal and rhombic, respectively). In addition, Raman spectroscopy can distinguish different numbers of hydration water as in the cases of gypsum ($CaSO_4{\cdot}2H_2O$), hemihydrate ($CaSO_4{\cdot}0.5H_2O$) and anhydrite ($CaSO_4$).

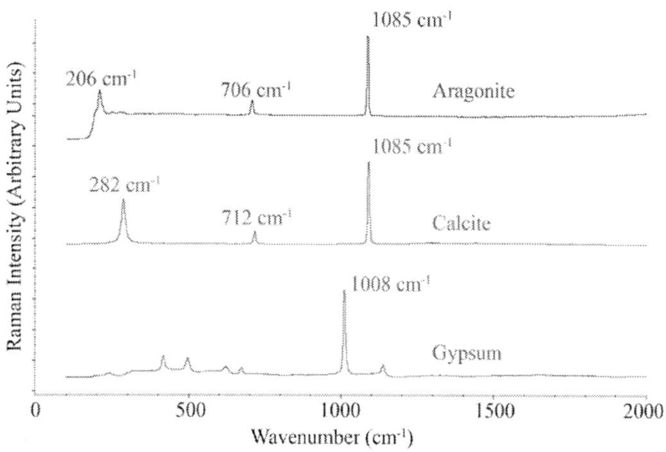

Figure 2. Raman spectra showing differences among polymorphous compounds (aragonite and calcite) and shift due to the presence of different functional groups, that is, carbonates versus sulphates (gypsum).

The applications of Raman spectroscopy are not only qualitative but also semiquantitative. Initially, it was used to examine inorganic compounds but its use is growing in organic and polymer analysis [7-9]. For example, it is used for bulk material characterization, online process analysis, remote sensing, microscopic analysis and chemical imaging [6, 10-16]. More recently, it is spreading in pharmaceutical applications, while other applications have been successfully established in characterization of pigments, semiconductors, art materials, archaeology and biotechnology [17].

In the particular case of building materials, Raman spectroscopy has demonstrated to be a powerful technique that allows to characterize both original and decaying compounds. The identification of decaying compounds formed as a consequence of the environmental impact (atmospheric pollution, infiltration waters or microorganism) is decisive to define the conservation state and the impacts suffered by building materials like bricks and mortars.

Generally, the alteration processes by environmental stressors is a natural and irreversible process of degradation that building material are destined to suffer. In this process, the crystallization of soluble salts into the building materials is considered one of the most destructive damage. The salts appear in the building surface and they become visible as efflorescences (Figure 3), but also they crystallize/dissolve within the pores and capillaries as subefflorescences which cause considerable damage. They are able to produce internal fractures due to volume changes when soluble ions recrystallize as salt in the pores or when the same compound changes its number of hydration water molecules. The most common salts are sulfates (SO_4^{-2}), nitrates (NO_3^{-}), and carbonates (CO_3^{-2}) [18-21]. These compounds are easily detectable by Raman spectroscopy as it has been already mentioned, so the characterization of salts found in the pores and surface of damaged building materials can be approached by this technique. The previous characterization is very important to identify the original components as well as the degradation compounds. This cataloging together with the experience of the researcher is critical to diagnose the chemical process leading to the deterioration.

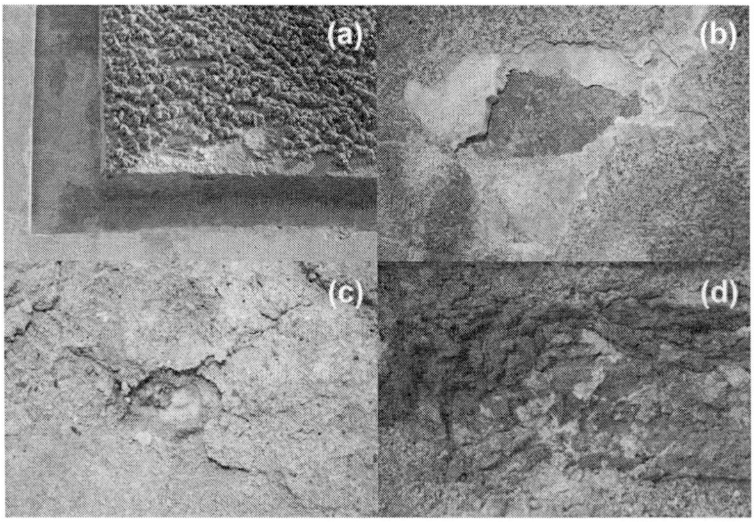

Figure 3. Deteriorations in buildings materials caused by soluble salts. (a) Efflorescence in an embellishing mortar, (b) detachments of the rendering mortar, (c-d) subefflorescence in a brick.

Besides, the major biodeterioration agents associated with decaying processes in bricks and mortars are lichens and mosses. They contribute in these processes not only mechanically and physically, but also chemically. The chemical damage has been widely studied by Raman Spectroscopy in last decades. This technique allows us to determine the original compounds of the support were microorganisms are settled together with decaying compounds related with their metabolic activity.

In the field of Raman characterization, the spectrometers have undergone a great development in last decades. Currently, several equipments offer different possibilities depending on the characteristics of the measurements or samples. Therefore, starting from most sophisticated laboratory non-portable equipments to easy to operate handheld spectrometers are available.

In this work, the usefulness of Raman spectroscopy on the characterization of building materials, in particular bricks and mortars, is demonstrated. Furthermore, the efficiency of portable spectrometers on the characterization of decayed building materials is stated. For this purpose some examples are presented of materials affected by different decaying mechanisms: environmental impact, infiltration waters and biodeterioration.

2. Performance of Commercially Available Raman Spectrometers

As mentioned above, nowadays different Raman spectrometers offering a wide range of characteristics are commercially available. The nature of the sample and the compounds to be determined establish the requirements of the instrument to be used. However, as it is going to be demonstrated, although spectra quality can be substantially improved by using laboratory equipments, it is not always necessary to perform laboratory analysis in order to get unambiguous results. In fact, portable instruments offer a good performance in the field of the analysis of construction materials. Moreover, when the materials to be analyzed belong to the built heritage, the use of such instruments allowing non-destructive in situ analyses are essential. The examples provided in this work were obtained by three Raman spectrometers that supply different possibilities. The first one was used for field measurements. This is an ultramobile (handheld) Raman spectrometer (InnoRam $BWTEK_{INC}$). It works with a 785 nm excitation laser and has a variable power range to control thermal decomposition of the measured surfaces. The probe offers also the possibility to perform microscopic analysis by using different optical lens (4x, 20x and 50x), that allow to measure areas of a diameter between 10 and 200 μm. For this purpose, the probe can be installed in a manually movable platform (see Figure 4a) to focus on the target area or it can be put up on a motorized tripod that allows to make microanalyses even in areas located higher than 2 meters from the floor.

Another portable instrument used in this work consists of a diode excitation laser of 785 nm and a CCD detector cooled by Peltier effect (Renishaw RA 100). Its working power is controlled by filters and a motorized tripod permits focusing the laser on the analysis area (Figure 4b).

Finally, depending on the complexity of the sample and the possibility of sampling, an inVia Renishaw confocal Raman microspectrometer (Renishaw, Gloucestershire, UK) coupled to a DMLM Leica microscope with long range lens (5×, 20×, 50×, and 100×) and a

Peltier cooled CCD detector was sometimes used. In addition, the instrument allows measurements with two excitation lasers (785 and 514 nm; nominal laser power 350 mW and 50 mW respectively) that offer the possibility of doing analysis under resonance conditions in some cases (Figure 4c).

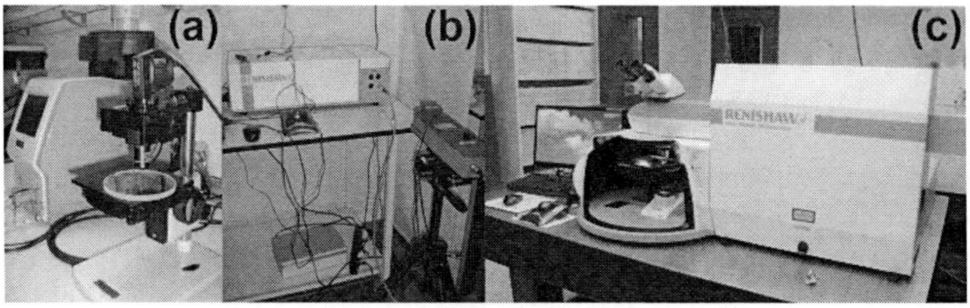

Figure 4. Some commercially available Raman spectrometers. (a) A InnoRam BWTEK$_{INC}$ handheld spectrometer, (b) a portable RA 100 (Renishaw) portable instrument and (c) non portable inVia Renishaw confocal spectrometer.

The efficiency of portable spectrometers still shows differences in comparison with laboratory equipments. However, in terms of identification of compounds, the great progress in this technique is noticeable if spectra obtained with handheld or portable spectrometers and those taken in the laboratory are compared (Figure 5). In situ measurements can be nowadays made with resolutions up to 3.5 cm^{-1}, which is good enough as to differentiate among the most common degradations compounds cited previously. The problem comes when the signal to noise ratio, which is better corrected in non movable equipments, is not good enough to distinguish low intensity Raman bands produced by poor scattering compounds or by substances at low concentration levels. This consequence is enhanced when compounds producing fluorescence effect are present in the sample.

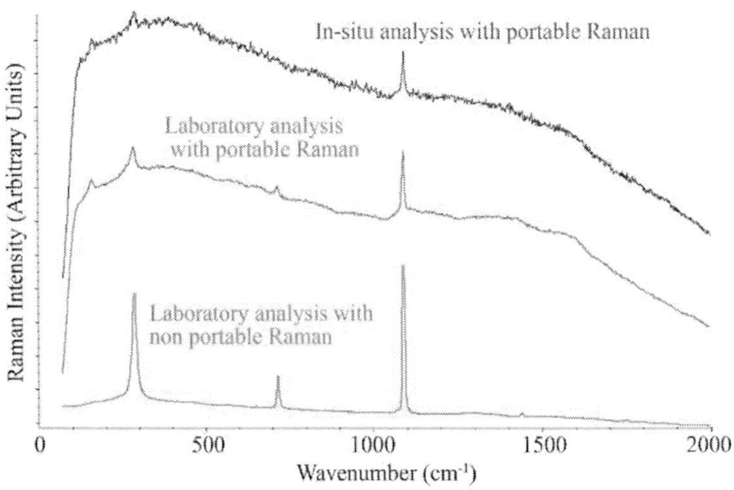

Figure 5. Comparison of the different spectra of calcite with portable and with a non portable spectrometer. The spectra are not treated or corrected.

On the other hand, the main disadvantages of in situ analysis are related with factors that not always can be controlled such as vibrations caused by traffic or wind, sunlight etc. According to our experience in field analyses, such factors can prevent correct spectral assignations or sometimes even disturb acquisition. For instance, the spectrum of the sunlight presents the most intense Raman bands in the spectral range where some sulfates are identified [22]. This is the case of coquimbite (see table 1), which formation has been associated with polluted environments rich on particulate iron [23]. With regard to vibrations, the difficulties come mainly when the microanalyses are performing, which can be sometimes even impossible.

The information obtained with Raman equipments is sometimes complemented with the use of other non-destructive techniques such as X-Ray spectroscopy or infrared spectroscopy (IR). Portable equipments implementing such spectroscopies have been developed as well. Thus, a combination of elemental and molecular information can be achieved by only carrying out in situ analysis. This way, a complete methodology that meets the desired requirements for almost all fields of Cultural Heritage can be designed [23-25].

All equipment described above has allowed to study many types of materials and matrices, from building materials (sandstone, limestone, mortar, bricks, cement and concrete) to pigments on different supports, slags, food or geological materials such as sediments, beachrock or meteorites [20, 23, 25-29]. However, Raman spectroscopy is the decisive technique during diagnosis since it is the molecular spectroscopy providing specific information that allows the identification of compounds as it has been already mentioned. Therefore, the results summarized in this work are focused on the Raman spectroscopic study of building materials and more specifically, on the assessment of the damage caused by the impact of atmospheric acid gases, water infiltration and biodeterioration in bricks and mortar of the built Heritage [23, 27, 30-33].

3. Deterioration Due to Atmospheric Pollution

In order to explain the deterioration of bricks and mortars caused by the impact of air pollutants, it is interesting to describe briefly the deposition mechanisms of pollutants on construction materials. Dry deposition is said to occur when a contaminant is transferred from the atmosphere to the material, in the absence of rain, being more common in sheltered areas of the building [27, 34]. However, wet deposition begins with the suspension of water in the atmosphere, where CO_2, SO_x and NO_x gases are converted into acid aerosols. For instance, the aerosol of sulfuric acid is present mainly in urban and industrial areas [35] and its formation is due to the dissolution of sulfur dioxide and subsequent oxidation to sulfuric acid, according to reaction 1.

The formation of nitric acid is a similar process as described above. In this case, nitrogen dioxide (NO_2) is emitted as a consequence of the fossil fuel use in industry or vehicles and NO_2 reacts with the water of the atmosphere to form nitric acid (HNO_3) (reaction 2) [36]. Carbon dioxide is other contaminant present in the atmosphere that in contact with water forms carbonic acid (reaction 3). This acid is corrosive and its action is important in densely populated industrial areas. Therefore, calcareous materials can be easily attacked and solubilized by these types of aerosols. Indeed, this problem is one of the main responsible for salts formation into construction materials. However, it should be noted that the effects from contaminants can be synergistic [33, 37].

$$SO_2 + H_2O \rightarrow H_2SO_3;\ H_2SO_3 + \tfrac{1}{2}O_2 \rightarrow H_2SO_4 \qquad (1)$$

$$NO_2 + H_2O \overset{Ox}{\rightarrow} HNO_3 + \tfrac{1}{2}H_2 \qquad (2)$$

$$CO_2 + H_2O \rightarrow H_2CO_3 \qquad (3)$$

Nevertheless, to study the influence of pollutants on materials, the chemistry between them should be clear because the effects observed are the result of their interaction. In this way, mortars are probably the most sensitive to degradation because of their calcareous nature and porosity. Calcareous mortars are composed by quartz together with carbonates (mainly calcite, $CaCO_3$) and diverse iron oxides such as hematite (Fe_2O_3). The presence of calcite is related with the addition of sandstone to the cement. Calcite appears in the cement when calcium oxide (CaO), belonging to the original composition, reacts exothermically with water to form calcium hydroxide ($Ca(OH)_2$), in a process called slaking. Subsequently, calcium hydroxide absorbs carbon dioxide to convert into calcite. This process is called curing [38].

Natural calcite can be dissolved by the attack of atmospheric carbonic acid to form hydrogencarbonate (HCO_3^-). This compound is more soluble and can migrate outward, landing on the surface like calcite efflorescence, which can be removed by rain washing with the consequence loss of material. The calcite in mortars can be presented like original or degradation component [38].

In environments with high atmospheric pollution, the most common degradation products of those original compounds, were gypsum ($CaSO_4{\cdot}2H_2O$) or nitrocalcite ($Ca(NO_3)_2{\cdot}\ 4H_2O$). Gypsum is formed by attack of the SO_x acid gases over calcite of the mortars [20]. However, the presence of gypsum can be also due to the use of additives in mortars to obtain certain mechanical properties.

Analogously, nitrocalcite is formed by the reaction of NO_x atmospheric acid gases respectively with calcite. Reactions 4-7 explain the formation of these types of salts, where M represents the different cations of the salts. In these reactions the formation of other compounds is expected, as CO_2, H_2 or H_2O. In this sense, the presence of water in the environment can produce different hydrated forms of the salts.

Formation of the hydroxide by hydration of the oxide:

$$M_2O_m + H_2O \rightarrow 2M(OH)_m \qquad (4)$$

Carbonation of the hydroxide:

$$2M(OH)_m + mH_2CO_3 \rightarrow M_2(CO_3)_m \cdot nH_2O \qquad (5)$$

Reaction of the carbonates:

$$M_2(CO_3)_m + mH_2SO_4 \rightarrow M_2(SO_4)_m \cdot nH_2O \quad (6)$$

$$M_2(CO_3)_m + mHNO_3 \rightarrow M(NO_3)_m \cdot nH_2O \quad (7)$$

Besides, other reactions that usually occur are hydration/dehydration processes of the salts, where crystallized salts change the hydration water molecule number depending on the environmental conditions. For example, gypsum may suffer dehydration reactions becoming hemihydrate ($CaSO_4 \cdot 0.5H_2O$) or anhydrite ($CaSO_4$) compounds. These changes produce an important deterioration due to the tension exerted in the volume changes. Field analyses have demonstrated that it is possible to distinguish among such similar compounds by using the handheld spectrometer as shown in figure 6. Note that not only the main band around 1000 cm^{-1} (pointed in the figure) is visible but also most of the middle intensity or minor bands corresponding to anhydrite and gypsum.

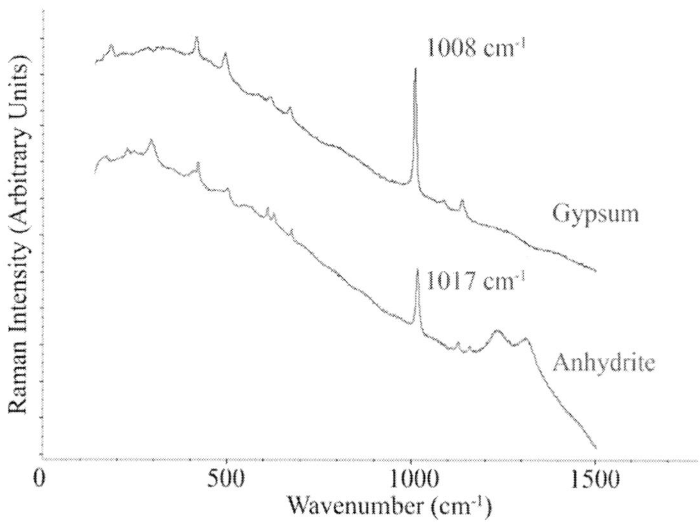

Figure 6. Comparison of the spectra collected by the InnoRam hand-held spectrometer ($BWTEK_{INC}$) on a brick sample. Gypsum and anhydrite are clearly identified thanks to the good quality of the measurements.

Bricks are likely the simplest construction materials to be diagnosed because it is easy to distinguish among original and degradation compounds [39]. Raw materials used for brick manufacturing are clays that are composed mainly of diverse silicates. After firing, the brick is composed mainly by oxides and silicate mixtures. For instance, silicates like amazonite ($KAlSi_3O_8$) and oxides like rutile (TiO_2) and hematite (Fe_2O_3) are common original compounds in bricks [27]. So that, if other compounds such as carbonates, sulfates or nitrates are identified in bricks they are always classified such as degradation compounds. Calcite often comes from a decaying due to the hydration and subsequent carbonation (by atmospheric CO_2) of the original calcium oxide (CaO) according to reactions 4 and 5. The Raman scattering of the cited compounds is high enough as to identify them even by handheld spectrometers. Unfortunately nitrates have not been often detected due to their high solubility. Such soluble salts are observed in zones protected from rain washing. This is also the case of chlorides. However, the acidic behavior of the pollutant precursor of nitrate formation, that is, nitric aerosol, allows to penetrate deeper in the material and it causes more destructive effects than chlorides do [10]. On the other hand, it has to be taken into account that the origin of such decaying compounds is not necessarily an atmospheric attack but also can be formed as well after infiltration of salt-rich waters.

Other original compounds can react analogously with atmospheric acid gases to form the respective carbonates, sulfates and/or nitrates, like natrite (Na_2CO_3), thenardite (Na_2SO_4) and nitratine ($NaNO_3$) that have been often referenced as soluble salts present in bricks [23, 31]. It is remarkable that other compounds successfully identified in the brick samples were limonite ($FeO(OH){\cdot}nH_2O$) and goethite (α-$FeO(OH)$) [23]. Its origin should be carefully studied, with special attention to when, where and how were manufactured.

The Raman features of the most usually compounds identified by Raman spectroscopy in mortar and bricks affected by atmospheric pollution are collected in the table 1.

Table 1. Raman bands of some of the different compounds formed in mortars and bricks by the action of atmospheric pollution. Results obtained using the 785 nm laser

Formula	Chemical name/Mineral	Raman bands ν (cm^{-1})
$CaCO_3$	Calcium carbonate/Calcite	281, 712, 1085
$CaSO_4$	Calcium sulphate/Anhydrite	169, 399, 417, 498, 594, 609, 628, 661, 675, 1017, 1111, 1129, 1161
$CaSO_4{\cdot}0.5H_2O$	Calcium sulphate hemihydrate/Bassanite	236, 322, 430, 487, 627, 668, 1015
$CaSO_4{\cdot}2H_2O$	Calcium sulphate dihydrate/Gypsum	414, 493, 619, 670, 1008, 1135
$Ca(NO_3)_2{\cdot}4H_2O$	Calcium nitrate tetrahydrate/Nitrocalcite	195, 719, 745, 1050
Fe_2O_3	Iron trioxide/Hematite	219, 237, 287, 403, 488, 606
α-$FeO(OH)$	Iron oxyhydroxide/Goethite	302, 384, 545
$FeO(OH){\cdot}nH_2O$	Iron oxyhydroxide/Limonite	240, 297, 394, 552
$Fe_2(SO_4)_3{\cdot}9H_2O$	Iron sulphate (III)/Coquimbite	497, 598, 1025, 1092, 1198
TiO_2	Titanium dioxide/Rutile	447, 608
SiO_2	Silicon dioxide/Quartz	263, 355, 395, 464, 695, 805, 1159
$KAlSi_3O_8$	Potasium feldspar/Amazonite	256, 283, 329, 355, 370, 403, 453, 474, 512, 650, 750, 814, 1124
Na_2CO_3	Sodium carbonate/ Natrite	193, 701, 1080
$Na_2SO_4{\cdot}10H_2O$	Sodium sulphate decahydrated/Mirabilite	167, 456, 617, 989, 1111
$NaNO_3$	Sodium nitrate/ Nitratine	190, 416, 724, 1067

4. Deterioration Due to Infiltration Waters

The effect of infiltration water is the second important impact affecting building materials. The water reaches the surface of the material by capillary action, reaching a certain height that depends on the porosity and water trapping capacity of the material, the evaporation rate as well as critical water content of the soil. The water coming from the soil is rich in carbonates, sulphates, chlorides and/or nitrates and calcium, magnesium, sodium, potassium and/or ammonium cations. In addition, in areas with high human settlements the soil water especially contains nitrates and chlorides. The later has been commonly related with coastal areas but, as it has been already mentioned, the current mechanism is likely the principal origin for these salts. High chloride concentrations have been determined in buildings located in estuaries suffering floods or close to roads or pavements were high amount of NaCl is uses during winter to avoid frost formation. However, most chlorides (being ionic compounds) are not Raman active so the complementary information of other techniques such as XRF spectroscopy is needed to characterize them.

In contrast, the presence of sulfate may be explained by the use of fertilizers in nearby fields, dedicated to agriculture, as well as, water contaminated nearby industries. Usually, the compounds are sodium or calcium sulfates with different number of hydration water as gypsum ($CaSO_4 \cdot 2H_2O$), bassanite ($CaSO_4 \cdot 0.5H_2O$), mirabilite ($Na_2SO_4 \cdot 10H_2O$) or thenardite (Na_2SO_4).

The formation of nitrate salts is mainly due to the impact coming from the organic matter degradation that produces ammonium nitrate (NH_4NO_3). Thanks to its acidity, this compound is very reactive and transforms the original compounds of the mortars and bricks like potassium feldspar ($KAlSi_3O_8$) or calcium oxide (CaO) producing frequently nitrates as for example niter (KNO_3), nitratine ($NaNO_3$), nitromagnesite ($Mg(NO_3)_2 \cdot 6H_2O$) and nitrocalcite ($Ca(NO_3)_2 \cdot 4H_2O$) [27], according to reactions 8 and 9. It is also remarkable that the composition of the salts appearing as efflorescences does not agree with the salt content of the samples. For instance, salts presence frequency order is Ca>Na>K>Mg in bricks whereas the frequency order for salts appearing as efflorescence is Na>K>Ca>Mg [40].

$$KAlSi_3O_8 + H_2O \rightarrow Al(OH)_3 + Si(OH)_4 + K^+ + OH^- \text{ (complete hydrolysis)}$$

$$KAlSi_3O_8 + H_2O \rightarrow SiAl_2O_5(OH)_4 + Si(OH)_4 + K^+ + 2OH^- \text{ (partial hydrolisis)}$$

$$K^+ + NH_4NO_3 \rightarrow KNO_3 + NH_4^+$$

$$NH_4^+ + H_2O \rightarrow NH_3 + H_3O^+ \qquad (8)$$

$$CaCO_3 + NH_4NO_3 + H_2O \rightarrow Ca(NO_3)_2 \cdot 4H_2O + CO_2 + NH_4^+$$

$$NH_4^+ + H_2O \rightarrow NH_3 + H_3O^+ \qquad (9)$$

Once water penetrates into the material, it migrates through capillary net. Then, a condensation process can occur at the surface of the material. In addition, the water will be removed by evaporation even transporting soluble salts [27]. The presence of one or another salt is directly related to environmental conditions such as humidity and temperature and thus, salt composition of facades affected by water infiltrations shows a clear seasonal variability [41]. Hence, Raman spectroscopy monitoring performed by portable instruments can be essential in order to determine the suitable treatment to remove the specific salts.

The common effects of the chemical degradation in buildings by atmospheric pollution or infiltration water are the alveoli, disaggregation, the typical disease plates and black crust formations. The latter is formed by the adhesion on the surface of wet material particles of dust, soot or smoke. Once deposited, it acts as a nucleus of reaction for chemical degradation, forming a permanent pollution patina. In the black crust can be found from common products of degradation to toxic particulate matter, heavy metals and PAHs [25].

The position of the bands of Raman spectra of the most common compounds found in mortars and bricks affected by infiltration water are in the table 2.

Table 2. Raman bands of some of the different compounds found usually in mortars and bricks by the action of infiltration waters measured with 785 nm laser

Formula	Chemical name/Mineralogical name	ν (cm^{-1}) (Main Raman bands)
$CaCO_3$	Calcium carbonate/Calcite	281, 712, 1085
$CaSO_4$	Calcium sulphate/Anhydrite	169, 399, 417, 498, 594, 609, 628, 661, 675, 1017, 1111, 1129, 1161
$CaSO_4·0.5H_2O$	Calcium sulphate hemihydrate/Bassanite	236, 322, 430, 487, 627, 668, 1015
$CaSO_4·2H_2O$	Calcium sulphate dihydrate/Gypsum	414, 493, 619, 670, 1008, 1135
$Ca(NO_3)_2·4H_2O$	Calcium nitrate tetrahydrate/Nitrocalcite	195, 719, 745, 1050
$KAlSi_3O_8$	Potassium feldspar/Amazonite	256, 283, 329, 355, 370, 403, 453, 474, 512, 650, 750, 814, 1124
Na_2CO_3	Sodium carbonate/ Natrite	193, 701, 1080
Na_2SO_4	Sodium sulphate/Thenardite	451, 464, 628, 631, 644, 992, 1100, 1128, 1148
$Na_2SO_4·10H_2O$	Sodium sulphate decahydrated/Mirabilite	167, 456, 617, 989, 1111
KNO_3	Potassium nitrate/Niter	198, 417, 715, 1050, 1344, 1359, 1777
NH_4NO_3	Ammonium nitrate/Nitrammite	193, 714, 1043, 1288, 1412, 1654, 1777

5. Biodeterioration Markers

Raman Spectroscopy is a very useful analytical technique in the characterization purposes of not only inorganic compounds but also in organic compounds present in bricks and mortars. Due to the usual exposure of these materials to open atmosphere, they are colonized by diverse kind of microorganisms.

Crustose lichens, which are extremely attached to the stone, develop physical structures called hyphae to penetrate into stone porous. These structures cause the cracking of the materials and promote physical decaying mechanisms [42]. In this sense, mortars are more susceptible than bricks to be attacked due to their higher porosity but also because of the composition. Lichens excrete oxalic acid in high quantities which reacts with calcite present in building materials and calcium oxalates appear following reactions 10 and 11:

$$H_2C_2O_4 + CaCO_3 \rightarrow CaC_2O_4 \cdot H_2O + CO_2 \quad (10)$$

$$2H_2C_2O_4 + CaCO_3 \rightarrow 2(CaC_2O_4 \cdot 2H_2O) + 2CO_2 \quad (11)$$

Calcium oxalates can appear in the monohydrate form or mineral whewellite ($CaC_2O_4·H_2O$) or in the dihydrate form or mineral weddellite ($CaC_2O_4·2H_2O$). Both compounds are clearly detected and differentiated by Raman Spectroscopy as can be seen in figure 7 [43, 44].

Lichens and mosses are photosynthetic organisms, and thus, they need of different pigments that can be distinguished by Raman Spectroscopy. The most common are chlorophyll and carotenoids. Both are present in photosynthetic reaction centre of the chloroplasts in cellular parts of these organisms. Other common pigment is phorphyrin which appears together with chlorophyll in the photosynthetic reaction centre [45].

When the carotenoid type molecules are present in the target sample which is going to be analyzed by Raman Spectroscopy, another excitation laser, which allows the resonance effect of these organic compounds, can be proposed. Carotenoid molecules are chromophoric compounds which present a maximum in the blue-green region of the absorption spectra. When the excitation laser of the Raman measures coincides with a maximum of the absorption spectra, carotenoids absorb this radiation in a resonance effect. Consequently the Raman signatures of these compounds are significantly enhanced respect to signatures of other compounds present jointly in the sample. Moreover, when the Raman spectra is extended until 3500-4000 cm^{-1} (depending on the spectral range of the equipment), new Raman signatures appear corresponding with Resonance Raman features of the carotenoid.

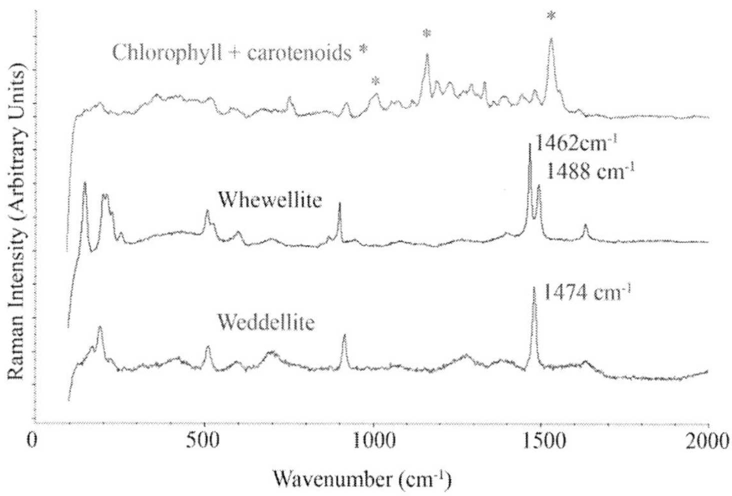

Figure 7. Different biodeterioration compounds found on the bricks and mortar. Spectra collected with 785 nm laser of the InnoRam ($BWTEK_{INC}$) handheld spectrometer. In the first spectrum a mixture of chlorophyll and carotenoids (peaks marked with *) was identified.

The global information that a Raman spectra gives, allow us to detect the presence of carotenoids as well as to distinguish among carotenes (carotenoids composed of a chain with certain quantity of polyenes) and xanthophylls (oxygen derivatives of carotenes). Figure 8 shows the evidence of the identification and consequent differentiation between carotenoids. The spectra obtained are collected with 514 nm laser of the InVia Raman spectrometer but other research studies used the 514 nm laser with handheld instrument in order to identify carotenoids in field measurements [46].

This differentiation is based on the Raman signals appeared in the range of 1500-1550 cm^{-1} coincident with C=C stretching vibration and the Resonance Raman features appeared between 2000 and 3500 cm^{-1}. On the other hand, in the case of chlorophyll and phorphyrin the use of 514 nm laser is not adequate due to the impossibility of determinate them. Consequently, the Raman signals of carotenoids that appear together with these pigments are significantly improved.

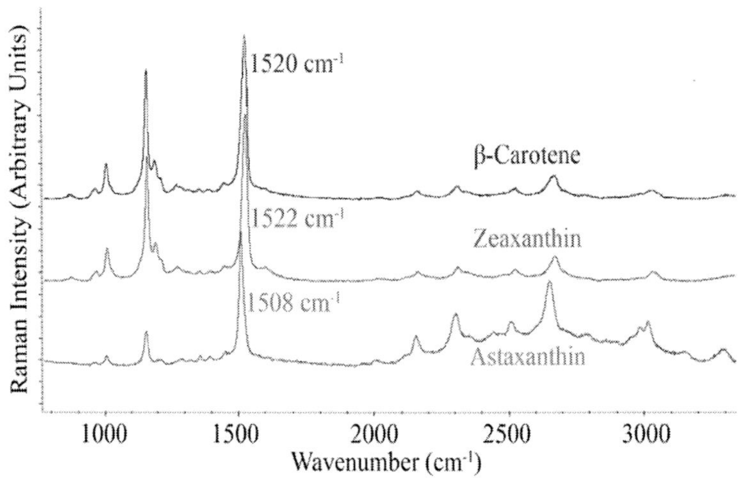

Figure 8. Carotenoid differentiation obtained with InVia Raman spectrometer using the laser of 514 nm which allows to achieve the Resonance effect of carotenoids.

Table 3. Raman bands of some of the different compounds found usually in mortars and bricks by biodeterioration measured with 785 or 514 nm laser

Formula	Laser (nm)	Chemical name/Mineral	Main Raman bands ν (cm $^{-1}$)
$CaC_2O_4 \cdot H_2O$	785	Calcium oxalate/ Whewellite	140, 195, 205, 221, 503, 520, 593, 863, 895, 937, 1391, 1462, 1488, 1627
$CaC_2O_4 \cdot 2H_2O$	785	Calcium oxalate dihydrated/ Weddellite	161, 188, 506, 589, 868, 910, 1410, 1474, 1629
$C_{55}H_{72}O_5N_4Mg$	785	Chlorophyll a	513, 744, 753, 919, 963, 986, 1048, 1068, 1110, 1185, 1221, 1286, 1326, 1385, 1438, 1550, 1606
Undetermined*	785	Carotenoids*	1003, 1156, 1524
$C_{36}H_{20}N_2O_4$	785	Scytonemin	574, 1167, 1330, 1385, 1508, 1558, 1592
$C_{36}H_{20}N_2O_4$	514	Scytonemin	439, 494, 575, 676, 1095, 1158, 1170, 1286, 1323, 1383, 1453, 1507, 1553, 1592, 1629, 1713
$C_{40}H_{52}O_4$	514	Astaxanthin	956, 1006, 1152, 1198, 1507, 2020, 2161, 2302, 2353, 2510, 2652, 2785, 2983, 3016, 3159, 3294, 3444
$C_{40}H_{56}O_2$	514	Zeaxanthin	867, 962, 1003, 1155, 1185, 1206, 1267, 1288, 1314, 1351, 1389, 1445, 1521, 2020, 2157, 2305, 2345, 2525, 2666, 2985, 3027, 3306, 3445
$C_{40}H_{56}$	514	β-carotene	959, 1003, 1155, 1186, 1207, 1520, 2156, 2309, 2521, 2672, 3030

* The carotenoid differentiation is not possible with the 785 nm laser.

Another molecule related with the metabolism of lichens is the UV screening pigment scytonemin that is unambiguously detected with Raman Spectroscopy. The position of its Raman bands does not differ essentially between two excitation lasers (785 or 514 nm), but the relative intensities of Raman bands change with the laser type. This fact makes necessary to establish a previous work to find the proper Raman signatures of each molecule with each laser or maybe in some cases it is necessary to measures pure standards of each compound [47].

The Raman features of each compound are collected in table 3, the identification of these compounds is crucial to diagnose the biodeterioration suffered by bricks and mortars.

CONCLUSION

Raman Spectroscopy is a very useful analytical technique to diagnose the conservation state of building materials, including bricks and mortars. It also has the possibility to carry out in situ analysis with a good quality results. Raman Spectroscopy can be classified as a non destructive technique, so the value, the importance and the total integrity of the samples are preserved.

Thanks to the possibility offered by this technique in distinguishing different types of molecules it is possible to establish deterioration reactions and mechanisms based only in Raman measures. But in the case of some compounds like calcite and gypsum, which can appear in mortars as original or decaying compounds, the criterion of the researcher is crucial to classify each one based on the shape, depth, location or even exposure suffered by the material used because Raman technique give the same response in both cases.

In any case, the possibility that Raman spectroscopy offers for field analysis is a clear advantage that makes this technique a suitable to diagnose construction materials and built heritage in particular. Even more, due to the development of new hand-held Raman spectrometers with high sensitivity and good spectral resolution, a fast diagnosis of the building materials (including bricks and mortars) can be performed. Hence, a restoration protocol can be more rapidly established from in situ analyses since decisions can be taken in the field.

ACKNOWLEDGMENTS

This work has been financially supported by the IMDICOGU project (ref.:BIA2008-06592) from the Spanish Ministry of Science and Innovation (MICINN). O. Gómez-Laserna and I. Ibarrondo acknowledge their grants from the University of the Basque Country and N. Prieto-Taboada acknowledges her grant from MICINN. Technical support provided by the Raman-LASPEA laboratory of the SGIker (UPV/EHU, MICINN, GV/EJ, ERDF and ESF) is gratefully acknowledged.

REFERENCES

[1] Chaplin, TD; Clark, RJH; Martinón-Torres, M; A combined Raman microscopy, XRF and SEM–EDX study of three valuable objects – A large painted leather screen and two illuminated title pages in 17th century books of ordinances of the Worshipful Company of Barbers, London. *Journal of Molecular Structure*, 2010 976, 350-359.

[2] Robinson, JW; Skelly Frame, EM; Frame II, GM; *Undergraduate instrumental analysis*. New York: Marcel Dekker; 2005.

[3] Smith, E; Dent, G; *Modern Raman Spectroscopy - A Practical Approach.* England: John Wiley and Sons; 2005.

[4] Fouad, E; Coherent anti-Stokes Raman scattering: Spectroscopy and microscopy. *Vibrational Spectroscopy*, 2011 55, 1-37.

[5] Gooijer, C; Mank, AJG; Laser spectroscopy in analytical chemistry: *Analytica Chimica Acta*, 1999 400, 281-295.

[6] Jimenez-Sandoval, S; Micro-Raman spectroscopy: a powerful technique for materials research. *Microelectronics Journal*, 2000 31, 419-427.

[7] Lutz, HD; Haeuseler, H; Infrared and Raman spectroscopy in inorganic solids research. *Journal of Molecular Structure*, 1999 511-512, 69-75.

[8] Moore, DS; McGrane, SD; Comparative infrared and Raman spectroscopy of energetic polymers. *Journal of Molecular Structure*, 2003 661-662, 561-566.

[9] Snively, CM; Koenig, JL; Polymer Applications of IR and Raman Spectroscopy. In: Lindon, J, editor. *Encyclopedia of Spectroscopy and Spectrometry*, Oxford: Academic Press; 1999; 2213-2220.

[10] Martínez-Arkarazo, I; Angulo, M; Zuloaga, O; Usobiaga, A; Madariaga, JM; Spectroscopic characterisation of moonmilk deposits in Pozalagua tourist Cave (Karrantza, Basque Country, North of Spain). *Spectrochimica Acta Part A: Molecular and Biomolecular Spectroscopy*, 2007 68, 1058-1064.

[11] Wachs, IE; Jehng, J; Deo, G; Weckhuysen, BM; Guliants, VV; Benziger, JB; In situ Raman spectroscopy studies of bulk and surface metal oxide phases during oxidation reactions. *Catalysis Today*, 1996 32, 47-55.

[12] Sahu, S; Exline, DL; Nelson, MP; Identification of thaumasite in concrete by Raman chemical imaging. *Cement and Concrete Composites,* 2002 24, 347-350.

[13] Brody, RH; Carter, EA; Edwards, HGM; Pollard, AM; FT-Raman Spectroscopy, Applications. In: Lindon, J, editor. *Encyclopedia of Spectroscopy and Spectrometry*, Oxford: Academic Press; 1999; 732-740.

[14] Lana-Villarreal, T; Pérez, JM; Gómez, R; Surface enhanced Raman spectroscopy for adsorption studies on semiconductor nanostructured films. *Surface Science*, 2004 572, 329-336.

[15] Herrero, AM; Cambero, MI; Ordóñez, JA; de la Hoz, L; Carmona, P; Raman spectroscopy study of the structural effect of microbial transglutaminase on meat systems and its relationship with textural characteristics. *Food Chemistry,* 2008 109, 25-32.

[16] Das, RS; Agrawal, YK; Raman spectroscopy: Recent advancements, techniques and applications. *Vibrational Spectroscopy*, 2011 57, 163-176.

[17] Madariaga, JM; Raman spectroscopy in art and archaeology. *Journal of Raman Spectroscopy*, 2010 41, 1389-1393.

[18] Parmrlee, CW; Soluble salts and clay wares. *Journal of the American Ceramic Society,* 1922 5, 538-553.

[19] Cardiano, P; Ioppolo, S; De Stefano, C; Pettignano, A; Sergi, S; Piraino, P; Study and characterization of the ancient bricks of monastery of "San Filippo di Fragalà" in Frazzanò (Sicily). *Analytica Chimica Acta,* 2004 519, 103-111.

[20] Charola, A; Pühringer, J; Steiger, M; Gypsum: a review of its role in the deterioration of building materials. *Environmental Geology*, 2007 52, 339-352.

[21] Charola, A; Salts in the Deterioration of Porous Materials: An Overview. *Journal of the American Institute for Conservation*, 2000 39, 327-343.

[22] Martinez-Arkarazo, I; Knuutinen, U; Maguregui, M; Castro, K; Madariaga, JM; Analytica Pompeiana Universitatis Vasconicae: field multianalytical analyses in Pompeii. *Book of Abstracts: 6th International Congress on the aplication of Raman spectroscopy in Art and Archaeology*, Italy: Timeo Editore Bologna; 2011; 50-51.

[23] Prieto-Taboada, N; Maguregui, M; Martinez-Arkarazo, I; Olazabal, M; Arana, G; Madariaga, J; Spectroscopic evaluation of the environmental impact on black crusted modern mortars in urban–industrial areas. *Analytical and Bioanalytical Chemistry*, 2010 399, 2949-2959.

[24] Castro, K; Sarmiento, A; Maguregui, M; Martínez-Arkarazo, I; Etxebarria, N; Angulo, M; Barrutia, M; González-Cembellín, J; Madariaga, J; Multianalytical approach to the analysis of English polychromed alabaster sculptures: μRaman, μEDXRF, and FTIR spectroscopies. *Analytical and Bioanalytical Chemistry*, 2008 392, 755-763.

[25] Martínez-Arkarazo, I; Angulo, M; Bartolomé, L; Etxebarria, N; Olazabal, MA; Madariaga, JM; An integrated analytical approach to diagnose the conservation state of building materials of a palace house in the metropolitan Bilbao (Basque Country, North of Spain). *Analytica Chimica Acta*, 2007 584, 350-359.

[26] Pérez-Alonso, M; Castro, K; Martínez-Arkarazo, I; Angulo, M; Olazabal, MA; Madariaga, JM; Analysis of bulk and inorganic degradation products of stones, mortars and wall paintings by portable Raman microprobe spectroscopy. *Analytical and Bioanalytical Chemistry*, 2004 379, 42-50.

[27] Maguregui, M; Sarmiento, A; Martínez-Arkarazo, I; Angulo, M; Castro, K; Arana, G; Etxebarria, N; Madariaga, JM; Analytical diagnosis methodology to evaluate nitrate impact on historical building materials. *Analytical and Bioanalytical Chemistry*, 2008 391, 1361-1370.

[28] Arrieta, N; Goienaga, N; Martínez-Arkarazo, I; Murelaga, X; Baceta, JI; Sarmiento, A; Madariaga, JM; Beachrock formation in temperate coastlines: Examples in sand-gravel beaches adjacent to the Nerbioi-Ibaizabal Estuary (Bilbao, Bay of Biscay, North of Spain). *Spectrochimica Acta Part A: Molecular and Biomolecular Spectroscopy*, 2011 80, 55-65.

[29] Aramendia, J; Gómez-Nubla, L; Fdez-Ortiz de Vallejuelo, S; Castro, K; Murelaga, X; Madariaga, JM; New Findings by Raman Microspectroscopy in the Bulk and Inclusions Trapped in Libyan Desert Glass. *Spectroscopy Letters*, *in press*.

[30] Ibarrondo, I; Prieto-Taboada, N; Martínez-Arkarazo, I; Madariaga, JM; The Suitable Carotene and Xanthophyll Identification in Lecanora Lichens: *Resonance Raman Spectroscopic Study*. In: Lunar and Planetary Institute, editor. Conference on Micro-

Raman and Luminiscence Studies in the earth and planetary sciences (CORALS II), Houston: Lunar and Planetary Institute (LPI) y Consejo Superior de Investigaciones Científicas (CSIC); 2011; 40-41.

[31] Gómez-Laserna, O; Morillas, H; Prieto-Taboada, N; Ibarrondo, I; Martínez-Arkarazo, I; Olazábal, MA; Madariaga, JM; Raman Spectroscopy study of a salt weathering process in mortars of a historical palace house. In: Sammartano, S and Crisponi, G, editor. *The acta of the International Symposia on Metal Complexes,* Messina, Italy: Ismec. group series; 2011; 125-126.

[32] Prieto-Taboada, N; Gómez-Laserna, O; Ibarrondo, I; Martinez-Arkarazo, I; Olazabal, MA; Madariaga, JM; *Cross-Section Analysis to Establish the Penetration Level of Atmospheric Pollution in Mortars.* In: Lunar and Planetary Institute, editor. Conference on Micro-Raman and Luminiscence Studies in the earth and planetary sciences (CORALS II), Houston: Lunar and Planetary Institute (LPI) y Consejo Superior de Investigaciones Científicas (CSIC); 2011; 66-67.

[33] Sarmiento, A; Maguregui, M; Martínez-Arkarazo, I; Angulo, M; Castro, K; Olazábal, MA; Fernández, LA; Rodríguez-Laso, MD; Mujika, AM; Gómez, J; Madariaga, JM; Raman spectroscopy as a tool to diagnose the impacts of combustion and greenhouse acid gases on properties of Built Heritage. *Journal of Raman Spectroscopy*, 2008 39, 1042-1049.

[34] Charola, AE; Review of the Literature on the Topic of Acidic Deposition on Stone. New York: *The National Center for Preservation Technology and Training*; 1998.

[35] Allen, GC; El-Turki, A; Hallam, KR; McLaughlin, D; Stacey, M; Role of NO2 and SO2 in degradation of limestone. *British Corrosion Journal,* 2000 35, 35-38.

[36] van Aardenne, JA; Carmichael, GR; Levy, H; Streets, D; Hordijk, L; Anthropogenic NOx emissions in Asia in the period 1990-2020. *Atmospheric Environment*, 1999 33, 633-646.

[37] Dotsika, E; Psomiadis, D; Poutoukis, D; Raco, B; Gamaletsos, P; Isotopic analysis for degradation diagnosis of calcite matrix in mortar. *Analytical and Bioanalytical Chemistry*, 2009 395, 2227-2234.

[38] Johansen, V; Klemm, W; Taylor, P; Why Chemistry Matters in Concrete. *Concrete International*, 2002 24, 84-89.

[39] Cultrone, G; Sebastián, E; Elert, K; de la Torre, MJ; Cazalla, O; Rodriguez–Navarro, C; Influence of mineralogy and firing temperature on the porosity of bricks. *Journal of the European Ceramic Society*, 2004 24, 547-564.

[40] Rincón, JM; Romero, M; Fundamentos y clasificación de las eflorescencias en ladrillos de construcción. *Materiales de Construcción*, 2000 50, 63-70.

[41] Matsukura, Y; Oguchi, CT; Kuchitsu, N; Salt damage to brick kiln walls in Japan: spatial and seasonal variation of efflorescence and moisture content. *Bulletin of Engineering Geology and the Environment,* 2004 63, 167-176.

[42] Jorge-Villar, SE; Edwards, HGM; Lichen colonization of an active volcanic environment: a Raman spectroscopic study of extremophile biomolecular protective strategies. *Journal of Raman Spectroscopy,* 2010 41, 63-67.

[43] Edwards, HGM; Seaward, MRD; Attwood, SJ; Little, S; de Oliveira, LFC; Tretiach, M; FT-Raman spectroscopy of lichens on dolomitic rocks: an assessment of metal oxalate formation. *The Analyst,* 2003 128, 1218-1121.

[44] Frost, RL; Raman spectroscopy of natural oxalates. *Analytica Chimica Acta*, 2004 517, 207-214.

[45] Withnall, R; Chowdhry, BZ; Silver, J; Edwards, HGM; de Oliveira, LFC; Raman spectra of carotenoids in natural products. *Spectrochimica Acta Part A: Molecular and Biomolecular Spectroscopy,* 2003 59, 2207-2212.

[46] Vítek, P; Jehlička, J; Edwards, H; Osterrothová, K; Identification of β-carotene in an evaporitic matrix—evaluation of Raman spectroscopic analysis for astrobiological research on Mars. *Analytical and Bioanalytical Chemistry,* 2009 393, 1967-1975.

[47] De Gelder, J; De Gussem, K; Vandenabeele, P; Moens, L; Reference database of Raman spectra of biological molecules. *Journal of Raman Spectroscopy,* 2007 38, 1133-1147.

In: Brick and Mortar Research
Editors: S. Manuel Rivera and A. L. Pena Diaz
ISBN: 978-1-61942-927-7

Chapter 7

PREPARATION OF COLOURED FACING BRICK FROM LOW MELTING CLAY UNDER A WATER VAPOUR ATMOSPHERE

Jadambaa Temuujin*[*a], Tsedev Jadambaa[b] and Shigeo Hayashi[c]

[a] Institute of Chemistry and Chemical Technology,
Mongolian Academy of Sciences, Mongolia
[b] Centre for New Materials, Mongolian
University of Science and Technology, Ulaanbaatar, Mongolia
[c] Graduate School of Engineering and Resource Science,
Akita University, Akita, Japan

ABSTRACT

Preparation of coloured facing brick by the traditional method requires high quality clay minerals, high temperature firing and occasionally colouring pigments. In this chapter we report our research on the preparation of coloured facing brick from locally available low melting clay (Tolgoit deposit, Mongolia) under a pressured water vapour atmosphere. It was found that colour change and mechanical properties of the brick samples strongly depends on chemical and mineralogical composition of the clay. The water vapour pressure and the chemical composition of the vapour also affects the properties of the brick after firing. The clay specimens were fired at 600, 800, 900 and 1000°C for various times. A water vapour atmosphere improved the structure formation of the ceramics and increased the crushing strength of the fired clay by up to three times. Ilmenite formed from the $FeO+TiO_2$ oxides of the original clay during firing under the water vapour atmosphere colours the brick a blue to grey colour, while $CaO+MgO+SiO_2$ oxides may form a green-coloured compound. Based on the experimental results, coloured facing bricks of acceptable quality were produced on a semi-industrial pilot plant scale.

---

* jtemuujin@yahoo.com

INTRODUCTION

Brick has been regarded throughout history as one of the longest lasting and strongest building materials. Bricks for building may be made from clay, shale, soft slate, calcium silicate, concrete, or shaped from quarried stone. However, true bricks are ceramics and are therefore created by the action of heating and cooling [1]. Buildings and various utensils made of clay have always been associated with Man's civilized development and history. For instance, sun-baked clay structures were built during the pre-European period of American history [2]. Bricks were commonly used in Europe at the time of the early European settlement of North America and fired bricks were produced at an early date [2]. The properties and colour of the fired clay bricks depend on many parameters, including the physical, chemical and mineralogical properties of the raw material, firing temperature and duration, method of production, firing atmosphere etc. During the firing of green clay bricks, chemical reactions produce structural and mineralogical composition changes of the clay constituents, resulting in hard and durable ceramic bodies that are highly resistant to weathering. The colour of the fired bricks depends not only on the chemical composition of the clay but also on the chemical composition of the compounds formed during firing. If the firing results in the incorporation of trivalent iron into mullite, metakaolinite or a fassaitic pyroxene, the colour changes to yellow or light brown. If after firing, the iron is present as free iron oxide (hematite), the colour of the brick becomes red [3]. The colour of ceramic bricks can be changed by incorporating manganese and iron ore and metallic slags and sludges from electroplating plants [4] or different pigments [5]. Generally the mineralogical composition of the clay determines the colour of the clay brick [6-8]. More than 800 different brick colours have been described, from the lightest of whites and creams to the darkest blues and purples, providing an unlimited opportunity to blend or accentuate [9].

Traditionally, brick was one of the most common building materials used in Mongolia since the Hunnu period 300-100 B.C. [10]. Archaeological study indicates that the production of quarrel-shaped red bricks in Mongolia started B.C. [11]. It is likely that the production of blue bricks began after the production of the red bricks. One of the earliest findings of blue brick related to the early Turks period in Mongolia. An archaeological survey of Kul-Tegin's tomb (died 731) revealed the presence of red and blue bricks and glazed tiles [12, 13]. The sintered blue bricks were used for floor and road paving, demonstrating their application as strong, durable and freeze-thaw resistant materials. From the 16th century in Mongolia, many Buddhist temples were built, mainly from blue coloured brick that produced from the nearby low melting-point clay deposits. We have carried out a comparative study of chemical composition of blue brick from the Amarbayasgalant temple (Bulgan aimag, central Mongolia) and the nearby low melting-point clay deposit. The use of these low melting clays by the master brick makers reflects the scarcity of high quality clay minerals in Mongolia. Table 1 shows the chemical composition of the blue brick and Amarbayasgalant low melting clay. The blue bricks were fired in a shaft-like kiln which allows the introduction of water during cooling, creating a pressurized water vapour atmosphere.

Table 1 clearly indicates that the low melting clay was used for the production of the blue brick. The similarity of the chemical compositions also suggests that these coloured bricks can be produced from the more abundant low melting clay deposit. Although clay brick is one of the best known building materials with a several thousand year old history, there is more to learn to improve the brick making process.

The influence of firing temperature and duration on the physical and mechanical properties of clay brick has been studied previously, indicating that the firing temperature is the most important factor [14, 15]. At higher firing temperatures, the physical and mechanical properties of the clay brick improve dramatically as the vitrification point of the clay is approached. On the other hand, the mechanical properties of clay samples fired at lower temperatures can be improved by firing under a water vapour atmosphere [16, 17]. Moreover, the colour of low-melting clay can be regulated by applying a water vapour atmosphere, as was used by the Mongolian masters in building the Buddhist temples. However, the colouration mechanism in low-melting clay was not determined scientifically. Therefore, it is important to determine this colour change mechanism in low melting clays fired under a water vapour atmosphere, to enable coloured facing bricks to be produced in a single firing without additives. It has been estimated that the use of facing brick in construction reduces the cost of a 1 m^2 of wall by 15% and the labour by 25%, and also reduces the expenditure of maintenance of the facade [4]. In this chapter we report our previous results on the preparation of coloured facing brick by firing under a pressurized water vapour atmosphere, as part of a project carried out by the authors [11, 18].

EXPERIMENTAL

The raw material was the low-melting clay from the Tolgoit deposit near Ulaanbaatar city. The total reserve of this clay is approx. 20 million m^3, and its chemical composition (in wt.%) is: SiO_2 58.7, Al_2O_3 16.06, TiO_2 0.66, Fe_2O_3 3.81, FeO 0.46, CaO 5.21, MgO 1.49, Na_2O 2.42, K_2O 2.59, MnO 0.08, LOI 7.01, H_2O 1.66. This chemical composition shows that Tolgoit clay contains a medium content of colored oxides (Fe_2O_3+FeO+TiO_2+MnO = 5.01%) and a high content alkali oxides (Na_2O+K_2O = 5.0%). The chemical composition of the clay was determined by traditional wet chemical analysis. Mineralogical analysis was carried out using a combination of XRD ("Dron 2" equipment) optical microscopy (MIN-8" microscope) and DTA/TG analyses ("MOM" derivatograph). The mineralogical composition consists of montmorillonite and hydromica minerals with impurities of kaolinite and halloysite. Thermal expansion or shrinkage was measured with a quartz dilatometer under water vapour and air atmospheres at a heating rate of 10°C/min, the water vapour being introduced into the dilatometer furnace from 400°C. Vapour pressure wasn't controlled during the dilatometer measurements. The technological properties of this clay were determined by standardized methods based on the Russian standards. The working moisture of the clay was 22.2% with an absolute working moisture of 28.5%. The upper border of plasticity was 47, the lower border was 28 and the plasticity number was 19, indicating that this clay is of medium plasticity. Granulometry distribution showed 27.1% of particles >0.05 mm, 33.05% between 0.05-0.01 mm, 26.65% between 0.005-0.001 mm and 4.35% <0.001mm.

Table 1. Chemical composition of blue brick and Amarbayasgalant clay

	SiO_2	Al_2O_3	Fe_2O_3	TiO_2	CaO	MgO	Na_2O	K_2O	SO_3	LOI
Blue brick	60	15.41	6.86	1.00	7.38	4.19	0.96	2.51	0.02	0.27
Amarbayasgalant clay	54.78	14.73	6.36	0.95	6.75	2.78	1.48	1.73	0.12	7.59

The raw clay was dried to a residual moisture of 3-5%, ground and passed through a 0.2 mm sieve. The clay powder was mixed with water and used to prepare cubes with 2 cm edges and 15 x 2 x 1 cm bars by the usual plastic method. These specimens were dried at ambient temperature to a residual moisture of 2-4%.

The samples were fired under a regulated water vapor atmosphere in a slightly modified autoclave furnace described elsewhere [16]. The furnace pressure was measured by a pressure gauge in terms of mm of water. The influence of firing temperature and dwell time on the technological and physico-mechanical characteristics of the clay specimens was determined by firing the cube samples at 600, 700, 800, 900 and 1000°C for 2, 4 and 8 hours under a constant pressure of 20 mm of water. The effect of the water vapour pressure on the technological and physico-mechanical characteristics of the samples was studied by firing the cube samples at 800 and 1000°C under pressures of 0, 10, 20, 100, 200 and 300 mm of water. Same samples showed a variation in the mechanical properties in different batches, but the general trends were similar. The furnace heating rate was 250°C/h and no water vapour was introduced to the furnace below 400°C in either the heating or cooling cycles. Compressive and bending strengths, open porosity and density values were used to characterize the physico-mechanical properties. At least 5 samples were used for each measurement. The degree of Fe_2O_3 reduction to FeO was determined volumetrically by titration with $K_2Cr_2O_7$. The samples were dissolved in a mixture of HF and H_2SO_4 under an inert atmosphere.

RESULTS AND DISCUSSION

Influence of Water Vapour Atmosphere on the Structure Formation of Ceramics

XRD analysis (not shown here) revealed that under water vapour atmosphere the decomposition of the minerals occurs faster than in ambient atmosphere, in good agreement with the previous results [11]. Moreover, crystallisation of hercynite occurred in this specimen but not in samples fired in ambient atmosphere [2]. Similar thermal behaviour of clay minerals under water vapour atmosphere was reported in our previous papers [16, 17]. However, in contrast to kaolinite, crystallisation of mullite was not detected [16, 19]. Although mullite was not detected in the sample fired at 1000 °C, a small amount, below the XRD detection limit, may form. Dilatometric curves determined under oxidising (ambient) and water vapour atmosphere are shown in Fig.1.

The dilatometer measurements indicate that the first and second shrinkage stages of the sample fired under water vapour starts 50-70 °C lower than in ambient atmosphere. The first shrinkage is caused by decomposition of minerals, release of chemically bound water and start of sintering, the second shrinkage is related to liquid phase sintering of the specimens. In other words, structure formation in the sample is facilitated by the water vapour atmosphere. Low temperature sintering of the samples fired in the water vapour atmosphere is confirmed by the bulk density and water absorption data. Fig.2 shows the change of bulk density and water absorption with the firing temperature under water vapour and ambient atmospheres.

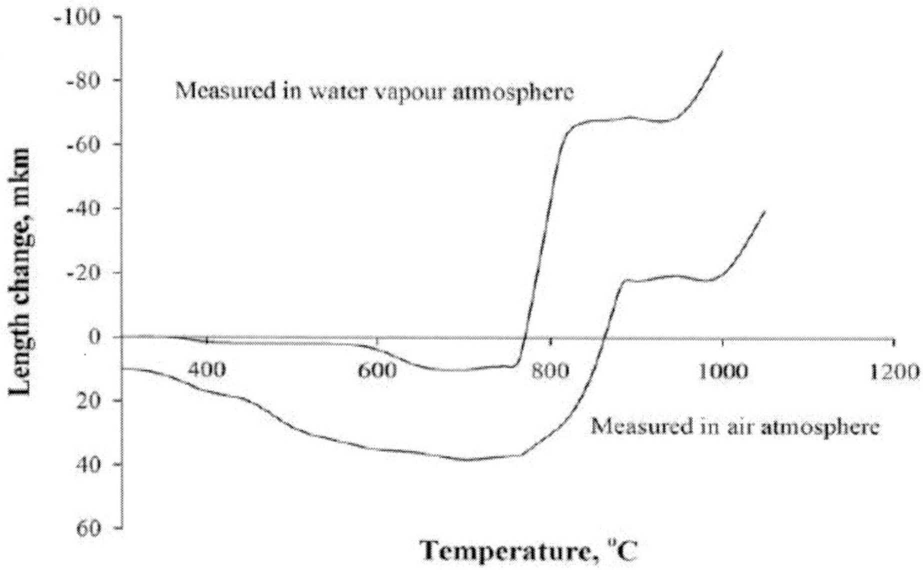

Figure 1. Dilatometer curves of samples fired under water vapour and oxidizing atmospheres.

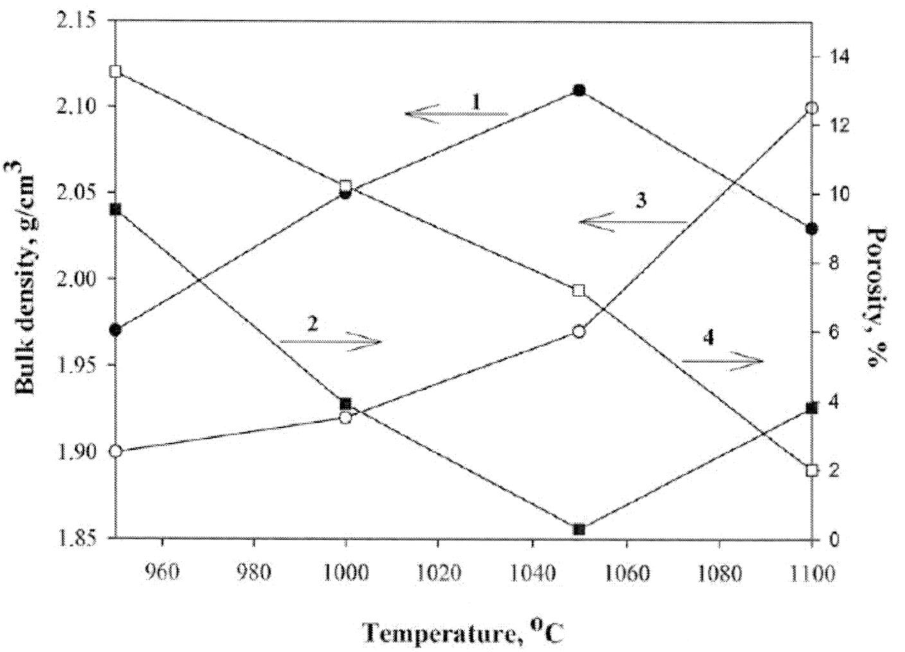

Figure 2. Bulk density and porosity curves of samples fired under water vapour (1, 2) and ambient (3, 4) atmosphere.

The density and porosity changes (reducing the bulk density and increasing porosity at 1100°C, curves 1 and 2) indicate that specimens fired in water vapour undergo partial melting. In other words, the water vapour atmosphere decreases the temperature of refractoriness by about 40-50°C. Previous authors have reported that a water vapour atmosphere decreases the sintering temperature by 170-250°C compared with samples fired in ambient atmosphere [20]. Possibly this large difference in sintering temperature is related to differences in the chemical and mineralogical composition of the raw material. We have previously suggested that the positive influence of a water vapour atmosphere on structure formation of the ceramic is a common phenomenon with breakage of Si-O bond and formation of Si-OH groups in kaolinite and montmorillonite clays [16, 17] and this may also be the case for low-temperature melting clay such as Tolgoit clay.

Firing of Tolgoit clay in a water vapour atmosphere accelerates formation of the glassy phase. The content of glassy phase increases with increasing firing temperature, as evidenced by the partial melting of the specimens. Therefore, it is expected that the dwell time should also influence the physico-mechanical properties of the clay specimens.

The influence of dwell time on the compressive strength of the clay brick is shown in Fig.3.

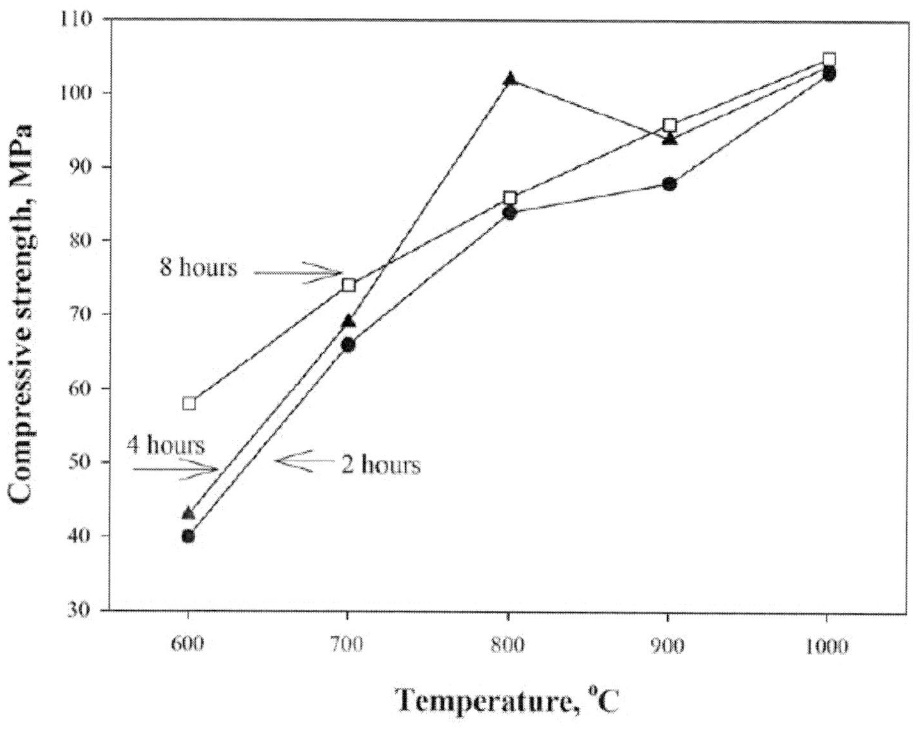

Figure 3. Change of the compressive strength with dwell time.

Fig. 3 shows that the compressive strength of the specimens increases almost linearly between 600 and 800°C, and also increases with increasing dwell time up to 8h. The change in compressive strength is negligible at 1000°C even when the dwell time is increased 4-fold. Colour changes occur in samples fired above 600°C, but the hue of these colours and the corresponding strength of the specimens did not warrant the use of firing temperatures below 800°C. Therefore, temperatures of 800 and 1000°C were chosen for further experiments. Fig.4 shows influence of water vapour pressure on the compressive strength of the samples fired at 800 and 1000°C.

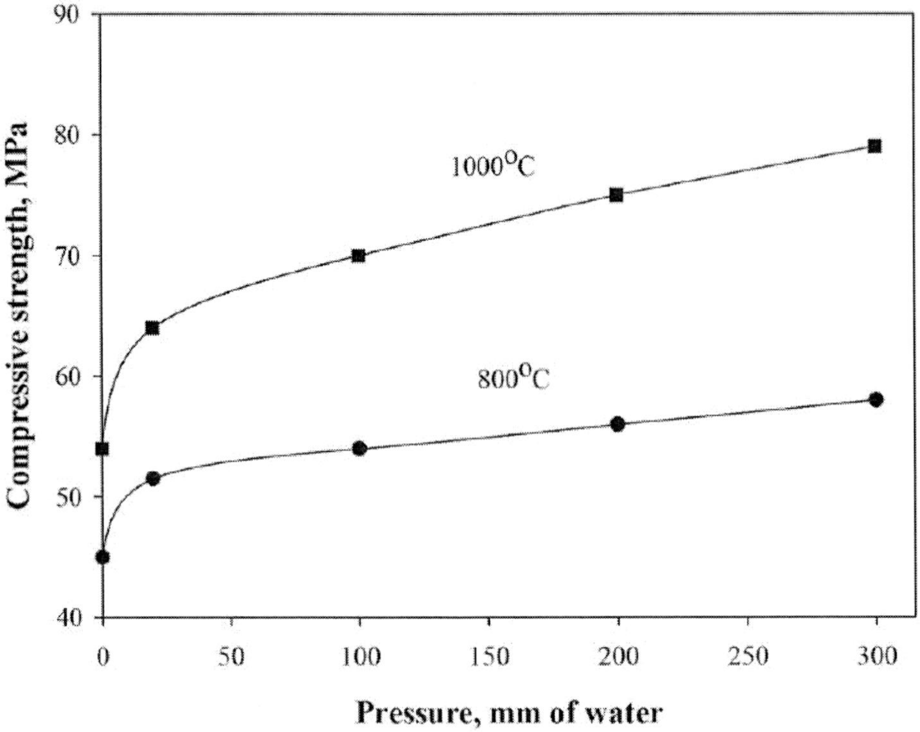

Figure 4. Influence of vapour pressure on the compressive strength of the samples fired at 800 and 1000 °C.

Fig.4 shows steep increases in compressive strength with increasing vapour pressure from 0 to 20 mm of water at 800 and 1000 °C. Further pressure increase to 300 mm of water results in a more gradual rate of strength increase. This graph shows that increasing the vapour pressure exerts a positive influence on the compressive strengths, which are quite high in the specimens fired under 20 mm of water pressure at 800 and 1000°C. Since 20 mm of water pressure is not very difficult to achieve under practical production conditions, this pressure was chosen as the working pressure. This is in agreement with previous results which showed the mechanical properties of kaolinite and montmorillonite clays can be improved substantially by firing under water vapour atmospheres [16, 17].

The influence of the water vapour atmosphere on the mechanical properties of the clay specimens is shown in Fig.5.

The compressive strengths of the specimens fired under water vapour atmosphere (curves 1 and 3) are higher than those fired under ambient (oxidising) atmosphere at all firing temperatures. The bending strengths of the same specimens (curves 2 and 4) are very similar at 600 and 700°C, but show considerably greater differences in samples fired at higher temperatures. The bending strength of the sample fired in water vapour atmosphere is higher than that fired in ambient atmosphere, except at 1000°C, where the bending strength of specimen fired under water vapour is lower than that fired at 900°C. This result appears to be related to the increased liquid phase content in the samples. We suggest that even low water vapour pressure (20 mm of water) accelerates the decomposition of the low-melting clay and improves the ceramic structure formation at lower temperatures compared with samples fired in ambient (oxidising) atmosphere.

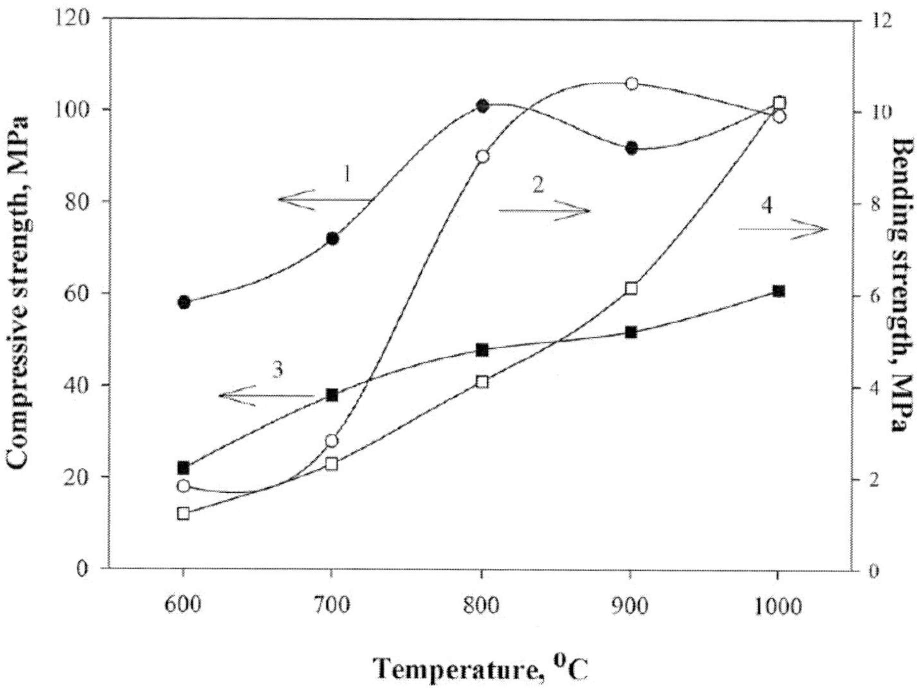

Figure 5. Influence of firing atmosphere on the mechanical properties of clay specimens, 1-Compressive strength of specimens fired under 20 mm of water vapour pressure; 2- Bending strength of specimens fired under 20 mm of water vapour pressure; 3-Compressive strength of specimens fired in ambient (oxidising) atmosphere; 4-Bending strength of specimens fired in ambient (oxidising) atmosphere.

Colour Characteristics of Samples Fired in Water Vapour and Ambient Atmosphere

The chemical and mineralogical composition of the clay raw materials and the formation of a new compound exerts a major influence on the colour of the fired products. Therefore, it is essential to regulate the mineralogical composition during firing to be able to control the colour of the bricks. Our experiments show that the colour of ceramics fired in water vapour depends on the firing conditions. Firing at 700, 800 and 900 °C with dwell times of 2 and 4 h produced bluish-grey colours, but a dwell time of 8h changed the colour to whitish-grey. Increasing the dwell time did not change the main colour but brightened it. Firing at 1000 °C produced a dark-grey colour. It was found that the heating rate has no influence on the colour; heating rates of 100 °C/h and 250 °C/h up to 800 °C produced similar bluish-grey colours.

The results of experiments showing the colours of specimens fired under water vapour atmosphere are shown in Table 2.

Table 2. Influence of vapour pressure on the colour of the ceramics

Firing temperature, °C	Pressure, mm of water						
	0	10	20	50	100	200	300
800	Bright-yellow	Bright-grey	Bluish-grey	Bluish-grey	Bluish-grey	Bluish-grey	Bluish-grey
1000	Reddish-yellow	Dark-grey with a bright tone	Dark-grey	Dark-grey	Dark-grey	Dark-grey	Dark-grey

Specimens fired under a water vapour pressure of 20 mm of water at 800 °C for 2h show a clear bluish-grey colour, becoming dark-grey with a better colour tone at 1000 °C. At high temperature liquid phase sintering occurs, involving the participation of gaseous components. The kinetics of solid state reactions and liquid phase sintering of clay minerals during firing depend on the rate of liquid phase formation and on the gaseous products of the reaction [21]. The rate of liquid phase formation depends on the concentration of fluxing agents and their efficiency which is also related to the characteristics of the gaseous medium. In reducing atmospheres, ferric oxide, Fe_2O_3, transforms to ferrous oxide (FeO) which in the presence of silica is an effective fluxing agent [22]. The reducing of iron oxide in clay minerals is not a simple reaction because of the presence of water vapour, CO_2, CO, H_2, N_2 and SO_2 gases. Reduction of ferric oxide to ferrous oxide during the clay firing can be written in terms of the following reactions:

$$3Fe_2O_3 + CO \leftrightarrow 2Fe_3O_4 + CO_2\uparrow \quad (1)$$
$$Fe_3O_4 + CO \leftrightarrow 3FeO + CO_2\uparrow \quad (2)$$
$$Fe_2O_3 + H_2 \rightarrow 2FeO + H_2O\uparrow \quad (3)$$
$$Fe_2O_3 + C \rightarrow 2FeO + CO\uparrow \quad (4)$$
$$CO + H_2O \rightarrow CO_2\uparrow + H_2\uparrow \quad (5)$$
$$C + 2H_2O \rightarrow CO_2\uparrow + H_2\uparrow \quad (6)$$
$$C + CO_2 \rightarrow 2CO\uparrow \quad (7)$$

The most effective reducing agents are hydrogen and carbon monoxide. The active iron compounds participating in ceramic structure formation from clay minerals during firing are the ferric-ferrous and ferrous oxides. When water vapour is admitted into the furnace under pressure, the water vapour is readily adsorbed on the surface of clay specimens because of its polarity [22]. Thermal diffusion of carbon oxides towards the structural water on the outer surface results in increased hydrogen concentration within the specimen, creating a local reducing atmosphere that reduces the ferric oxide to ferrous oxide. The increase in the ferrous oxide concentration with dwell times during firing at different temperatures is shown in Fig.6.

Fig. 6 shows that the content of ferrous oxide in the specimens depends on the firing temperature and dwell time. Ferrous oxide itself exerts a considerable influence on the formation of new phases, their structure and the ratio between these phases. Fig.7 shows the XRD diffraction patterns of the specimens fired in water vapour atmosphere, which indicate that hercynite ($FeAl_2O_4$) with lattice spacings of 1.43, 1.55 and 2.018 Å was formed at 800, 900 and 1000 °C.

The hercynite ($FeAl_2O_4$) could be formed by a gas-transport mechanism as previously described [11], where Fe^{2+} is transported in the gas phase as $FeO{\cdot}H_2O$ to an aluminium compound where it condenses with the formation of hercynite. Although Si^{4+} can also be transported as SiO in the gas phase, mullite was not detected in the sample fired under water vapour at 1000 °C, possibly due to the higher mobility of Fe^{2+} than Si^{4+}.

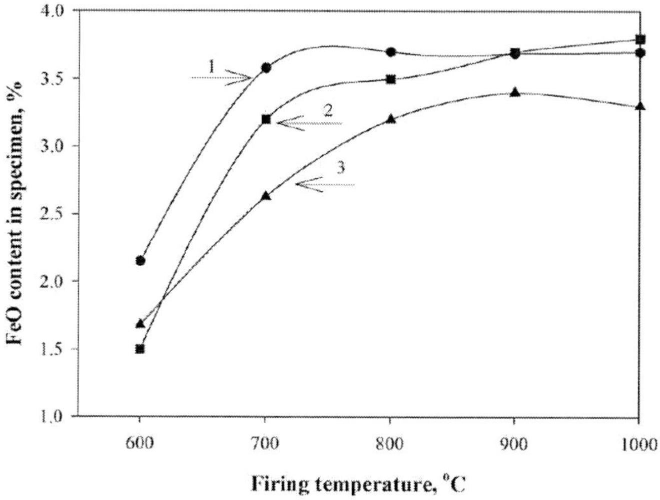

Figure 6. Change of the FeO content with dwell time in the clay specimen, 1- 2h, 2- 4h, 3- 8h.

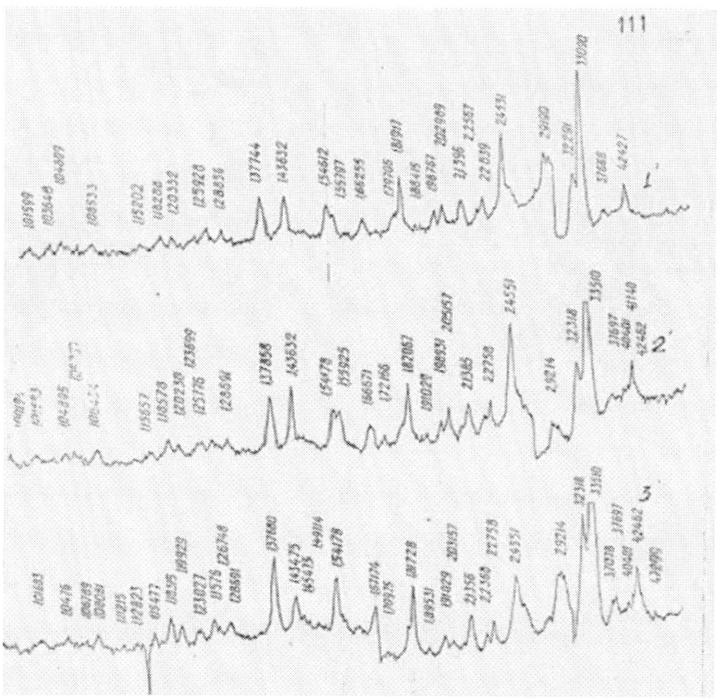

Figure 7. XRD patterns of samples fired in water vapour, taken from reference [11]. 1- 900 °C for 4 hr, 2- 1000 °C for 4h, 3- 1000 °C for 8 h.

The present clay contains colouring oxides including Fe_2O_3, FeO, TiO_2, and MnO. Therefore these oxides, or minerals formed by a combination of these oxides, are likely to be responsible for the colour of the ceramic. The colours of these oxides are different; ferric oxide (hematite) is red, ferrous oxide (wustite) is black, MnO is grey-green, while TiO_2 is white. The colours of these oxides are different from those observed in the water vapour atmospheres fired specimens, possibly due to the formation of differently coloured compounds during firing. In order to verify this proposal, two different specimens were prepared from the pure oxides and fired under a water vapour atmosphere:

1) TiO_2-Fe_2O_3-organic
2) SiO_2-Al_2O_3-Fe_2O_3-$CaCO_3$-MgO-TiO_2-organic

After firing these mixtures at 950 °C, ilmenite ($FeTiO_3$) was formed in the first mixture and diopside ($MgCaSi_2O_6$) was formed in the second mixture. The ilmenite-containing specimen was bluish-dark grey in color, becoming darker with increasing vapour pressure. The colour of the second mixture was whitish-green, becoming greener with increasing vapour pressure. This suggests that ilmenite is the mineral responsible for the colour of the present clay fired under water vapour. The colour of the second mixture also suggests that the addition of the diopside-forming compound into a clay mixture could produce facing bricks of green colour. The experiments reveal that a gas-transport reaction mechanism is occurring under a pressured water vapour atmosphere firing, resulting in the recrystallization of coloured minerals from the gas phase [11].

The colour stability of specimens fired in 20 mm of water vapour pressure at 800, 850, 900, 950 and 1000 °C was determined with a mercury-quartz lamp [23]. The distance between the lamp and the specimens was 55 cm and specimens were irradiated for 500 h. Visual examination after 500 h exposure times did not reveal any colour change, indicating the stability of the ceramic colours.

Technological Scheme for Production of Coloured Bricks from Low Melting Clay

The experimental results carried out on the laboratory scale suggest the following technological scheme for the production of coloured face bricks:

Exploration of the clay deposit
↓
Storage of the clay
↓
Processing of the clay
↓
Preparation of a plastic mass
↓
Extruding of the brick mass
↓
Drying

↓

Firing under controlled water vapour atmosphere

Exploring and processing of the clay should proceed by the usual method used by brick manufacturers worldwide. Figs. 8 and 9 show the firing schemes used in the present experiments.

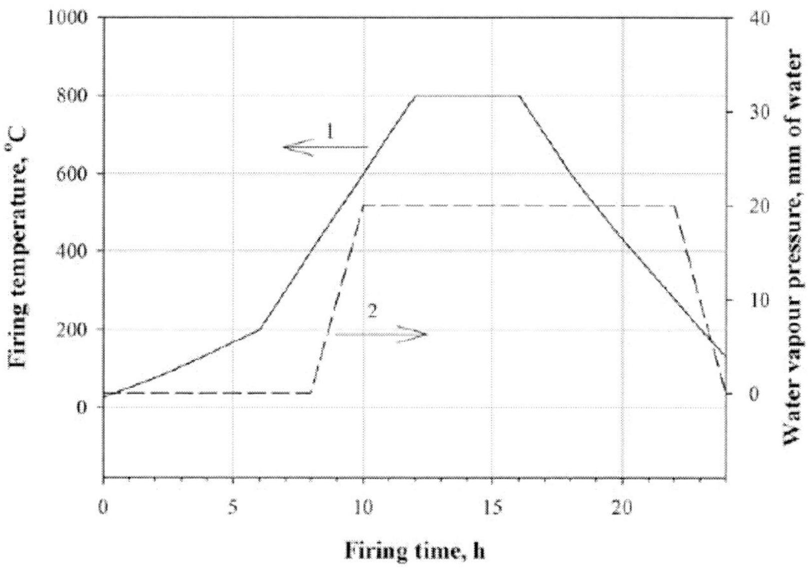

Figure 8. Firing scheme of specimens fired at 800 °C. 1- temperature, 2-pressure.

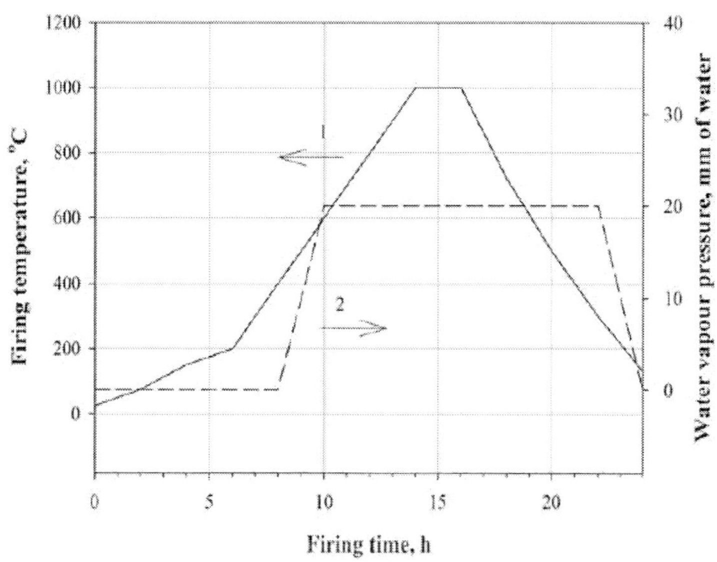

Figure 9. Firing scheme of specimens fired at 1000 °C. 1- temperature, 2- pressure.

Specimens fired at 800 and 1000 °C showed less water absorption than specimens fired under ambient atmosphere (Table 3).

Table 3. Water absorption of samples fired in different atmospheres

Firing condition	Water absorption, %
800°C, water vapour atmosphere, 4 h	11.13 (0.24)
1000°C, water vapour atmosphere, 2 h	3.24 (0.12)
800°C, ambient (oxidising) atmosphere, 4 h	16 (0.31)
1000°C, ambient (oxidising) atmosphere, 2 h	12.87 (0.27)

Samples fired under water vapour atmosphere at both 800 and 1000 °C show a good softening coefficient almost equal to 1. Both samples also show acceptable mechanical properties up to 50 freeze-thaw cycles and can be classified as being resistant to 50 freeze-thaw cycles.

Based on the laboratory scale experiments, a firing scheme for the preparation of coloured bricks on a semi-industrial pilot plant scale under pressure of 20 mm of water at 800 and 1000 °C is shown in Fig.10.

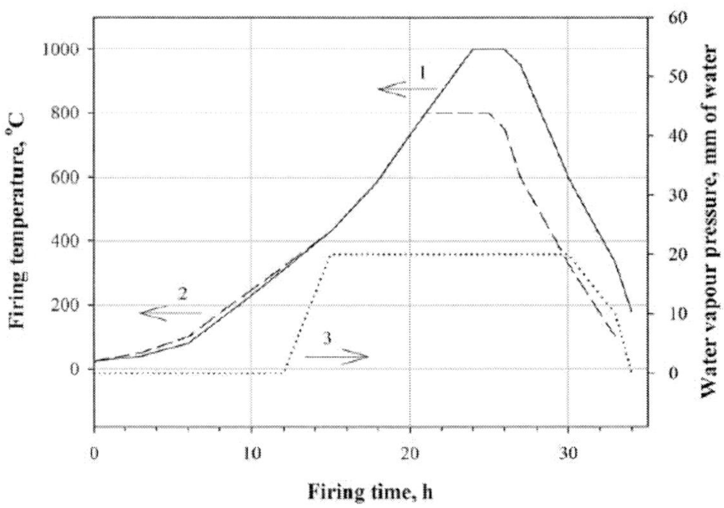

Figure 10. Firing scheme for coloured bricks on the industrial pilot plant scale, 1-1000 °C, 2-800 °C, 3-water vapour pressure.

For semi-industrial pilot scale firing, a newly designed discontinuous type camera kiln was used [24, 25]. The physico-mechanical properties of coloured bricks obtained on the semi-industrial pilot plant scale are shown below:

1. Firing temperature: 800 °C

Compressive strength: 22.1 MPa
Bending strength: 8.6 MPa

Water absorption: 12.6%
Volume weight: 1.92 g/cm^3

2. Firing temperature: 1000 °C

Compressive strength: 24.7 MPa
Bending strength: 7.9 MPa
Water absorption: 6.8%
Volume weight: 1.98 g/cm^3

During the industrial-scale experiment, about 20,000 coloured bricks were prepared and were used to face a civil building.

CONCLUSION

Dark blue-grey coloured facing bricks have been prepared by firing low-melting clay-based brick under 20 mm of water pressure at 800 and 1000 °C for 4 and 2 h, respectively. Firing under a pressurized water vapour atmosphere produced a locally reducing atmosphere within the brick mass, reducing ferric oxide to ferrous oxide with the formation of a liquid phase. The dark blue-grey colour of the low-melting clay is related to the formation of ilmenite ($FeTiO_3$) under the water vapour atmosphere firing, which also brought about an improvement in the mechanical properties. The compressive strength of clay specimens fired under a water vapour atmosphere improved 2 to 3-fold by comparison with samples fired under ambient (oxidising) conditions.

ACKNOWLEDGMENT

Authors wish to express thank to Prof. K.J.D. MacKenzie of the MacDiarmid Institute for Advanced Materials and Nanotechnology, Victoria University of Wellington (New Zealand) for critical reading and editing of this chapter.

REFERENCES

http://en.wikipedia.org/wiki/Brick, cited on 12th September, 2011

Gillott, J.E.(1962). Clay mineralogy in building research, *Clays Clay Min.* 11, 296-298

Kreimeyer, R. (1987). Some notes on the firing colour of clay bricks, *Appl. Clay Sci.* 2, 175-183

Pisch, I.V., Maslennikova, G.N., Gvozdeva, N.A., Klimosh, Yu.A., Baranovskaya, E.I. (2007). Methods of dyeing ceramic brick, *Glass and Ceram*, 64, 270-273

http://www.southerncolor.com/brick-pigments.html, cited on 13th September, 2011

Valanciene, V., Siauciunas, R., Baltusnikaite, J. J. (2010). The influence of mineralogical composition on the colour of clay body, *J. Eur. Ceram. Soc.* 30, 1609-1617

Karaman, S., Gunal, H., Ersahin, S. (2006). Assesment of clay bricks compressive strength using quantitative values of colour components, *Const. Build Mat.* 20, 348-354

Rusovich-Yugai, N.S., Neklyudova, T., L. (2007). Color characteristics of low-melting clays, *Glass and Ceram*, 64, 232-234

http://www.thinkbrick.com.au/brick-colours-mortar-and-profiles/, cited on 13th September, 2011

Maidar, D. Arkhitektura i gradostroitelstvo Mongolia; State Publishing House Mongolia, 1968

Jadambaa, Ts. Production method of facing coloured ceramics under regulated vapour-gas atmosphere /with examples of low melting clay from Tolgoit deposit from MPR. PhD dissertation. Moscow Institute of Engineers for Land Arrangements, Moscow, 1979. (in Russian)

Ser-Odjav, N. Shinjlekh ukhaan. 1957, vol 2 (in Mongolian)

Ser-Odjav, N. Shinjlekh ukhaan. 1958, vol 5-6 (in Mongolian)

Litvan, G.G. (1984). Determination of the firing temperature of clay brick, *Am. Ceram. Soc. Bull.*, 63, 617-627

Karaman, S., Ersahin, S., Gunal, H. J. (2006). Firing temperature and firing time influence on mechanical and physical properties of clay bricks. *Sci. Indust. Res.* 65, 153-189

Temuujin, J., Okada, K., MacKenzie, K.J.D., Jadambaa, Ts. (1999). The Effect of water vapour atmospheres on the thermal transformation of kaolinite investigated by XRD, FTIR and Solid State MAS NMR, *J. Eur. Ceram. Soc*, 19, 105-112.

Temuujin, J., Jadambaa, Ts., MacKenzie, K.J.D. (2000). Effect of water vapour atmospheres on thermal transformations and mechanical strength of montmorillonite clay compacts. *Brit. Ceram. Trans*, 99, 63-65.

Nekhoroshev, A.V., Tsitelauri, G.I., Khlebionek, E., Jadambaa, Ts. Resource saving technologies of ceramics, silicate and concrete production', 481, Moscow, Stroiizdat, 1991 (in Russian)

MacKenzie, K.J.D. (1969). The effects of impurities on the formation of mullite from kaolinite-type minerals. III. The effect of the firing atmosphere. *Trans. Brit. Ceram Soc*, 69, 103-109

Pavlov, B.O, Nazarov, A.P., Influence of gaseous atmosphere on the properties of ceramics, *Glass and Ceram,* 1973, 2 (in Russian)

Mchedlov-Petrosyan, O.P. Chemistry of inorganic building materials, Moscow, *Stroiizdat,* 1971 (in Russian)

Rogovoi, M.I. Technology of synthetic porous fillers and ceramics, Moscow, *Stroiizdat,* 1974 (in Russian)

Kholopova, L.I. Decorative synthetic stone and its application to building construction, Leningrad, *Stroiizdat,* 1976 (in Russian)

Jadambaa, Ts., Nekhoroshev, A.B., Method for production of coloured bricks, Certificate of Inventions, *MPR,* N35, 1982

Jadambaa, Ts., Nekhoroshev, A.B., Camera and tunnel kilns design for firing of coloured bricks, Certificate of Inventions, *MPR*, N34, 1982

Reviewed at Nova Science Publisher.

In: Brick and Mortar Research
Editors: S. Manuel Rivera and A. L. Pena Diaz
ISBN: 978-1-61942-927-7

Chapter 8

NEGATIVE EFFECTS OF THE USE OF WHITE PORTLAND CEMENT AS ADDITIVE TO AERIAL LIME MORTARS SET AT ATMOSPHERIC CONDITIONS: A CHEMICAL, MINERALOGICAL AND PHYSICAL-MECHANICAL INVESTIGATION

A. Arizzi and G. Cultrone*
Universidad de Granada, Departamento de Mineralogía
y Petrología, Campus Fuentenueva s/n, Granada, España

ABSTRACT

In recent years the use and study of lime mortars are being improved because lime shows a high compatibility with the ancient materials of the Architectural Heritage. On the other hand, the scarce knowledge of the fabrication techniques, the application methods used in the past and the long time of setting have driven scientists and restorers to add other binders (i.e. cement, pozzolans) to the lime, which claim to obviate the inconveniences of the use of this material alone. The use of Portland cement as additive in lime-based mortars seems to be an easy and diffused solution for the industrial production of repair mortars. In this work, we have compared the chemical-mineralogical and physical-mechanical properties of mortars prepared with calcitic lime and a calcareous aggregate, with mortars in which a white Portland cement (WPC) has been used in substitution of a 25 wt.% of the lime. This chapter shows that mortars prepared with Portland cement do not provide satisfactory results after two months since their elaboration. From mineralogic data and porosimetric analyses no improvements were observed by the addition of cement. In fact, even if this percentage of cement produced lower shrinkage than lime mortars, it did not improve the hydric and mechanical properties. Most importantly, the WPC caused the formation of ettringite that provoked fissures in mortars samples. This chapter demonstrates that the use of cement, even if only as additive in lime mortars, would impair the longevity of the masonry that encloses it.

* Email: arizzina@ugr.es; Tel.: +34958246614; Fax: +34958243368.

1. INTRODUCTION

After its discovery in 1824, Portland cement, a binder of high mechanical resistances and workability, has replaced lime in the great majority of new construction and restoration works. The rapid diffusion of this material all over the world was promoted by the low costs and times of production that, in turn, were favoured by the standardization of the cement properties. Due to these economic advantages, in the last century Portland cement has been used in many historic buildings as repair material, where it has caused serious and irreversible damages. The reasons of that stay in the high stiffness, low permeability and content of soluble salts of this binder, which make it non-compatible with the traditional construction materials [1,2]. Because the application of Portland cement in the Architectural Heritage has demonstrated to be counterproductive, its use in this field has been drastically reduced in the last decades whilst, in contrast, the production of lime mortars is being improved.

On the one hand, lime is the binder that better performs in ancient buildings, because of its high permeability and porosity that allow the materials to "transpire", its low mechanical strengths that allow the materials to "move", and its high reversibility that allows the repair materials to be removed without damaging the original ones. On the other hand, there exists a scarce knowledge of the methodologies used in the past for its production and application, as well as there is a lack of standards that prevents its production at an industrial level, especially if ones consider the wide functional and constructive versatility of lime mortar (i.e. renders, plasters, structural and pointing mortars, etc.).

These problems have pushed manufacturers and architects to the use of other materials blended to the lime in the mortar mix, with the function of second binders or additives. This halfway solution is being achieved with the use, for example, of the Portland cement. It has been found that, by blending cement to the lime, it is possible to reduce cracking, to quicken the application and hardening of the mortar, thus providing protection from rain before carbonation has been completed, and to ensure reliability and predictability of its properties [3]. Moreover, new types of cement have been developed to obtain better performances with the lime, such as the white Portland cement (WPC). This material is manufactured from materials virtually free of iron oxide and other impurities, which impart the grey colour to the normal Portland cement. However, because of the different manufacturing processes (the absence of iron causes a temperature increase), it is much more expensive (i.e. twice the price) than the equivalent grey product. Typical applications of WPC are architectonical elements, such as rendering, cast stone, precast and in-situ structural concrete and pointing, and ornamental elements that require a white colour for eventual pigment addition.

The main interest of this work is to find out whether and how the addition of white Portland cement improves the physic-mechanical properties of an aerial lime-based mortar cured at atmospheric conditions, by checking the mineralogical and textural changes induced by it. In this research, the WPC is used as an additive rather than as a real binder, therefore it seems reasonable to expose the mortars to a non-moist environment, at which aerial lime mortars are normally cured. In this way, it is possible to know if the cement is still effective in the mortar hardening and, consequently, in the improvement of its physical characteristics (i.e. water absorption, shrinkage), without developing excessive high strength and low porosity. This study claims to verify if the use of cement as additive in repair lime mortars is a really good halfway point or, instead, represents a solution that has to be definitively discarded in restoration works.

2. Materials and Methods

2.1. Mortars Elaboration

CC mortar was prepared with a calcitic dry hydrated lime, (named CL, CL90-S [4]) produced by ANCASA (Seville, Spain) and a calcareous aggregate (named CA) with a continuous grading from 0.063 to 1.5 mm. A white Portland cement (named WPC, BL II/A-LL 52.5 R [5]), produced by CEMEX commercial (Valencia, Spain), was used in substitution of 25 wt.% of lime in the elaboration of a new mortar named CPC. The binder-to-aggregate (B/A) ratio used in both mortars is 1:3 by weight. The amount of water, established in order to obtain mortars with a plastic consistency, was 31% and 27% of total mass, for CC and CPC mortars, respectively. The flow was calculated as the average of the spread diameter (in mm) measured on three fresh samples during the flowability assay [6], in the range of 140-160 mm. After preparation [7], mortars were cured for 7 days in standard steel moulds (4×4×16 cm) at RH = 60±5% and T = 20±5 °C and, after removal from the moulds, were cured for a further 21 days in the same laboratory following the modification of the UNE-EN 1015-2 standard [7] proposed by Cazalla [8]. Mortars were conserved at the same conditions throughout the study.

2.2. Characterization of the Mortars Components

The chemical composition (major and minor elements) of the components of mortars was studied using a Bruker S4 Pioneer X-ray fluorescence spectrometer (XRF), with wavelength dispersion equipped with a goniometer that analyzes crystals (LIF200/PET/OVO-55) and a Rh X-ray tube (60 kV, 150 mA). Semiquantitative scanning spectra were obtained using Spectraplus software. Mortar powders (ca. 5g) were dispersed in KBr, deposited in an aluminium cup and then pressed at 10 ton to obtain a pressed pellet (40 mm sample disc). Measurements were performed in a vacuum with a rotating sample.

The mineralogical phases of binders and aggregate were determined by X-ray diffraction (XRD) and thermogravimetric (TGA) analyses. In the first case, a Philips PW-1710 (disoriented powder method) was used. Analysis conditions were: radiation CuK_{α} (λ=1.5405 Å), 3 to 60 °2θ explored area, 40 kV voltage, 40 mA current intensity and goniometer speed of 0.1 °2θ/s. The interpretation and quantification of the mineral phases was performed using the X-Powder software package [9]. Thermal measurements were made with a Shimadzu TGA-50H analyzer. Approximately 50 mg of sample was heated in an alumina crucible, in a flushed-air atmosphere (100 mL/min), at a heating rate of 5°/min over a range of 25-950 °C. TGA data treatment was carried out according to the UNE-EN ISO 11358 [10] standard.

2.3. Mortars Characterization

Both inner (core) and outer (1 cm from the core to the surface) parts of the mortars samples were analyzed after 28 and 60 days since mortars elaboration.

The mineralogical phases of the mortars were determined by TGA using the same analytical device described above. Mortar phases were determined by measuring the weight loss resulting from the stoichiometric reactions of:

1. gypsum decomposition ($110<T<170$ °C);
2. calcium silicate hydrates decomposition ($100<T<125$ °C);
3. portlandite dehydroxylation ($450<T<550$ °C);
4. calcite decomposition ($700<T<900$ °C).

The decrease in portlandite content was taken as reference to estimate the carbonation degree of mortars (I_{CD}, %).

The textural characteristics of mortars were studied by means of field emission scanning electron microscopy (FESEM), using a Carl Zeiss Leo-Gemini 1530 microscope on previously dried, carbon-coated mortar fragments.

Open porosity (P_o, %) and pore size distribution (PSD, in a range of $0.002<r<200$ µm) were determined using a Micrometics Autopore III 9410 mercury injection porosimeter (MIP). Mortar fragments of ca. 2 cm^3 were oven-dried for 24 h at 60 °C before the analysis.

The capillary imbibition test was performed only after 60 days since mortars elaboration, on three samples per mortar type of 4×4×16 cm that were oven-dried at 100 °C for 24 h before measurements were taken. Two imbibition coefficients (A and B) were determined from the mass uptake per surface unit and the height over time, according to the Beck et al. [11] procedure.

Flexural and compressive strength (R_f and R_c) were measured by means of a hydraulic press INCOTECNIC-Matest. According to the UNE-EN 1015-11 [12] standard, flexural assays were carried out at 28 days of carbonation, on three samples per mortar (of 4×4×16 cm). The six samples obtained after the flexural rupture were used for the compressive assays.

Shrinkage of hardened mortars was determined after 60 days measuring the variation of the dimensions (expressed as percentage) on three perpendicular directions: a, parallel to the compaction plane; b, parallel to the compaction plane along the largest face of the sample; c, perpendicular to the compaction plane. Measurements were carried out on three samples per mortars.

3. Results and Discussion

3.1. Characterization of Mortars Components

The chemical analysis (XRF, Table 1) of the white Portland cement shows high contents of CaO and SiO_2. These elements (Ca and Si) correspond to the calcium silicates C_3S (tri-calcium silicate, or alite) and C_2S (di-calcium silicate, or belite), found by XRD (Table 1).

The TiO_2 proceeds from a pigment added to white Portland cement to further enhance the whiteness.

The presence of Al in WPC was detected only by XRF and this indicates that aluminate phases are present in amounts not detectable by XRD (Figure 1 and Table 1).

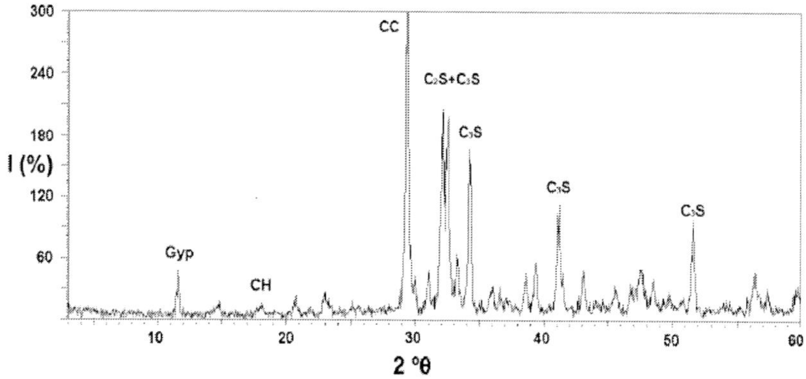

Figure 1. X-ray diffraction pattern of white Portland cement (WPC). CC = calcite; CH = portlandite; Gyp = gypsum; C_2S = dicalcium silicates; C_3S = tricalcium silicates.

By means of thermogravimetric analysis only calcite (CC), portlandite (CH) and gypsum (Gyp) phases were identified (Figure 2 and Table 1).

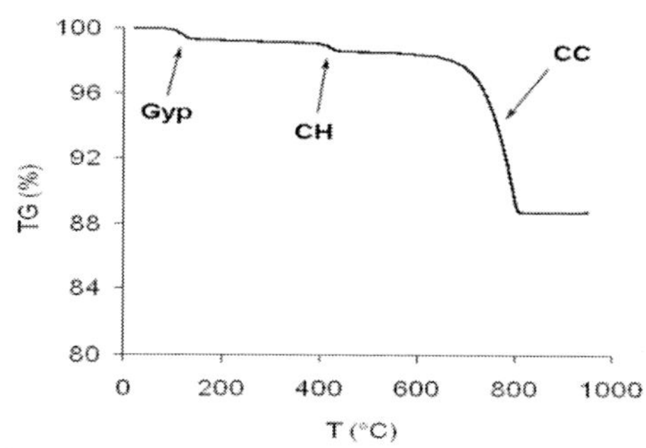

Figure 2. Thermogravimetric curve of white Portland cement (WPC) where the weight loss (TGA, expressed in %) is represented as function of temperature (T, in °C). CC = calcite; CH = portlandite; Gyp = gypsum.

3.2. Evaluation of the Hardening Processes in Lime and Cement-Lime Mortars

3.2.1. Carbonation Degree

An important question to consider is whether and how the cement takes part in the hardening of mortars at these curing conditions. According to Cizer et al. [13], in cement-lime mortars there is no competition between hydration and carbonation under atmospheric conditions because cement hydration always takes place initially and evolves before carbonation becomes effective. This means that in CPC mortars, which have been cured at atmospheric conditions (see section 2.1), both processes took place, although it is likely that this happens at different degrees in the surface and in the core of the mortars.

Table 1. Chemical (X-ray fluorescence analysis, XRF) and mineralogical (X-ray diffraction, XRD and thermogravimetric analysis, TGA) composition of white Portland cement (WPC). CC=calcite; CH=portlandite; Gyp=gypsum; C_2S=dicalcium silicates; C_3S=tricalcium silicates

Chemical composition (XRF)	
(Oxides %)	WPC
SiO_2	14.98
CaO	70.49
SO_3	5.58
MgO	0.61
Fe_2O_3	0.26
Al_2O_3	4.22
TiO2	0.08
Mineralogical composition (XRD)	
Mineral phases (%)	WPC
CC	20.9
CH	16.8
Gyp	2.3
C_2S	29.9
C_3S	30.1
Mineralogical composition (TGA)	
Mineral phases (%)	WPC
CC	22.2
CH	2.2
Gyp	2.4

After 2 months since the elaboration, CC and CPC mortars show the same carbonation degree (expressed as I_{CD}, in %, Table 2) in the external part (-ex) of the samples, whilst the interior of mortars (-in) presents always a lower degree of carbonation than the surface. Moreover, this difference in the degree of carbonation between the surface and the core is much higher in CPC mortars (about 55%) than in CC ones (about 23%).

We remember that the carbonation degree index (I_{CD}) is calculated for each mortar considering the variation in the amount of portlandite obtained between the beginning of mortar life and the end of the study. This determination does not consider the amount of portlandite liberated during the hydration reactions in cement-lime mortars, so that the I_{CD} is not an accurate value in this case, although it is useful for estimating the progress of carbonation and hydration of mortars. Thus, a lower I_{CD} value, which indicates a higher amount of calcium hydroxide, is due to a slower consumption of the lime time (i.e. formation of a smaller amount of calcite) in CC mortars; whilst, in CPC mortars, it indicates the formation of more CSH phase (see Table 2) that, during the hydration reaction, liberates $Ca(OH)_2$ [14], thus increasing its amount in the internal samples. Both options can be attributed to the fact that the interior of mortars dries slower than the exterior and this might result in a more moist ambient in the core than in the surface. In the smallest pores, water condensation is likely to occur and this would hinder the diffusion of CO_2 [15], thus slowing down carbonation in this zone. Moreover, the formation of calcium silicate hydrates would be enhanced by the alkaline ambient created due to the dissolution of calcium hydroxide in the pore water [16].

Table 2. Physical parameters of CC (CL+CA) and CPC (CL+WPC+CA) mortars: Carbonation degree index (I_{CD}, in %) of the surface (ex) and the core (in) after 60 days since the elaboration; Open porosity (P_o, in %) values and standard deviation after 28 and 60 days since the elaboration; Imbibition coefficients (A, in g/cm^2 min$^{1/2}$ and B, in cm/min$^{1/2}$) after 60 days since the elaboration; Flexural (R_f, in Mpa) and compressive (R_c, in Mpa) resistances and standard deviation after 28 days, according to the UNE-EN 1015-11 (2000); Shrinkage values (in %) and standard deviation along a, b and c directions of mortars samples after 60 days since their elaboration

Carbonation degree index (I_{CD})			
Mortar sample	Zone		
	Ex	In	
CC	71.67	49.04	
CPC	71.77	17.12	
Open porosity (P_o)			
Mortar sample	Time (days)		
	28	60	
CC	32.5±2.74	35.0±1.10	
CPC	36.2±1.80	37.3±1.00	
Imbibition coefficients			
Mortar sample	A	B	
CC	0.08	9.95	
CPC	0.09	10.79	
Mechanical resistances			
Mortar sample	R_f	R_c	
CC	1.03±0.03	1.43±0.16	
CPC	2.59±0.09	1.62±0.23	
Shrinkage			
Mortar sample	a	b	c
CC	3.6±1.41	3.8±0.03	6.8±0.00
CPC	3.3±0.71	0.0±0.00	6.0±0.35

This result confirms our initial assumption that carbonation and hydration happens at different degrees in the exterior and the interior of cement-lime mortars: in the surface, the carbonation process takes place at the same degree than in lime mortars whilst, in the core, carbonation is less favoured and hydration reaction enhanced.

3.2.2. Identification of the Hydrated Phases in Cement-Lime Mortar

The CSH phase in cement paste is amorphous or semicrystalline and it gives powder patterns very similar to those of the calcium silicates present in cement (C_2S and C_3S) [17]. However, the X-ray diffraction analysis was not able to detect the presence of hydrated phases in CPC mortars, because their peaks were marked by other mineral phases (calcite and portlandite) with higher cristallinity and reflecting power. The thermogravimetric analysis, in turn, allows the identification of CSH that decomposes at a temperature between 100 and 125 ºC. However, this range of temperature coincides partially with that in which gypsum decomposes (see section 2.3 and Figure 2), so the quantification of CSH and gypsum phases in the carbonated mortars by means of TGA can be a difficult task.

The electron microscopy, in turn, is effective in the identification and characterisation of the hydrated phases morphologies [18] although not in their quantification. Both upper and inner zones of mortars were observed by FESEM after two months since their elaboration. After this period of time, the CPC external samples presented some CSH phases in form of small fibres (only 200 nm long, Figure 3a), which have been recognised elsewhere as an early hydration structure [17], and many unhydrated or partially hydrated phases (Figure 3b) in form of particles with a rough surface. Internal samples of CPC showed some flattened particles of CSH (Figure 3c), which are typically present in samples with twenty-eight days of hydration [17].

The morphologies observed in Figure 3 suggest that a poor hydration takes place in the exterior of CPC mortars, whilst this process is more extended in the interior. Once again, it is shown that carbonation and hydration processes take place at different degrees in the mortar zones.

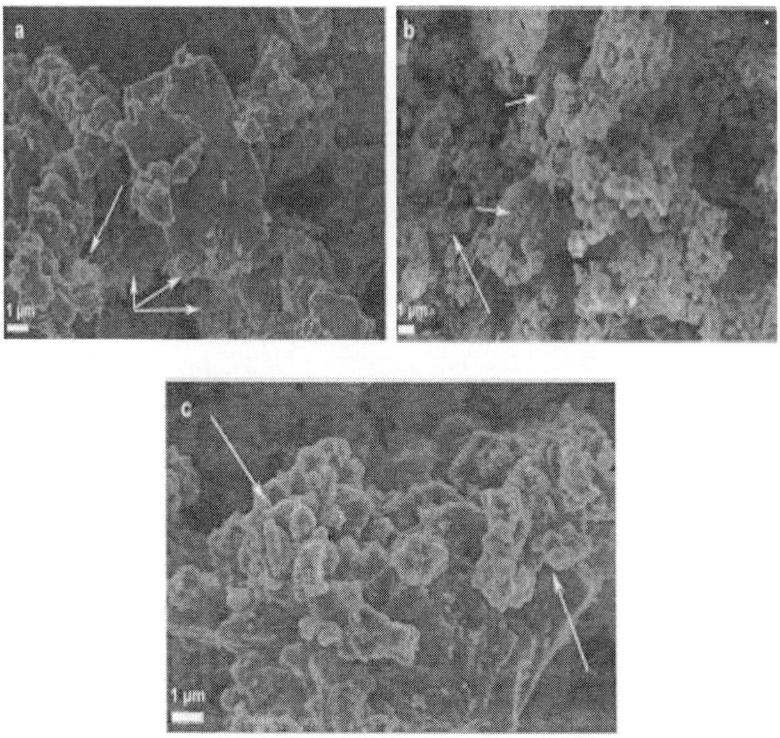

Figure 3. FESEM images of CPC mortar: upper part of the mortar, in which there are small fibres of CSH (a) and partially hydrated or unhydrated particles (CS_2 and CS_3) with rough surface (b); internal part of mortar (c), in which small and flattened particles of CSH are observed.

3.2.3. Pore System

The pore system is another textural aspect that helps to understand the progress of the hydration and carbonation processes. Two main types of pores typical of cement mortars have been obtained here (Figure 4): the gel pores, of radius comprised between around 0.01 and 0.05 μm, and the capillary pores, in the range of 0.1-1 μm. The former do not depend to the water amount but to the production of hydrated phases, as also observed by other authors in lime [19] and cement [20] mortars. The latter depend to the water-to-cement ratio and decrease as the hydration reaction progresses; they are responsible for the water transfert properties and the mechanical behaviour of cement mortars [21]. One of the advantages of adding cement to lime-based mortars is that the hydrated phases give place to a segmentation of the capillary pores, which reduces importantly the water absorption in mortar [21]. Here, this effect was not obtained, since CPC mortar presents only slight lower volume of capillary pores compared to CC mortar (Figure 4), and this little difference has to be attributed only to the lower amount of water used for the preparation of CPC mortar (see section 2.1).

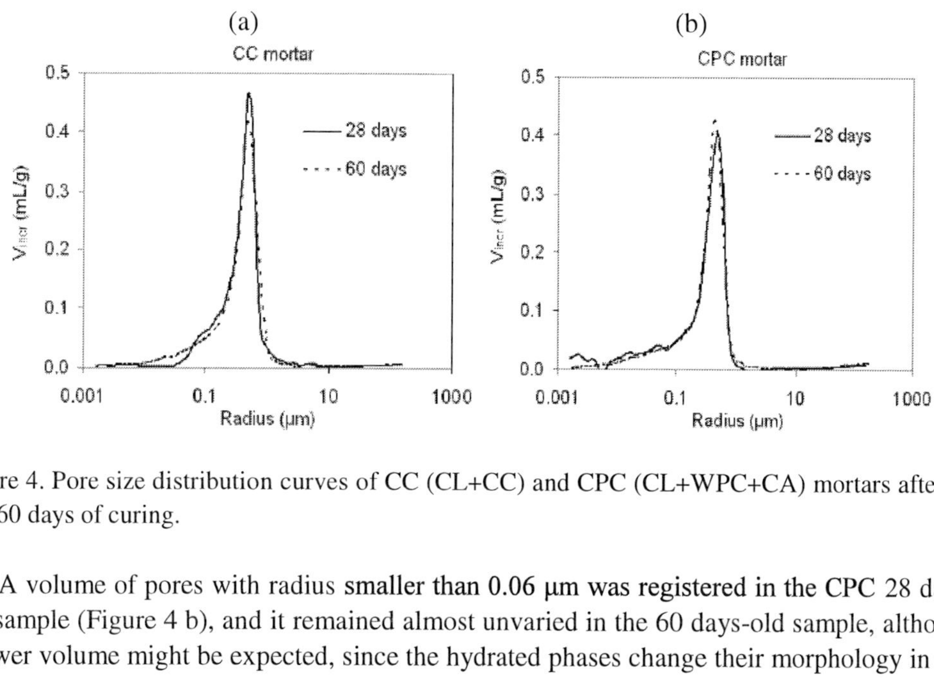

Figure 4. Pore size distribution curves of CC (CL+CC) and CPC (CL+WPC+CA) mortars after 28 and 60 days of curing.

A volume of pores with radius smaller than 0.06 μm was registered in the CPC 28 days-old sample (Figure 4 b), and it remained almost unvaried in the 60 days-old sample, although a lower volume might be expected, since the hydrated phases change their morphology in this period of time and occupy the pores left by the cement phases and the CSH gel [17].

In turn, in CC mortar, an increase of the volume of pores in the 0.01 - 0.06 μm size range was registered from 28 to 60 days (Figure 4a), and it is responsible for the porosity increase registered after 2 months (Table 2). This phenomenon is attributed to the formation of agglomerates of small calcite particles at the interfacial transition zone (ITZ) between aggregate grains and matrix [22], which is also expected to occur in presence of cement.

The final consequence of these two opposite mechanisms (decrease of the pore gel volume and increase of the pores at the ITZ) is that the volume of small pores remains unvaried in CPC mortars, as well as the open porosity value determined at 28 and 60 days (Table 2).

3.3. Physic-Mechanical Properties of Mortars

3.3.1. Capillary Uptake

Figure 5 shows the capillarity curves obtained for CC and CPC mortars after 2 months of curing. It can be noticed that, although CPC mortar presents slightly higher values of open porosity (Table 2), it does not absorb the highest amount of water. This fact, related with the pore size distribution of mortars, indicates that only pores of radius ranging between 0.1 and 1 μm, whose volume is bigger in CC samples at every curing age, are responsible for the water capillary uptake in mortars. However, this difference in the amount of water absorbed and the rate of absorption is not big between the two mortars, as reflected in the similar values of open porosity and imbibition coefficients (P_o, A and B, Table 2).

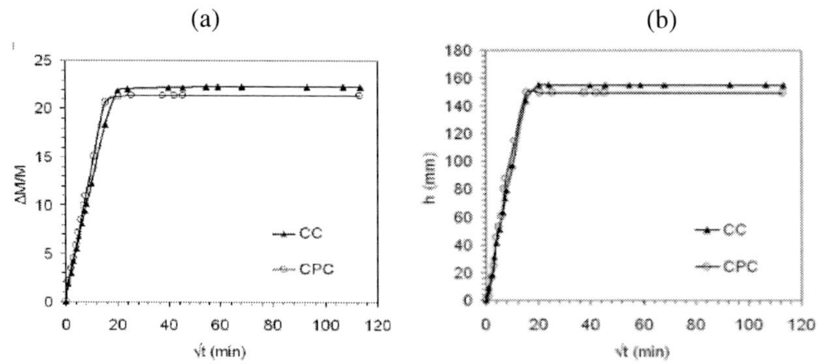

Figure 5. Capillarity curves of CC (CL+CA) and CPC (CL+WPC+CA) mortars after 60 days since their elaboration. Time (in minutes) is plotted versus the weight increase (ΔM/M, in %) (a) and the water uptake level (h, in mm) (b).

During the capillary test, some fissures appeared on the three samples of CPC, mainly between the dry and the wet zones of samples, and spread out, causing even important fractures in some points, as shown in Figure 6.

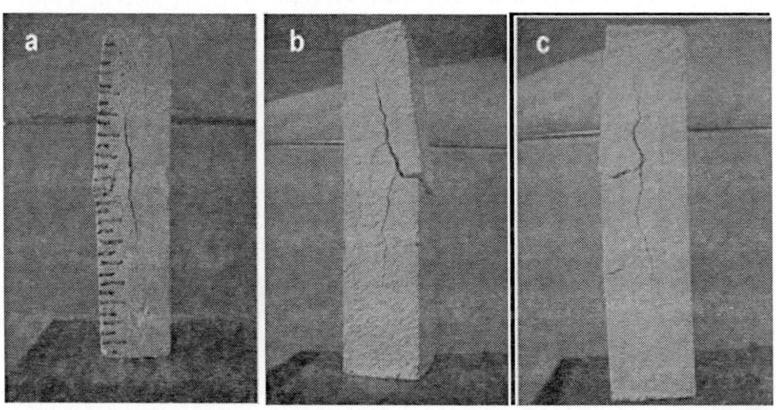

Figure 6. Development of fissures on the three samples of CPC (CL+WPC+CA) mortar after the capillary assay.

Once the capillary assay was finished, CPC samples were broken and the internal parts, in which the fissures had propagated, were analysed by means of X-ray diffraction to find out the cause of these fractures. XRD pattern (Figure 7) shows the presence of small amount of ettringite, in addition to the original phases initially detected in this mortar.

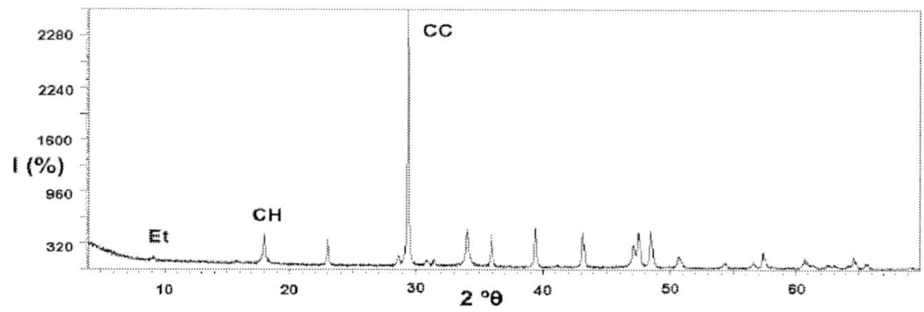

Figure 7. X-ray diffraction pattern of CPC mortar after the water capillary uptake. Et = ettringite; CH = portlandite; CC = calcite.

Ettringite ($Ca_6Al_2(SO_4)_3(OH)_{12}{\cdot}26H_2O$) derives from the reaction between the aluminates phases and the SO_4^{2-} ions present in cement.

The moist conditions created during the water imbibition test may have induced this reaction between the gypsum and the few aluminate phases of cement (see section 3.1), which had remained unhydrated in the cement matrix until this moment, given that no traces of aluminate hydrates were found during this study. Then the ettringite precipitated at the interface between the dry and wet zones of the samples, where the fissures had been initially observed. In fact, drying-wetting cycles contribute to a slight acceleration of the reaction although they are not necessary for the occurrence of ettringite [23].

3.3.2. Mechanical Strength

CPC mortar shows the highest values of mechanical resistances (especially R_f, Table 2), as expected for the presence of cement, although this increase was not high, due to the low percentage of cement replacing the lime [24] as well as to the atmospheric conditions to which the CPC mortar was exposed during the curing [21].

3.3.3. Shrinkage

The shrinkage values (in %) measured on the hardened samples (after 60 days) in three perpendicular directions (a, 4 cm; b, 16 cm; c, 4 cm) are shown in Table 2.

In CPC mortar, shrinkage is smaller than in CC, especially in the longest direction (b), where is equal to zero. The highest value of shrinkage has been measured in both mortars along the c direction. This is caused by the fact that the surface perpendicular to direction c is the most exposed to the air when the sample is in the mould (first 7 days of curing) and, consequently, faster water evaporation occurs in this face at early age.

Conclusion

This study shows that at the curing conditions applied for aerial lime mortars (T=20 ºC and RH=60%), the phases of Portland cement do not undergo to complete hydration so that only few changes in the microstructure and physical-mechanical properties are produced in the blended mortars. It was observed that the progress of the carbonation and hydration processes occurs at different degrees in the interior and the surface of mortars but few hydrated phases are formed in both cases. This is the reason of the scarce reduction of the capillary pores volume, which has resulted in no improvement of the water transfer properties (i.e. no reduction of the capillary uptake).

On the contrary, the little hydration occurred gave place to similar mechanical resistances between the lime and cement-lime mortars, which may represent a positive aspect for the whole masonry. The addition of cement presented also the advantage of reducing importantly the shrinkage in one of the three dimension of mortar sample. Notwithstanding, these benefits of using white Portland cement in lime-based mortars are totally countered by the occurrence of soluble salts, such as the ettringite found here.

In summary, the white Portland cement has demonstrated to be harmful and not very effective when mixed to lime mortars, even if only as additive. In view of these results, we recommend using and producing repair mortars without cement and, at the same time, we encourage researchers to improve the study of other alternative and more effective materials to be added to the lime.

Acknowledgment

This study was financially supported by Research Group RNM179 of the Junta de Andalucía and by Research Project P09-RNM-4905. We are grateful to Argos Derivados Del Cemento, S.L. for providing the raw materials used in this study.

References

[1] Veniale, F., Setti, M., Rodriguez-Navarro, C., Lodola, S., Palestra, W., Busetto, *A. Cem. Concr. Compos*. 2003, 25, 1123-1129.

[2] Callebaut, K., Elsen, J., Van Balen, K., Viaene, W. *Cem. Concr. Res*. 2001, 31, 397-403.

[3] O'Hare, G. (1995) Lime mortars and renders: the relative merits of adding cement. www.buildingconservation.com/articles/cement/cement.htm.

[4] UNE-EN 459-1. Cales para la construcción. Parte 1: Definiciones, especificaciones y criterios de conformidad. *AENOR*, Madrid; 2002.

[5] UNE-EN 197-1. Cemento. Parte 1: composición, especificaciones y criterios de conformidad para los cementos comunes. *AENOR*, Madrid; 2000.

[6] UNE EN 1015-3. Métodos de ensayo de los morteros para albañilería. Parte 3: Determinación de la consistencia del mortero fresco (por la mesa de sacudidas). *AENOR,* Madrid; 1998.

[7] UNE EN 1015-2. Métodos de ensayo de los morteros para albañilería. Parte 2: Toma de muestra total de morteros y preparación de los morteros para ensayo. *AENOR*, Madrid; 1999.

[8] Cazalla, O. Morteros de cal. Aplicación en el Patrimonio Histórico. *Ph.D. Thesis*, Universidad de Granada; 2002.

[9] J. D. Martín Ramos, XPowder. A software package for powder X-ray diffraction analysis. *Lgl. Dep. GR* 1001/04; 2004.

[10] UNE-EN ISO 11358. Plásticos. Termogravimetría (TG) de polimeros. Principios generales. *AENOR*, Madrid; 1997.

[11] Beck, K., Al-Mukhatar, M., Rozenbaum, O., Rautureau, M. *Build. Environ.* 2003, 38, 1151-1162.

[12] UNE-EN 1015-11. Metodos de ensayo de los morteros para albañileria. Parte 11: Determinación de la resistencia a flexión y a compresión del mortero endurecido. *AENOR*, Madrid; 2000.

[13] Cizer, O., Van Balen, K., Van Gemert, D., Elsen, J. Building materials and building technology to preserve the built heritage. ISBN: 978-3-937066-15-8; WTA-Schriftenreihe Heft 33 VII-XIV: *Leuven (BE),* 2009; Vol. 2, pp. 353-368.

[14] Giraldo, M., Manuel Alejandro. *Dyna. Rev. fac. Nac. Minas.* 2006, 73 (148), 69-81. ISSN 0012-7353.

[15] Houst, Y. F.; Wittmann, F. H. *Cem. Concr. Res.* 1994, 24 (6), 1165-1176.

[16] Cizer, O. Competition between carbonation and hydration on the hardening of calcium hydroxide and calcium silicate binders. *PhD Thesis*; Katholieke Universiteit Leuven, 2009.

[17] Ramachandran, V. S., Feldman, R. F. *Concrete Admixtures Handbook, Properties, Science and Technology, Noyes Publications*, New Jersey (US), 1995; Ch. 1, pp.1-36.

[18] Varas, M. J., Alvarez de Buergo, M., Fort, R. *Cem. Concr. Res.* 2005, 35, 2055-2065.

[19] Arandigoyen, M., Alvarez, J. I. *Appl. Surf. Sci.* 2006, 252, 8077-8085.

[20] Pandey, S. P., Sharma, R. L. *Cem. Concr. Res.* 2000, 30, 19-23.

[21] Fernandez Luco, L. Valoración de técnicas no destructivas para el control de la eficiencia de curado del hormigón. *PhD Thesis*, Universidad de Alicante; 2008, pp. 33-64.

[22] Lawrence, R. M., Mays, T. J., Rigby, S. P. Walker, P., D'Ayala, D. *Cem. Concr. Res.* 2007, 37, 1059-1069.

[23] Escadeillas, G., Aubert, J. E., Segerer, M., Price, W. *Cem. Concr. Res.* 2007, 37, 1445-1452.

[24] Arandigoyen, M., Alvarez, J. I. *Cem. Concr. Res.* 2007, 37, 767-775.

In: Brick and Mortar Research
Editors: S. Manuel Rivera and A. L. Pena Diaz
ISBN: 978-1-61942-927-7

Chapter 9

INDUSTRIAL COATINGS FOR HIGH PERFORMANCE APPLICATION: PHYSICOCHEMICAL CHARACTERISTICS AND ANTI-CORROSIVE BEHAVIOR

Th. Zafeiropoulou*, E. Rakanta and G. Batis
Department of Materials Science and Engineering,
School of Chemical Engineering, National Technical University
of Athens, Athens, Greece

ABSTRACT

The anti-corrosive behavior of four different types of organic coatings was examined in the present study. High performance application coatings including epoxy, polyurethane, chlorinated rubber and solvent based coatings were exposed to highly corrosive environment and their behavior against induced chloride ion corrosion was evaluated with electrochemical measurements such as Linear Polarization Technique (LPR), Half-cell Potential and Electrochemical Impedance Spectroscopy (EIS). The durability of the organic coatings was evaluated by assessing basic physicochemical properties including water vapor and liquid water absorption and chloride ions permeability. From the results obtained it is shown that cement mortar specimens that were covered with the polyurethane coating perform better, in both electrochemical and physicochemical measurements, compared to the other three high performance application coatings. Nevertheless, all the coating tested in the present study, are able to protect steel rebars from corrosion more effectively that conventional coatings.

Keywords: polyurethane; chlorinated rubber; organic coatings; corrosion; electrochemical measurements

* Fax:+30 210 772 3184; Tel:+30 210 772 3186; E-mail: dia_zaf@mail.ntua.gr

1. INTRODUCTION

The corrosion of reinforcing steel in cement-based materials is a worldwide problem with high economic costs related to maintenance and repairs of concrete structures damaged by corrosion. Organic coatings applied on the surface of concrete structures are one of the most widely used methods for the corrosion control of embedded steel rebars, in both conventional and high performance applications. High-performance anticorrosion coatings are applied to concrete structures to provide protection from corrosive environments. Generally, the coatings are separated into two major categories:

- Protective coatings applied to structures in oil, gas, petrochemical and power generation industries, as well as to bridges and water and waste treatment plants.
- Marine coatings applied to commercial ships, including freight carriers, tows, cruise ships and others.

Modified polyurethane paints are widely used as topcoats due to their good anti-corrosion properties and weather resistance. Particularly, acrylic polyurethane and aliphatic polyurethane coatings are the most widely used types. The aliphatic polyurethane is based on aliphatic isocyanate, which reacts with acrylic ester containing hydroxyl and has strong adhesion, good chemical resistance and aging resistance [1–2]. But aliphatic polyurethane is easy to become yellow because of the decomposition of urethane bonds. It has been found that the acrylic polyurethane composite coatings showed better performance than aliphatic polyurethane composite coatings and under the testing condition, for aliphatic polyurethane composite coating the coating resistance decreased below 10^6 Ωcm^2 after 28 days, while for acrylic polyurethane composite coatings it was 35 days [3]. Chlorinated rubber coatings are widely used for corrosion protection of surfaces. The material has excellent chemical and water resistance [4]. The coatings have, for a long time, been applied to steel, concrete and other building surfaces which are regularly exposed to water or chemicals as in swimming pools, road markings, marine building and bridge construction etc. A large number of pigments can be used in this type of paint, similar to most other oleo-resin type paints. Among these are titanium dioxide, micaceous iron oxide, zinc powder, red lead and metallic lead [5]. Both chlorinated rubber and epoxy coatings have been applied on the footings and piers, to avoid concrete deterioration due to sulfate attack. However, concrete coatings of several generic types are now marketed for protecting concrete at both above and below ground levels. Studies [6-8] to date have shown encouraging results on the performance of vapor barriers, vapor permeable coatings, surface penetrating sealers and protective coatings.

Dulaijan et al. [9, 10] evaluated the performance of cement- and epoxy-based coatings in protecting concrete. Results of that study [9] indicated that epoxy modified cement based coatings provide adequate protection to concrete.

However, the crack bridging capacity of the polymer modified cementitious coating was reported to be better than that of other cement-based coatings. The adhesion of all the epoxy resin-based coatings, to the concrete substrate, was noted to be better than that of the acrylic resin-based surface coatings. The water permeability in the concrete specimens coated with the selected resin based surface coatings was reported to be very low and they exhibited good crack bridging ability.

Further, all the coatings were noted to considerably reduce the diffusion of carbon dioxide into the concrete matrix. However, not all the coatings were able to withstand acidic exposure.

The chemical-resistance of epoxy resin based surface coatings was shown to be better than that of acrylic resin-based coatings [10]. In the present study four types of high performance organic coatings were evaluated electrochemically and physicochemically, regarding the protection they offer to embedded steel rebar when cement mortar specimens are exposed to highly corrosive environments under the presence of chloride ions. The electrochemical techniques include Linear Polarization Technique and Half-cell Potential measurements which are usually used for corrosion monitoring in coated specimens and also Electrochemical Impedance Spectroscopy which is an often used method when rebars are coated but is not widely adopted for coated cement mortar specimens.

2. Experimental

Reinforced cement mortar specimens were constructed and four different types of organic coatings were investigated, regarding the protection they offer against chloride ion corrosion. The specimens were partially immersed in 3.5 wt% NaCl solution for a period of four months. Reinforced cement mortar specimens were electrochemically investigated regarding their corrosion behavior and for every coating system physicochemical properties were also determined.

2.1. Materials

Cement II 32.5N, Greek quarry sand of 250 μm - 4 mm maximum grain dimension and drinkable water from NTUA water supply network, appropriate for preparing specimens according to ELOT 452 [11] were used for the construction of the specimens. Steel rebars type B500C were used, according to Hellenic Organization for Standardization ELOT 1421-3[12], having 12 mm nominal diameter and 100 mm length. The w/c ratio was 0.55. The specimens were cylindrical formed in 40 mm × 100 mm. Steel rebars were axially embedded in each one, located as shown in Figure 1 and were cleaned prior their installation into the mortars according to ISO/DIS 8407.3[13] and weighted to a 0.1 mg accuracy. The rebars were embedded 80 mm deep into the mortar and consequently 20 mm exerted. Reinforced cement mortar specimens were cast and stored at ambient conditions in the laboratory for 24 h. After being demoulded, the specimens were placed in a curing room (RH > 98%, T=20±1.5°C) for 24 h and then kept 1 day in the laboratory environment to stabilize internal humidity. Prior coating application specimens were left to dry for 8 days under laboratory conditions. In order to receive electrochemical measurements, a copper wire was enwrapped to each steel rebar. A mixture of two different epoxy resins was used aiming to protect the upper part of the rebar from atmospheric corrosion.

Organic coatings were applied by brush on the dried surface of the specimens at two layers, the second layer 24h after the first one. The composition of the coatings used is given in Table 1. Dry film thickness of all coatings was measured using ultrasonic thickness gauge meter according to ASTM D 6132-08 [14] Standard Test Method.

Table 1. Coatings applied on specimens

Nomeclature	Description	Thickness (μm)
Ref	Specimens without coating application	-
Polyurethane	Two-component, anticorrosive epoxy primer (solids 58%). Two component pure epoxy paint, hardened with amine. Solids b.w.: 95%. Two-component polyurethane paint with aliphatic isocyanate. Solids b.v.: 54%. Spreading rate: 9-11 m^2/l.	450
Epoxy	Two-component, anticorrosive epoxy primer (solids 58%). Two component pure epoxy paint, hardened with amine. Solids b.w.: 95%. Spreading rate: $6m^2/kg$.	300
Chlorinated rubber	Primer with special resins. Liquid rubber acrylic based sealant. Density: 1.37±0.5 g/cm^3.	250
Solvent based	Solvent based primer for exterior use (solids b.w.: 26±2%).Solvent-based paint for cement surfaces, styrene copolymers and acrylic resins, solids b.w.: 73.5±2%. Diluted 4% v/v with white spirit. Density: 1.48±0.04 g/ml, viscosity: 9.5±1.5 poise, spreading rate: 7.3±0.8m^2/l	250

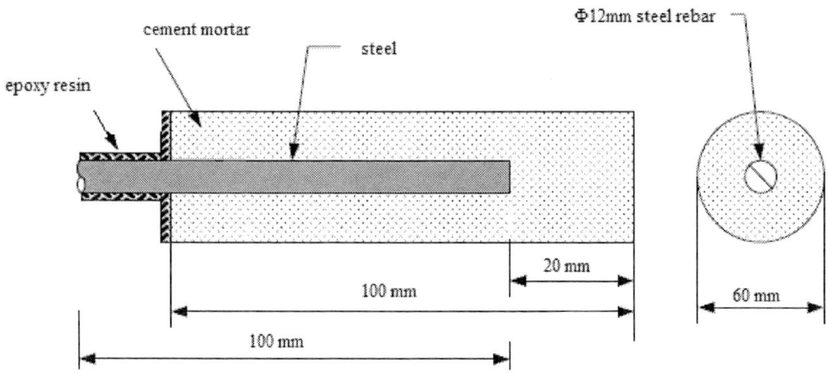

Figure 1. Cement mortar specimens.

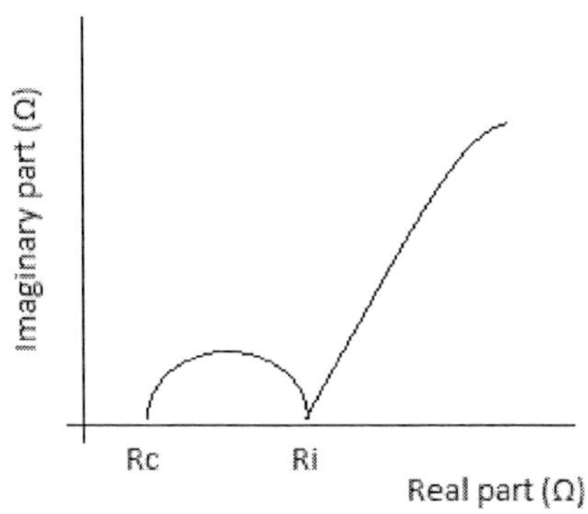

Figure 2. Typical Nyquist diagram used in the present study.

2.2. Methods

In this study, the following techniques were performed on all reinforced cement mortar specimens, exposed to chloride environment in order to study the corrosion process induced by chloride ion and the effect of the coating systems as a protection method.

Half-Cell Potential Measurements

For a period of four months steels' half-cell potential was periodically measured versus a Saturated Calomel Electrode (SCE), according to ASTM C876-87, using a high impedance voltmeter (10MΩ) [15].

Linear Polarization Technique (LPR)

Linear polarization technique experiments were conducted using a Potentiostat / Galvanostat Model 263A from EGandG Princeton Applied Research and an associated software package Power Suite to analyze the obtained data.

The electrochemical parameter of polarization resistance (R_p), was defined, as described in ASTM G59-97 (2009) [16].

The experimental set up was consisted of three electrodes where steel rebars represented the working electrode, saturated calomel electrode (SCE) the reference electrode and a carbon bar served as the counter electrode. The potential scan range was ± 10mV from Open Circuit Potential (OCP) and the scan rate was 0.160 mV/s.

Electrochemical Impedance Spectroscopy (EIS)

In the present study the impedance was measured using a Solarton SI 1260, Impedance/Gain-phase analyzer. The measurements were performed at room temperature using a three-electrode electrochemical cell, consisting of the working electrode, saturated calomel electrode (SCE) as reference and carbon bar as counter electrode. The measuring frequency ranged from 105 Hz down to 10-2 Hz. Spectra were treated using the Software Zplot / Z-view (Scribner Assoc. Inc.).

Transmission Rates

Water vapor transmission rate (WVTR) and liquid water transmission rate (w_t) were determined according to DIN EN 7783-2:1999 [17] and DIN EN 1062-3 [18] respectively. For the evaluation of chloride ions penetration the experimental set up was consisted of two permeation cells in which the tested sample acted as a membrane between two solutions. The first cell was filled with sodium chloride solution (2.0% w/v NaCl), whereas distilled water was used in the second cell.

The painted side of the membrane was facing the salt solution. Chlorides will traverse the membrane from the concentrated solution to the diluted one under natural diffusion driving force. For the determination of the total amount of chlorides passed through the coated membrane, conductivity measurements were periodically performed with a Sension7 conductivity meter by Hach, which were afterwards converted into concentration values. Figure 3 represents the experimental set up that was used in this study.

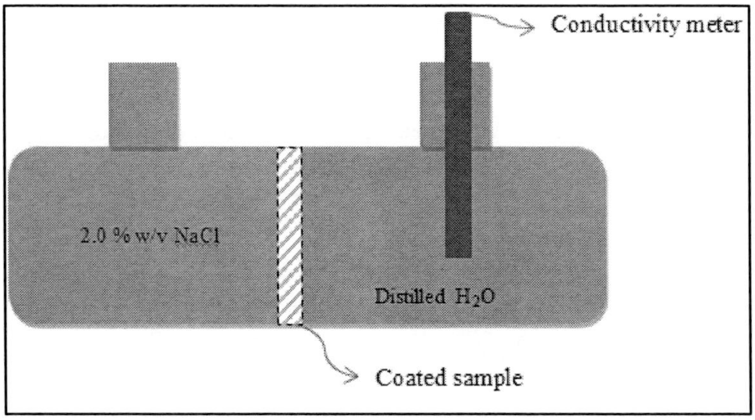

Figure 3. Experimental set up for the evaluation of Cl^- penetration.

3. Results and Discussion

Half-Cell Potential Measurements

Half – cell potential measurements of steel rebars is the most typical procedure to the routine inspection of reinforced concrete structures regarding the corrosion trend of the samples. Potential readings, however, are not sufficient as criterion, since they are affected by number of factors, which include polarization by limited diffusion of oxygen, concrete porosity and the presence of highly resistive layer. According to Table 2 [19], potentials more negative than -426mV, with respect to SCE, indicate severe corrosion of the reinforcement, whereas values less than -276 mV present high risk of corrosion with 90% probability. The area between -276 to -126 mV is indicative of intermediate corrosion risk and lower than -126 mV the risk of active corrosion is below 10%.

The results from half-cell potential measurements are presented in Figure 4, were the corrosion potential is depicted as a function of exposure period for cement mortars specimens covered with organic coatings and partially immersed in 3.5 wt.% NaCl solution.

Polyurethane and chlorinated rubber coatings, as depicted, present the highest corrosion potential values among all the coatings during the entire immersion time. The lowest value attained is approximately -200mV which may suggest passivation of the rebar, although this cannot be taken as a conclusive evidence of the passive state of the steel. According to Table 2 [19] an intermediate risk of corrosion is suggested for both coatings.

Table 2. Corrosion potential and corrosion condition

Steel's corrosion potential vs SCE (mV)	Corrosion condition
> -126	Low (10% risk of corrosion)
-126 to -276	Intermediate corrosion risk
< -276	High (90% risk of corrosion)
< -426	Severe corrosion

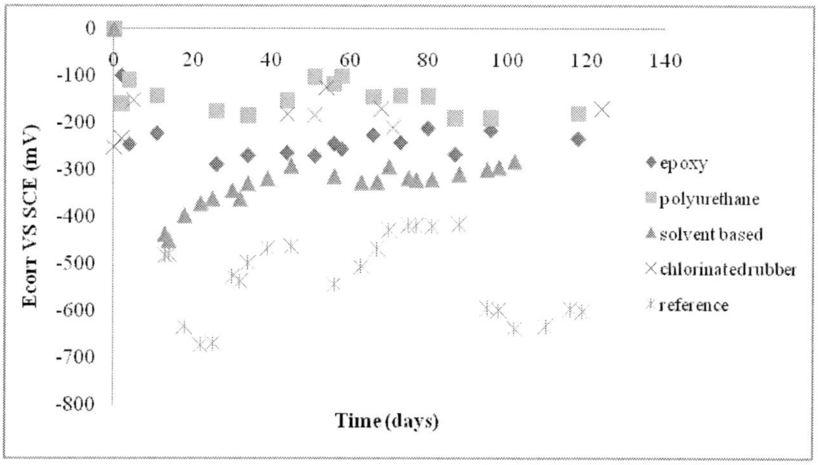

Figure 4. Half-cell potential as a function of exposure time for reinforced cement mortar specimens partially immersed in 3.5 wt.% NaCl solution.

A slightly higher corrosion tendency is observed for the epoxy coating, where corrosion potential is moving to – 290 mV, but the risk of corrosion remains intermediate.

During the first 20 days of exposure the corrosion potential for solvent based coating is moving to more electronegative direction, reaching -450 mV, but progressively after 45 days of immersion remains stable around -300mV, value which indicates high risk of corrosion. Reference specimens that remained uncoated, exhibit the worst behavior, since they demonstrate corrosion potential values up to -700 mV from the beginning of the experiment. The behavior exhibited from all the coatings can be considered as satisfied, since relatively noble corrosion potentials were attained, compared to severe corrosion state that is observed from other authors regarding conventional coatings [20-22].

Linear Polarization Technique

Linear Polarization technique is a corrosion monitoring method that allows corrosion rates to be measured directly, in real time. Although is limited to conducting liquids, the response time and data quality of this technique make it very useful, where applicable, to all other forms of corrosion monitoring. The technique is rapid and non-intrusive, requiring only a connection to the reinforcing steel. The data provides a valuable insight into the instantaneous corrosion rate of the steel reinforcement, giving more detailed information than a simple potential survey.

The LPR data enables a more detailed assessment of the structural condition and is a major tool in deciding upon the optimum remedial strategy to be adopted. It is thus imperative that the LPR measurements obtained are accurate. In LPR measurements the reinforcing steel is perturbed by a small amount from its equilibrium potential. This can be accomplished potentiostatically by changing the potential of the reinforcing steel by a fixed amount, ΔE, and monitoring the current decay, ΔI, after a fixed time. The polarization resistance, R_P, of the steel is then calculated from the equation

$$R_p = \frac{\Delta E}{\Delta I} \quad (1)$$

From which the corrosion rate, *I*corr, can then be calculated

$$I_{corr} = \frac{\beta_\alpha \cdot \beta_c}{2.303 \cdot (\beta_\alpha + \beta_c)} \cdot \frac{1}{R_p} = \frac{B}{R_p} \quad (2)$$

where β_α β_c are the anodic and cathodic Tafel slopes respectively and R_p is the polarization resistance (Ohm). For Stern–Geary constant B a value of 26mV has been adopted for active corroding steel bars and 52mV for passive conditions. In order to determine the corrosion current density, *i*corr, the surface area, *A*, of steel that has been polarized needs to be accurately known:

$$i_{corr} = \frac{I_{corr}}{A} \quad (3)$$

In the present study the electrochemical parameter of polarization resistance (R_p) for steel rebars was obtained using linear polarization technique. Periodical measurements were performed to 6 specimens for each coating and the results are shown in Figure 5 where is depicted the polarization resistance as a function of immersion time and in Table 3, where analytical R_p values are depicted.

Polyurethane coating demonstrates an increasing tendency regarding R_p values, until the 40th day of immersion. Between the 70th and the 100th day of the experimental procedure R_p values remain stable and a reduction is observed afterwards. R_p values for the Polyurethane coating are the highest among all the tested coatings but a stable condition is not yet reached.

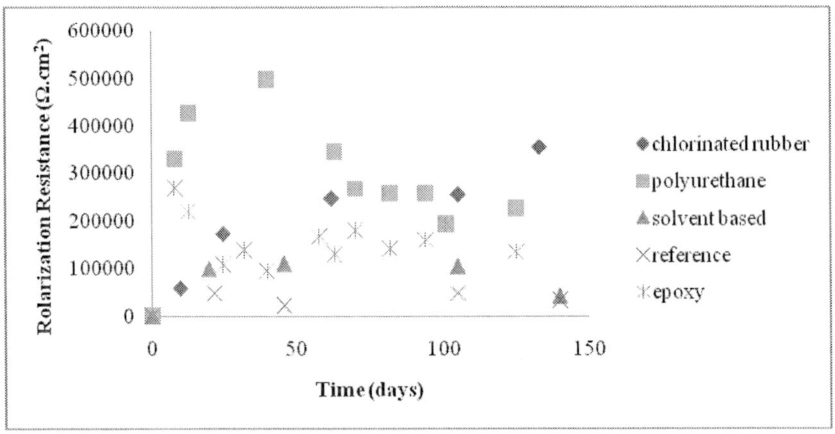

Figure 5. Polarization Resistance after 4 months of exposure to 3.5 wt.% NaCl solution.

Regarding chlorinated rubber coating, progressively increasing R_p values are depicted which are demonstrating a reduction to the rate of the corrosion process and are indicative of the protection that the coating is offering against corrosion. Cement mortar specimens the were covered with the epoxy coating exhibit a sudden reduction in R_p values after the 10th day and they remain more or less stable in lower values compared to the previous coatings. Solvent based coating presents the lowest R_p values and at the end of the immersion period the electrochemical decomposition of the metal surface appears to become quicker.

Table 3. Average values of polarization resistance R_p and standard deviations

Time (days)	Rp (Ω)				
	Epoxy	Polyurethane	Chlorinated rubber	Solvent based	Reference
0	0	0	0	0	0
8	268967±563	331240±229	58519±128	58519±128	
13	221289±712	427939±709			
25	108767±590		173176±454	173176±454	48260±221
32	139573±991				
40	95518±362	498795±821			22509±285
58	168140±321				
63	129888±856	346374±483	247519±559	247519±559	
70	180720±703	267742±634			
82	142413±257	258270±456			
94	160112±399	258474±657			
101	193131±755	193780±533	255874±732	255874±732	48003±331
125	136159±722	227601±240	355370±611	355370±611	33032±129

Electrochemical Impedance Spectroscopy

EIS technique was chosen based on the fact that provides information on the electrochemical process that occurs on the steel mortars interface, in addition to the kinetics of the reaction. In general, impedance consists of complex resistance when current flows through a circuit made of capacitors, resistors, or insulators, or any combination of these [23] and is commonly depicted as a complex function, having both real and imaginary components. It is usually plotted in Nyquist diagram, as imaginary part versus real part, proving a convenient method for the determination of important electrical parameters. In the present study a typical plot for all measurements conducted is given in Figure 2. The salient features of the spectrum are labeled as follows:

- The cement mortar resistance (R_c) which is highly related to the microstructure and pores of the cement based material and is given by the high frequency limit of the diagram.
- The steel-cement mortar interface film resistance (R_i) which is given by the diameter of the high frequency loop. The interfacial film actually represents the layer formed on the steel surface and its impedance properties are affected by cement hydration

products deposited in the layer and/or corrosion products of steel diffused into the layer [24].

Electrochemical impedance spectroscopy is an alternative tool to evaluate the activity difference between the tested coatings. The high frequency loop varies, depending on the porosity of the cement paste and consequently in the present study R_c values were determined for all the coatings. The results of EIS are depicted in Figure 6, where is shown a comparative Nyquist plot for all coating systems. In Figure 7 average R_c values are presented for 6 mortar specimens from each coating system. From the results it can be observed that polyurethane coating is demonstrating the best anticorrosive behavior among the other coatings, since it presents the highest R_c values. Chlorinated rubber and epoxy coatings are exhibit similar behavior which appears to be better than the solvent based coating.

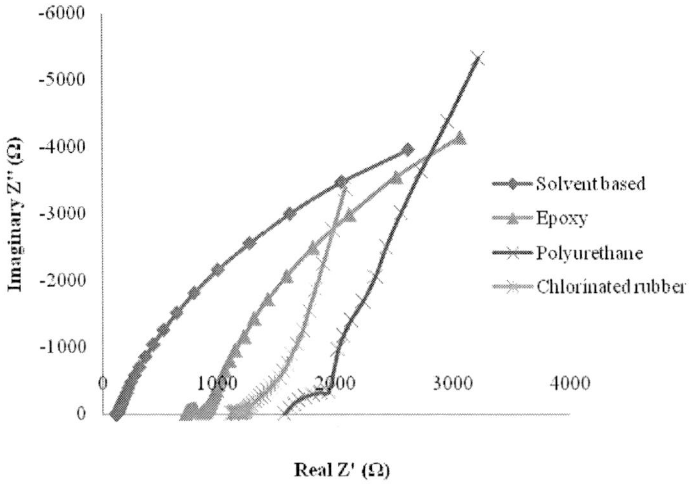

Figure 6. Nyquist plot after 40 days of immersion.

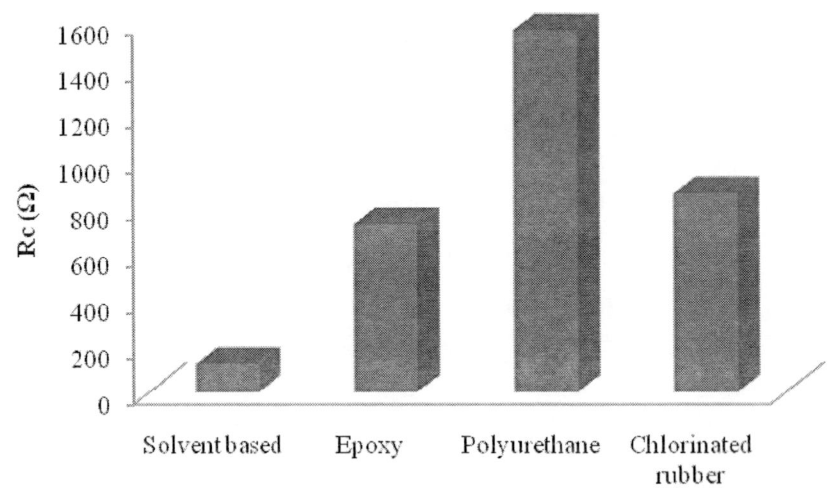

Figure 7. Average R_c values for all coatings.

Transmission Rates

The results for water vapor transmission rate (*WVTR*) and liquid water transmission rate (w_t) of all types of coating systems are presented in Figure 8 and Figure 9 respectively. The best combination for an effective coating is to exhibit high *WVTR* and low w_t values. Polyurethane, epoxy and chlorinated rubber coatings can be considered as water-repellent coatings, since they present very low liquid water permeability. According to DIN standards they can be subsumed to Category III (low liquid water permeability).

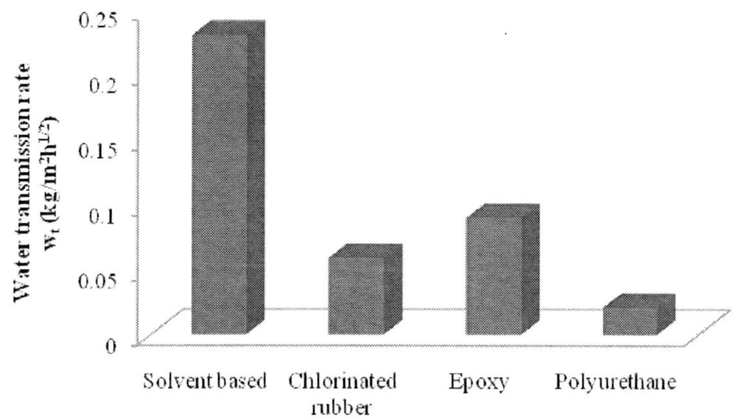

Figure 8. Liquid water transmission rate for every coating system.

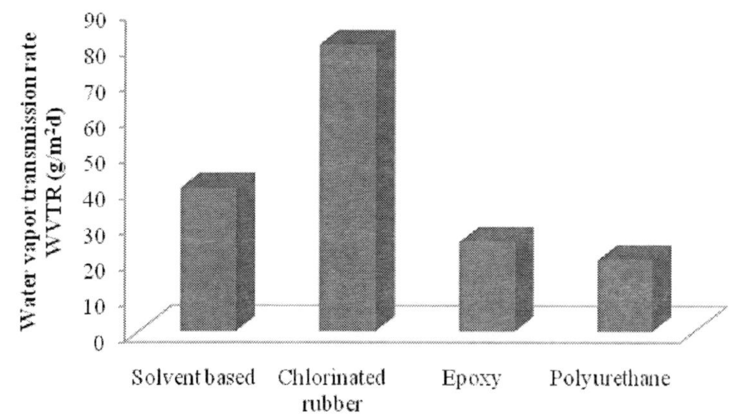

Figure 9. Water vapor transmission rate for coatings.

In contrast solvent based coating is assigned to Category II (medium liquid water permeability). Regarding *WVTR*, all coatings are exhibiting similar behavior and are classified to Category II (medium water vapor transmission rate).

Chloride permeability results are presented in Figure 10. It can be observed that all coating systems exhibited very low permeation values and especially the polyurethane coating had the minimum chloride permeability versus time from all the other coating systems tested. It must also be pointed out that the results from chloride permeability are in agreement with liquid water transmission rates for all the coatings.

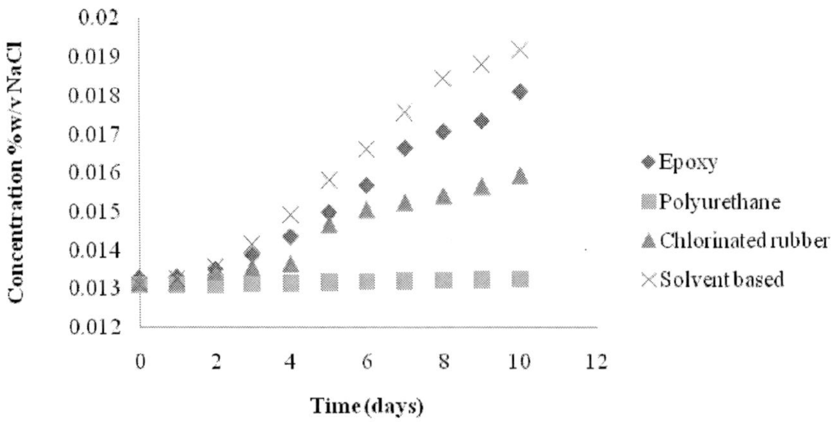

Figure 10. % w/v NaCl concentration versus time for all coatings.

CONCLUSION

From the results of all electrochemical measurements that were used in the present study it can be observed that polyurethane coating is offering the highest resistance to the corrosion process of the embedded steel rebar. Nevertheless, chlorinated rubber coating, demonstrates an increasing tendency regarding R_p values which is indicative of the effectiveness of the coating through time. The results from the electrochemical properties are in accordance with corrosion durability measurements since, according to DIN standards [19], polyurethane, epoxy and chlorinated rubber coatings are subsumed in Category III (low liquid water permeability). Moreover chloride ions permeability results are in agreement with the aforementioned classification.

REFERENCES

[1] Chattopadhyay, D.K.; Raju, K.V.S.N. Structural engineering of polyurethane coatings for high performance applications *Prog. Polym. Sci.* 2007, 2, 352–418.

[2] Fratricova, M.; Simon, P.; Schwarzer, P. Residual stability of polyurethane automotive coatings measured by chemiluminescence and equivalence of Xenotest and Solisi ageing tests *Polym. Degrad. Stabil.* 2006, 91, 94–100.

[3] Zhu, Y.; Xiong, J.; Tang, Y.; Zuo, Y. EIS study on failure process of two polyurethane composite coatings *Progr. Org. Coat*, 2010, 69, 7–11.

[4] Lambourne R. Paint and surface coatings: theory and practice London: Ellis Horwood, 1987.

[5] Chandler K.A.; Bayliss, D.A. Corrosion protection of steel structures *Elsevier App. Scien. Pubs.* 1985.

[6] Swamy, R.N.; Suryavanshi, A.K.; Tanikawa, S. Protective ability of an acrylic based surface coating system against chloride and carbonation penetration into concrete *ACI Mater. J.* 1998, 95, 101–112.

[7] Kamal, M.M.; Salama, A.E. Protection of reinforced concrete elements against corrosion by polymer coatings In: Dhir RK, Green JW, editors, Proceedings of International Conference on Protection of Concrete, London: E and FN Spon; 1991, 281–291

[8] Ibrahim, M.; Al-Gahtani, A.S.; Maslehuddin. M. Use of surface treatment materials to improve concrete durability, *ACI Mater. J.* 1999, 36–40.

[9] Dulaijan, S.U.; Maslehuddin, M.; Al-Zahrani, M.M.; Al-Juraifani, E.A.; Alidi, S.A.; Al-Meththel, M. Performance evaluation of cement based surface coatings In: Proceedings of 2000 International Conference, Repair, Rehabilitation and Maintenance of Concrete Structures and Innovations in Design and Construction, Seoul, Korea, September 19–22, 2000, 321–338.

[10] Dulaijan, S.U.; Maslehuddin, M.; Al-Zahrani, M.M.; Sharif, A.M.; Al- Juraifani, E.A.; Al-Idi, S.H. Performance evaluation of resin based surface coatings, In: Proceedings of 6th International Conference in Deterioration and Repair of Reinforced Concrete in the Arabian Gulf, Bahrain, November 20–22, 2000, 345–362.

[11] Hellenic Organization for Standardization ELOT 452, Determination of total Hg content to water with Atomic Absorption Spectroscopy.

[12] Hellenic Organization for Standardization ELOT 1421-3, Steel for the reinforcement of concrete – Weldable reinforcing steel – Part 3: Technical class B500C.

[13] ISO/DIS 8407.3, Procedures for removal of corrosion products from corrosion test specimen.

[14] ASTM D 6132-08 Standard Test Method for Nondestructive Measurement of Dry Film Thickness of Applied Organic Coatings Using an Ultrasonic Gage.

[15] ASTMC 876-87, Standard test method for half-cell potentials of reinforcing steel in concrete.

[16] ASTM G59-97(2009), Standard test method for conducting potentiodynamic polarization resistance measurements.

[17] DIN EN 7783-2:1999, Determination and classification of water-vapour transmission rate (permeability).

[18] DIN EN 1062-3, Determination and classification of liquid-water transmission rate (permeability).

[19] Broomfield, J. Corrosion of steel in concrete understanding, investigation and repair; 1st Edition, E and FN SPOK, UK, 1997, 5, 16-21.

[20] Batis, G.; Pantazopoulou, P.; Routoulas, A. Corrosion protection investigation of reinforcement by inorganic coating in the presence of alkanolamine-based inhibitor *Cem. Concr. Comp.* 2003, 25, 371–377.

[21] Kouloumbi, N.; Batis, G. The anticorrosive effect of fly ash, slag and a Greek pozzolan in reinforced concrete *Cem. Concr. Comp.* 1994, 16, 253–260.

[22] Zafeiropoulou, Th.; Rakanta, E.; Batis, G. Performance evaluation of organic coatings against corrosion in reinforced cement mortars *Progr. Org. Coat.* 2011, 72, 175-180.

[23] Park, S. M.; Yoo, J.S. Electrochemical impedance spectroscopy for better electrochemical measurements *Anal Chem*. 2003, 75, 455-461.

[24] Gu, P.; Beaudoin J. Estimation of steel corrosion rate in reinforced concrete by means of equivalent fittings of impedance spectra *Adv. Cem. Res.*, 1998, 10, 43-56.

In: Brick and Mortar Research
Editors: S. Manuel Rivera and A. L. Pena Diaz
ISBN: 978-1-61942-927-7

Chapter 10

On the Modeling of Brick-Mortar Interface

Fazia Fouchal,[1] Frédéric Lebon,[2] Amna Rekik[3] and Isabella Titeux[4]
[1]GEMH, Limoges University, IUT Egletons, France
[2]LMA, CNRS UPR7051, Aix-Marseille University, France
[3]PRISME, Polytech Orléans, France
[4]GRESPI, LGC EA4694 Reims University, France

1. Introduction

Traditionally, for thousand of years, masonry techniques were commonly used for the building of homes, monuments, walls, and retaining walls. Masonry is one of the oldest construction materials. Since masonry is a composite structure, however, failure of these structures will depend on the properties of the materials (mortar, bricks, etc.), as well as on the characteristics of the bonding between the various components. Two main methods of modeling masonry structures have been developed in the literature. The first method involves macroscopic models and homogenization techniques: the wall is assumed to be a single structural element characterized by a non-linear response when it is exposed to external forces. The second method has been developed for predicting the evolution of cracks and damage at the interface brick-mortar or brick-brick. This chapter deals with two families for the modeling of brick-mortar interface. The first part of this chapter deals with the experimental characterization of the materials (bricks and mortar) and the brick-mortar interface. Hereafter, we describe experimental studies on the shear behavior of masonry on the local scale, in the case of two different assemblies composed of two and three full/hollow bricks, with or without confinement. Two other structures are presented: a small wall made by seven bricks and the classical problem of a small wall under diagonal compression. The second part of this chapter deals with a phenomenological model of interface. The mechanical modeling approach (and in particular the RCCM [10] adhesive model adopted) is first presented. The model is based on the concept of adhesion variable due to Frémond [12]. A variation equation of this variable is introduced and the coefficients are identified experimentally. The numerical procedure used is then described. Lastly, some numerical examples are given and compared with the experimental data. The third part presents a multi-scale model of interface. The method is based on homogenization theories and asymptotic analysis. In the present work, we assume the existence of a third material: the brick-mortar interface, which is considered as a mixture of brick and mortar with a crack density. To obtain the effective properties of the damaged intermediate material, three aim steps are performed. First, we calculate the exact effective properties of the crack-free material using homogenization techniques for laminate composites, and thus define a first homogeneous equivalent medium. In the second step, we assign a crack density to the material. To model the macroscopic behavior of the cracked material, we use the Kachanov model [18] and then define a new homogeneous equivalent medium. Finally, in order to be sandwiched between the brick and mortar, this material is given a small thickness, and its mechanical behavior is derived using asymptotic techniques to shift from the micro to the macro level. A variation law of the crack length is introduced. The numerical procedure used is then described. Lastly, some numerical examples are given and compared with the experimental data.

2. Experimental Study of The Masonry

2.1. Introduction

The first part of this paper is dedicated to the experimental study of the mechanical behavior of the masonry. The experimental studies were performed on the materials and masonry specimens (single, couplet, triplet, wall) composed of full and hollow bricks in order to identify the mechanical behavior of the bricks, mortar and bricks /mortar joint, in particular under shear tests with and without confinement pressure.

2.2. Characterization of Materials

2.2.1. Mortar

The mortar used in this study is a ready-to-use mortar from "Weber and Boutrin" based on sand and cement (ciment is Portland type 52.5). The maximum size of the sand grains is 5 mm. The mortar characteristics are given in Table 1.

Table 1. Characteristics of the mortar

Granulometry	$0-5\ mm$
Dosing of bond / m^3 dry sand	350 kg
Water for mixing	12.5% or: • 5 L/ sack of 40 kg • 3 L/ sack of 25 kg
Density after 28 days	2.25
Contraction after 28 days	0.65

The standard tests following the EN 196-1 norm [5] were performed to determine the traction and compression strength [1].

The results concerning the flexion and compression limits are summarized in Table 2. The flexion strength is around $4, 26MPa$ and the compression strength is around $23MPa$ [1]. The elastic modulus given in [3] is equal to $(4000MPa)$.

Table 2. Mechanical characteristics of the mortar

Nb.	Flexural test		Compression test			
	Traction loading (kN)	Flexion strength (MPa)	Compression loading (kN)		Compression strength (MPa)	
			1st half prism	2nd half prism	1st half prism	2nd half prism
1	1.95	4.56	39.5	36.9	24.7	23.1
2	1.87	4.39	39.2	34.4	24.5	21.5
3	1.69	3.95	33.6	30.7	21.0	19.1
4	1.91	4.47	37.7	44.7	23.5	27.9
5	1.65	3.86	37.3	35.8	23.3	22.4
6	1.85	4.33	31.8	39.5	19.8	24.7
Mean value	1.82	4.26	36.5	37.0	22.8	23.1

2.2.2. Bricks

Full and hollow cooked earth bricks were used for the mechanical tests. The dimensions of each brick are 210 mm (length), 100 mm (width), 50 mm (height). Hollow Bricks are composed with ten holes ($25 \times 25\ mm^2$ in size) arranged in two rows. A compressive standard tests using NF EN 772-1 norm [6] were carried out on a series of three units full/hollow bricks in order to identify the compressive strength, elastic modulus and Poisson coefficient of the masonry units. Bricks are equipped with strain gauges glued in transverse and longitudinal directions (unidirectional compression test) and an extensometric sensor. Compressive load was applied on the upper face of the brick up to failure in two steps: the load was gradually increased as shown in Table 3; half-way to maximum pressure within 1 min. The elastic modulus is determined from the strain-stress curve at 30 % of the ultimate load.

Table 3. Loading rate

Expected Strength (N/mm^2)	Loading rate $(N/mm^2)/s$
< 10	0.05
from 11 to 20	0.15
from 21 to 40	0.3
from 41 to 80	0.6
> 80	1.0

Three tests are carried out on each kind of brick [1]. The curve giving the longitudinal deformation with respect to the normal stress and transversal deformation with respect to the normal stress are shown in Figures 1 and 2.

The quantitative data obtained on these mechanical parameters are summarized in Tables 4 and 5.

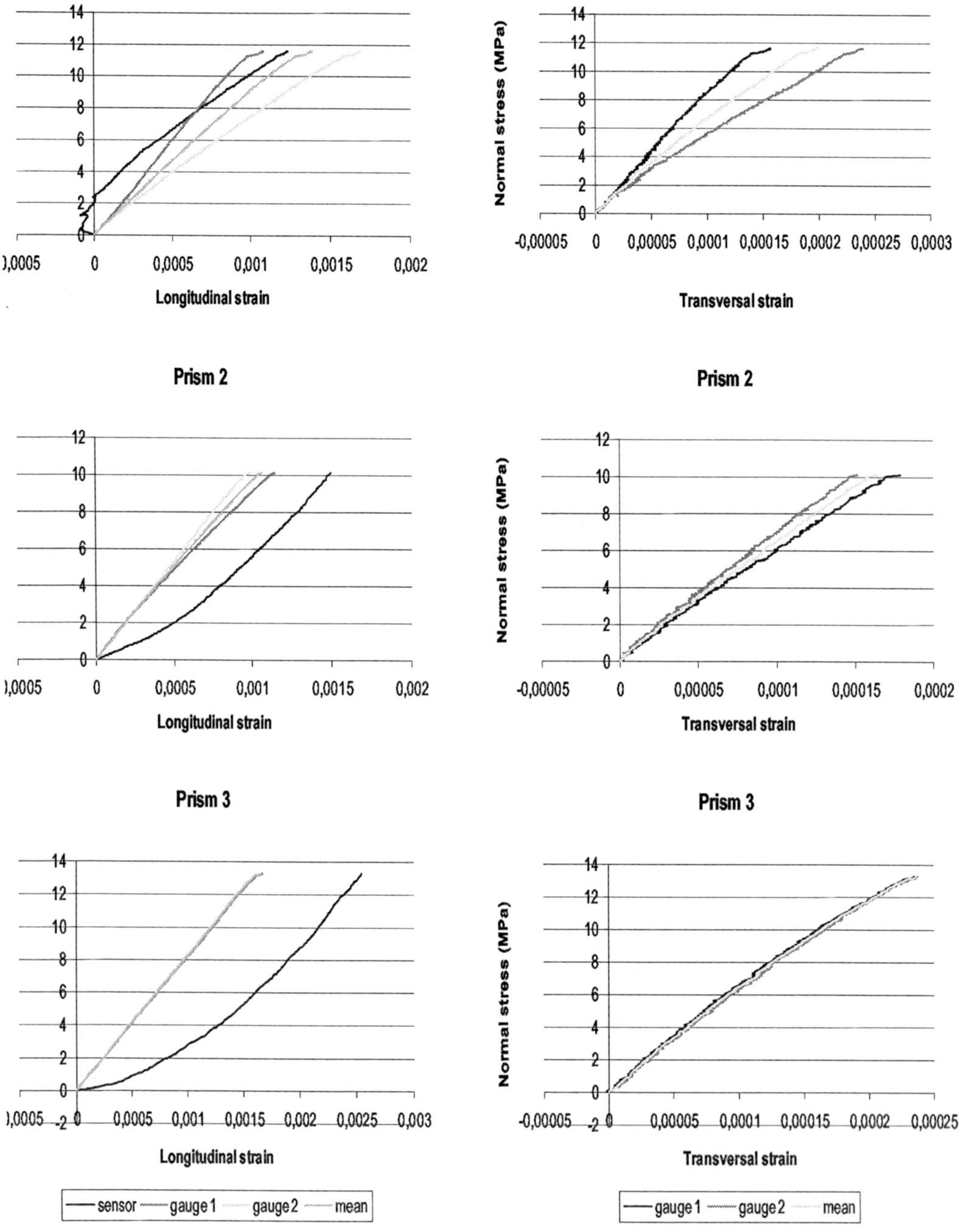

Figure 1. Compression test on full bricks: Elastic modulus and Poisson's coefficient

The experimental results obtained were similar to the mean value recorded by two gauges. The dispersion of the results was very small.

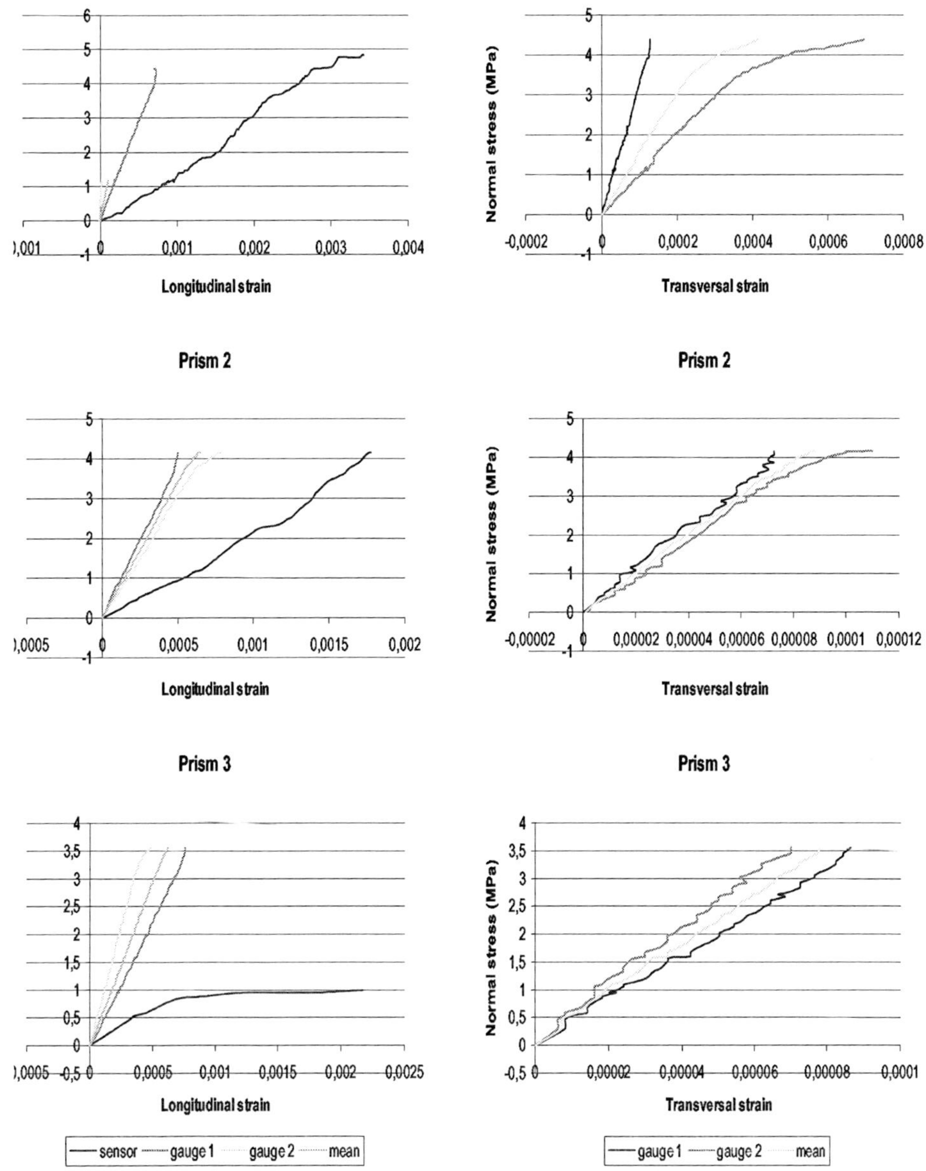

Figure 2. Compression test on hollow bricks: Elastic modulus and Poisson's coefficient

Table 4. Mechanical properties for full bricks

Nb	compression strength (MPa)	Elastic Modulus (MPa)	Poisson Coefficient
1	11.7	9 439	0.13
2	10.2	10 447	0.14
3	13.3	8 429	0.13
Mean value	11.7	9 438	0.13

Table 5. Mechanical properties for hollow bricks

Nb	Compression strength (MPa)	Elastic modulus (MPa)	Poisson coefficient
1	4.4	5 934	0.14
2	4.4	6 737	0.14
3	4.6	5 505	0.12
Mean value	4.46	6 059	0.13

2.3. Mechanical Behavior of Small Masonry Structures Under Shear Loading Without Confinement

2.3.1. Experimental Procedure

In order to determine the influence of the unit volume, two kinds of prism have been tested. The first kind is composed of two bricks, the second one is composed of three bricks (Figure 3). For each kind of unit volume, full or hollow bricks have been used. Mortar joint thickness used for the samples is equal to 10 mm. The aim of this experimental study is to determine the shear behavior and the rupture modes at the interface brick/mortar.

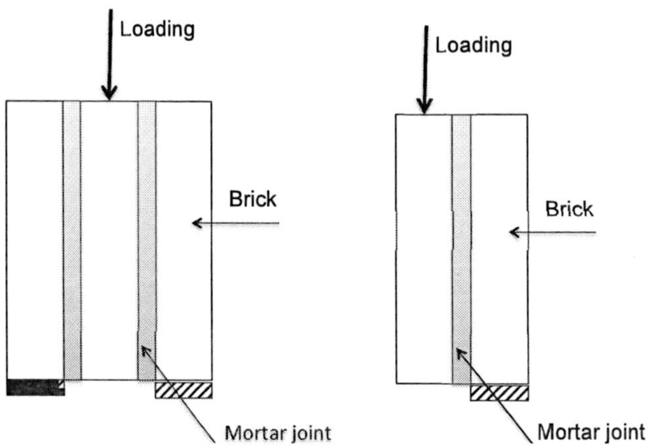

Figure 3. Shear test: (a) Three bricks unit (b)Two bricks unit

Prisms were prepared following the standard NF EN 1052-3 norm [7].

The samples were subjected to a monotonously increasing load until damage occurred [2].

Relative displacement between the loading brick and the unloading brick is measured by extensometric LVDT captors. Test machine and captors are connected with a PC in order to record data during the test notably the longitudinal strain and the applied load data.

Curves giving the shear stress with respect to the relative displacement were plotted that will determine the shear behavior law. The results are presented and commented in the following paragraphs.

2.3.2. Full Bricks

Figure 4 represents the recorded displacements with respect to the shear load $\tau = \frac{F}{A}$ where F is the imposed load on the upper face of the prism and A is the shear surface ($A = hight \times length = 210 \times 100\ mm^2$).

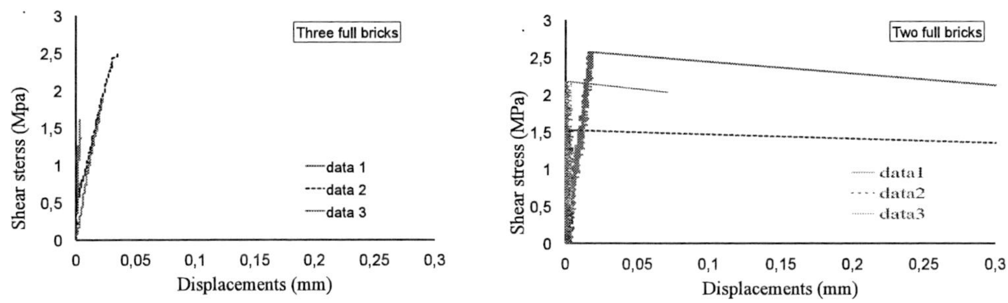

Figure 4. Shear test for full brick prisms

Figure 5 shows the different failurc modes obtained under shear test for the full brick prisms.

Figure 5. Failure modes - full brick prisms

2.3.3. Hollow Bricks

Figure 6 represents the recorded displacements with respect to the shear load $\tau = \frac{F}{A}$ where F is the imposed load on the upper face of the prism and A is the shear surface ($A = hight \times length = 210 \times 100\ mm^2$).

Figure 7 shows the different failure modes obtained under shear test for the Hollow brick prisms.

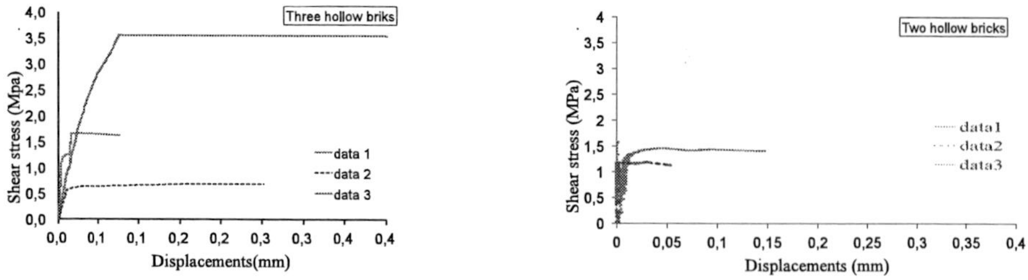

Figure 6. Shear test for Hollow brick prisms

Figure 7. Failure modes - Hollow brick prisms

2.3.4. Discussions

The experimental studies of different masonry samples presented have shown a nonlinear behavior on the local scale. We have observed a rigid elastic behavior to the rupture. The behavior of the specimen in full bricks is fragile in contrast to the response of the specimen in hollow bricks which is characterized by a softening behavior followed by the sliding movement between the adjacent bricks. The obtained results on hollow bricks samples have shown a wide dispersion. This is due essentially to the nonuniform distribution of the mortars spikes and local defects of the components of bricks masonry. We have noted that the couplet and the triplet samples made by a hollow bricks possess the same behavior. Consequently, the choices of basic cell have no influence on local scale. The rupture modes analysis confirms heterogeneity of masonry structures. This study allows us to identify phenomena that govern the rupture of the masonry at a local scale and to characterize the mecha! nical properties and the behavior law to take into account in finite elements computations [1] presented in the next section of this chapter.

2.4. Mechanical Behavior of Small Masonry Structures Under Shear Loading With Confinement

Small masonry structures subjected to both shear load and lateral confinement were study in [4]. In this section, we will focus on mechanical tests data obtained in [3, 4]. For this case, the dimensions of hollow ceramic bricks used are $(210 \times 100 \times 50mm^3)$, composed of ten holes. The holes sizes are $25 \times 25\ mm^2$ arranged in two rows. Mean value of compression strength and elastic modulus obtained by the standard compressive test on ten hollow bricks are $13,7MPa$ and $5820MPa$, respectively. Mortar used in this study is a ready-to-use mortar from "Lafarge-LANKO151" based on sand and cement (ciment is Portland type 52.5). The maximum size of the sand grains is 5 mm. The mortar characteristics are given in Table 1. Figure 8 presents the samples that were used for the mechanical tests performed according to RILEM recommendations [8]. Eighteen prisms were tested under loads confinement with various values between $0MPa$ and $1.6MPa$. The mechanical behavior of prisms showed in Figure 9 is characterized by a very rigid behavior for elastic domain, with little displacement (order of microns). The softening behavior is observed after the maximum stress followed by sliding movement between adjacent bricks. The confinement stress led to softening behavior in the masonry, when the confinement stress $\sigma = 0MPa$ the failure appears immediately. Identical tests give the same curves but with dispersion on the value of shear stress limit. This difference can be explained by the presence of spikes in the brick holes. Figure 10 show than spicks in the holes prevents brutal failure of a shearing prism and conditioned the shear stress limit. Analysis of the interface brick-mortar damage can identify two phenomena which are responsible for the prisms failure.

- spicks shearing
- rupture of the inner walls of the brick

The stress confinement σ has no significant influence on the evolution of limit shear stresses. The experimental data of these tests identifies three main values: $1.63MPa$, $1.85MPa$ and $2.2MPa$, respectively, as the lower, middle, and maximal value.

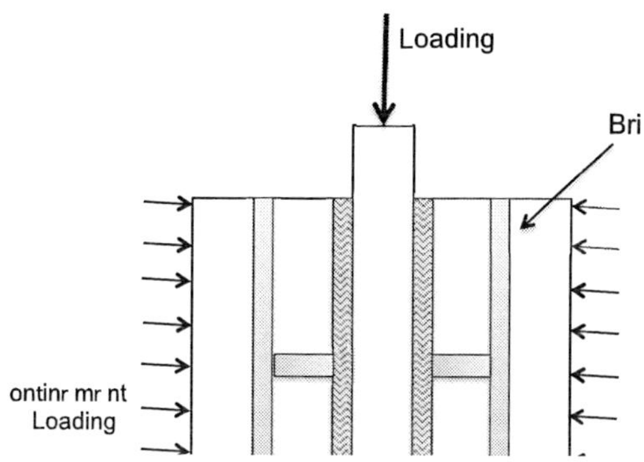

Figure 8. Experimental shear test: Hollow bricks [3]

2.5. Mechanical Behavior of Small Wall in Diagonal Compression

Now we present the mechanical results of the compression diagonal test obtained on the small wall masonry performed by Gabor [3]. This test can produce a stress and strain state where shear and compression are predominant. It is also possible to reproduce the failure by traction diagonal. Such loads are similar to loading developed in masonry construction subjected to earthquake. Figure 11 shows the experimental device used for this study. Test

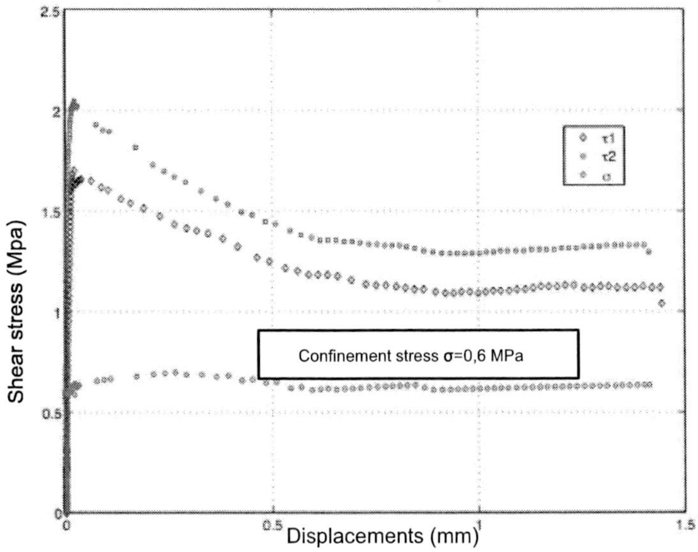

Figure 9. Measured Shear-displacement diagram for test $\sigma = 0,6MPa$ confinement stress

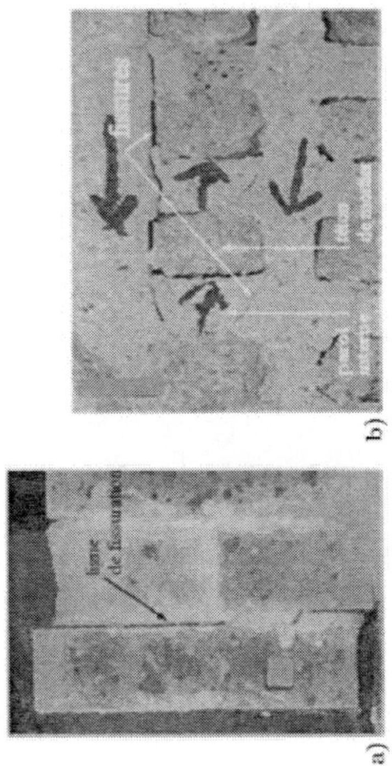

Figure 10. Failure modes - Hollow brick prisms

is monotonous with controlled loading. Wall dimensions are $870 \times 840 \times 100mm^3$; it is made with the same materials presented in prevoius paragraphs as recommended by RILEM [9]. Walls were tested after 28 days after manufacture. Mortar joint thickness is $10mm$. Three walls were created in this work.

To obtain a good load level during a test and produce a global stress - strains state on masonry in the aim to produce a failure on the wall diagonal corners have been set by a metal shoe filled with concrete. Two different values for metal shoe length were considered in mechanical test. The first is equal to $L/10$ the second one is equal to $L/6$. The first wall was tested with the first value in the second and third walls were tested with the second value.

State strain was measured by the compression (resp. tense) diagonal shortening (resp. elongation). Extensometric LVDT captors were fixed on the twice diagonal directions in order to measure the displacement between two points in the same diagonal.

Curves load-strain Figures 12 obtained experimentally can evaluate global response of masonry wall.

In the first test with $L/10$, a failure was produced by the sliding and crushing of the joints in support zone. The maximum load is equal to $160kN$ and failure is located on the three rows. In this case, the embedding conditions of corners wall " metal shoe length " cannot distribute the load in wall satisfactorily.

Cracks did not appear in the central zone of wall after test. Evolution load with respect to the strains Figure 12 identifies a elastic linear behavior of wall except zone support.

In order to avoid these failure modes, the boundary conditions were changed to a new length support equal to $L/10$ was considered for next tests. In this case, failure was observed on central zone wall following the diagonals. Cracks were crossing both bricks and mortar joint, see Figure (13).

Elastic linear behavior was observed in analysis of the curves r_1 and r_2 (Figure 12). The first wall is characterized by an elastic post level which appears at 90% of the limit load that predicts a degradation of mechanical properties. Maximal load for the first wall is equal to $215,3kN$ and $251,8kN$, for the second. Strains and loading measured are summarized in Table 6.

Table 6. Compression diagonal test data $L/6$

Wall and diagonal		Load (kN)		Strains (mm/mm)	
Test	Diagonal	Elastic	Maximal	Elastic	Maximal
r1	Compressive	188,3	215,3	415	520
	tense			98	190
r2	Compressive	251,8	215,8	670	700
	tense			170	220

Results shows a gap of 17% between maximal load obtained on the last two walls tested. This difference can be explained by the internal structure wall. The strength is governed essentially by the brick-mortar interfaces. That is confirmed by the experimental study presented in the first section of this work.

Figure 11. Experimental device for compression diagonal test

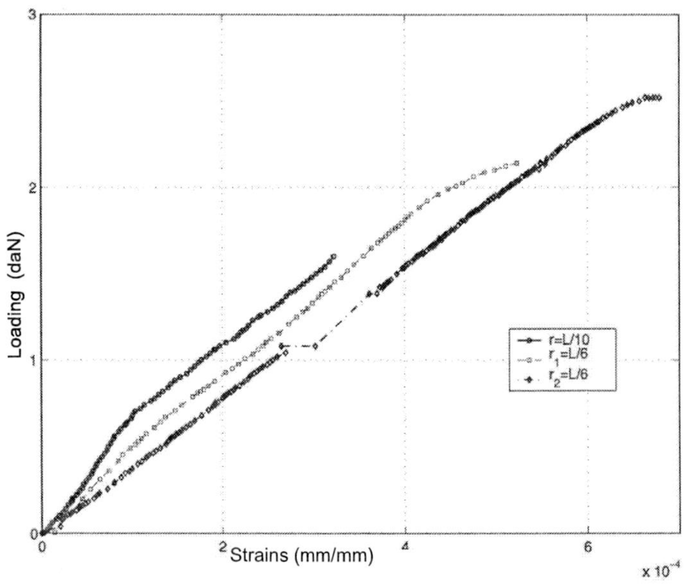

Figure 12. Compression diagonal test - Strain of compressive diagonal

Figure 13. Cracking and failure of the wall

3. Phenomenological Model of Interface

3.1. Introduction

In this section, we are interested in the mechanical behavior of the interfaces for masonry modeling structures when using the materials and interfaces characteristics that were determined experimentally in the previous section [2, 1]. First, RCCM model [10, 11] that is based on the adhesion intensity is presented. Second, this model was adopted to describe the interfaces between mortar and full or hollow bricks in particular; in this case, triplet assemblies subjected to shear loading. The mechanical behavior predicted by this model is compared with the previously obtained in the experimental data.

3.2. Presentation of RCCM Model

RCCM adhesion model was adopted to model the mortar/brick interfaces [10, 11]. This model combines the unilateral contact conditions (non-penetration between the mortar and the brick) with friction and adhesion between two deformable solids. The local constitutive equations required for this model are deduced from thermodynamic considerations and based on a material surface hypothesis for the contact zone. The basic idea underlying this model is to introduce a new state variable describing the contact state. This adhesion intensity variable, denoted β, was initially introduced by [12]. This variable gives the relative proportion of the active links between two bodies in contact. This variable is chosen as:

- $\beta = 1$, total adhesion
- $0 < \beta < 1$, partial adhesion
- $\beta = 0$, no adhesion

We are working in the framework of the contact between two deformable solids. The contact could be defined by a punctual correspondence between two surfaces in contact Γ_c^1 and Γ_c^2 of the domains Ω^1 and Ω^2 of R^d ($d = 2, 3$), respectively. At initial time $t = t_0$, we assume that $\Gamma_c = \Gamma_c^1 = \Gamma_c^2$. The relative displacement between two points located on the two surfaces in contact is denoted by $[u]$ with $[u] = u^1 - u^2$. Let R be the density of the contact forces. We take n^1 and n^2 to denote the external unit normal vectors to the boundaries of the two domains. The decomposition into normal and tangential parts is written:

$$[u] = [u_N]n^1 + [u_T] \quad with \quad [u_N] = [u].n^1 \tag{1}$$

$$R = R_N n^1 + R_T \quad with \quad R_N = R.n^1 \tag{2}$$

The state variables for the thermodynamical description of this model are:

- Strain tensor $e = (e_{ij})$
- Displacement jump $[u]$
- Intensity of adhesion β

b is taken to denote the adhesion viscosity coefficient, w to denote the Dupré energy, μ to denote the friction coefficient and C_N (resp. C_T) to denote the initial normal (resp. tangential) stiffness of the interface. In what follows, $\dot{f}$ denotes the rate of the function f. The constitutive equations of the interface are given by the following equations, as based on state laws and complementarity laws:

Unilateral Contact With Adhesion

$$R_N - C_N[u_N]\beta^2 \geq 0; \quad [u_N] \geq 0; \quad (R_N - C_N[u_N]\beta^2)[u_N] = 0 \tag{3}$$

Friction With Adhesion

$$\begin{array}{lcl} ||R_T - C_T[u_T]\beta^2|| & \leq & \mu|R_N - C_N[u_N]\beta^2| \\ ||R_T - C_T[u_T]\beta^2|| & < & \mu|R_N - C_N[u_N]\beta^2| \quad \Rightarrow \quad [\dot{u_T}] = 0 \\ ||R_T - C_T[u_T]\beta^2|| & = & \mu|R_N - C_N[u_N]\beta^2| \quad \Rightarrow \quad \exists\lambda \geq 0, \\ & & [\dot{u_T}] = \lambda(R_T - C_T[u_T]\beta^2) \end{array} \tag{4}$$

Evolution of The Intensity of Adhesion

$$\begin{array}{rcll} b\dot{\beta} & = & -(w-(C_N[u_T]^2+C_T\|[u_T]\|^2)\beta)^- & if \quad \beta\in[0,1[\\ b\dot{\beta} & \geq & -(w-(C_N[u_T]^2+C_T\|[u_T]\|^2)\beta)^- & if \quad \beta=1 \end{array} \tag{5}$$

Note that if there is no adhesion ($\beta = 0$), this model involves the classical Signorini-Coulomb problem.

A graphic interpretation of the tangential part of this RCCM model given in figure 14 shows the changes with time in the tangential forces depending on the tangential displacement jump. The changes in β lead to irreversible effects. If β decreases, the adhesive forces will decrease and eventually disappear. In the case of pure traction ($[u_N] > 0$), the adhesion resistance ($R_N = C_N[u_N]\beta^2$) is activated (elasticity without damage). β decreases when the displacement becomes sufficiently large for the elastic energy to become larger than the adhesion limit w.

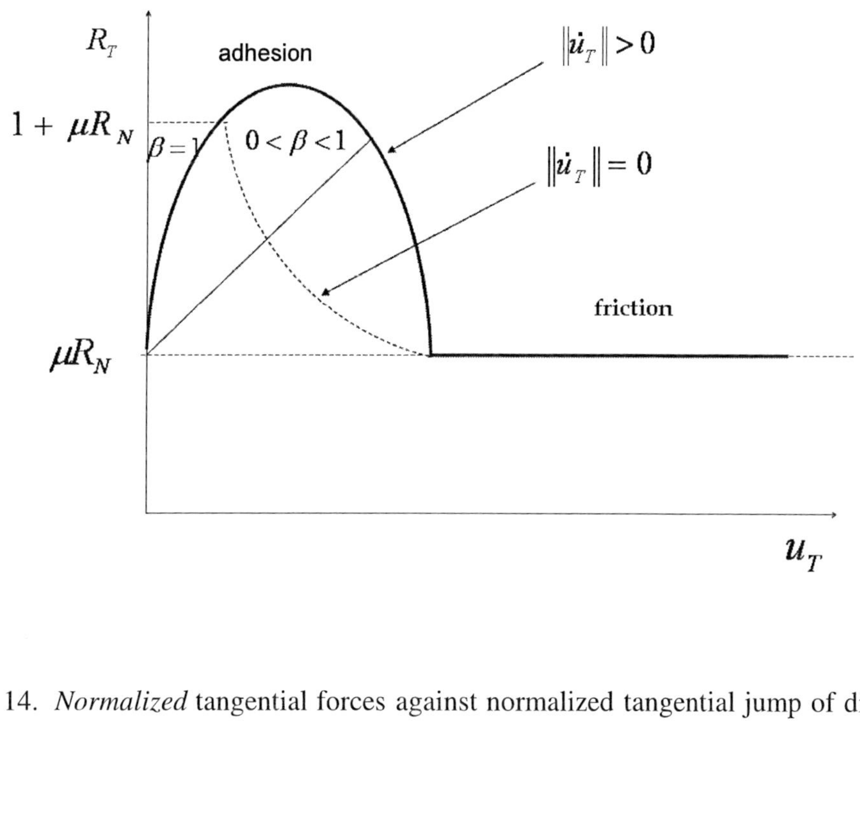

Figure 14. *Normalized* tangential forces against normalized tangential jump of displacement

3.3. Implementation of RCCM Model

The numerical problem was solved using the open computer code LMGC90 (http://www.lmgc.univ-montp2.fr/ dubois/LMGC90/). This code is a numerical platform dedicated to the modeling and simulation of dynamic multibody problems. Problems were approached in the general framework of dynamics. Discretized equations of the problem P are written as:

$$\begin{aligned} M\ddot{q} &= F(q,\dot{q}) + P(t) + r, \\ &+ \text{ interface conditions} \\ &+ \text{ initial conditions and boundary conditions} \end{aligned} \tag{6}$$

with q is a parametrization of the system (degrees of freedom), M is the mass matrix, $F(q,\dot{q})+P(t)$ are the internal and external loading vectors, r is the vector of contact forces.

Problems were usually approached using θ-methods and the Non-Smooth Contact Dynamics (NSCD) method [13, 14, 15]. Due to the contact conditions, a fairly small time-step is chosen and the problem is condensed in the local frame associated with the contact nodes. The local problem is solved using a non-linear Gauss-Seidel method.

The interface is governed by the RCCM law as presented earlier in this chapter. Contact between bodies is defined by contact nodes. Contact nodes are located between two nodes in the mesh of an element in contact at distances of 0.2 and 0.8, respectively, along each segment in contact (figure 15).

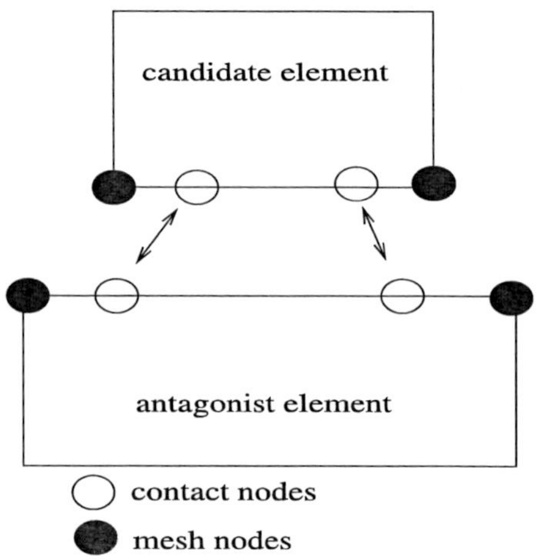

Figure 15. Position of contact nodes between two bodies

In this procedure, the bodies (bricks and mortar) are modeled using $Q4$ quadrangular finite elements. The numerical tests are performed with a constant time step equal to $\Delta t = 0.5 * 10^{-3}\ s$. The computations require 2000 increments in order to reach values resembling the experimental data, and θ (in the time integration method) is fixed and taken to be equal to 0.55.

3.4. Numerical Results: Mechanical Behavior of Small Masonry Structures Under Shear Loading

In this section, numerical simulations carried out using LMGC90 are presented. In particular, this study deals with various structures (triplets and couplets consisting of full or hollow bricks) previously studied [2, 1] and presented in the first section. It is attempted to model the same conditions as those pertaining in the experimental tests, in terms of the geometries, mechanical characteristics, boundary conditions, and loading conditions. The problems are treated under quasi-static conditions, adopting the plane strain hypothesis. For the sake of simplicity, the parameters used for the computations are: $\mu = 0.2$, $b \approx 0$, $w = 0.9\ J/mm^2$. Coefficients C_N and C_T, the normal and tangential stiffness of the interface, will be determined in the following sections. These parameters will be obtained by comparing the experimental and numerical data.

The mechanical characteristics of the materials are as follows Table 7:

Table 7. Characteristics of Mortar and Bricks

Materials	Young's modulus MPa	Poisson's ratio
Mortar	4 000	0.3
Full bricks	9 439	0.13
Hollow bricks	6 058	0.13

3.4.1. Full Brick Triplets

In [2], we noted the occurrence of two kinds of fracture processes:

1. The fracture occurs along the interface (tests N1 and N4)
2. The fracture begins along the interface and propagates into the mortar (tests N2 and N3).

Fracture along the interface The experimental data used here was published in [2]. The mesh consists of 130 $Q4$ finite elements. For the sake of symmetry, only the half-structure is considered. The loading on the upper part ranges from 0 to 53 kN. The first step here consists in determining the stiffness values of the interface, C_N and C_T. Various results are presented in figure 16 on pairs of C_N and C_T. The global behavior can be seen to depend strongly on C_N and C_T. The rigidity of the assembly depends mainly on the interface stiffness. Since the problem is highly nonlinear, even small perturbations in the stiffness coefficients can greatly affect the numerical responses.

The results shown in figure 17 were obtained in tests numbers 1 and 4, and are most encouraging, since the general shape of the curves is similar to that of the experimental ones. This confirms that the model accounts accurately for the mechanical behavior of the interface. This is also confirmed by the rupture modes obtained numerically (see figure 18). The variations in the shear stresses σ_{xy} show that the stress concentration develops in the regions containing the discontinuities, or more specifically, at the level of the interface.

Cracks developing into the mortar We now take the case where the interface and the mortar are fractured together. The results of the tests performed in this case are not as satisfactory as the previous ones. In this experimental part of this study, we note that the cracks started to develop at the brick-mortar interfaces but continued into the mortar itself (tests 2 and 3).

It is therefore proposed to introduce the damage to the mortar, using a cohesive zone model. The mortar is split into sub-bodies. Each sub-body is meshed using a quadrangular

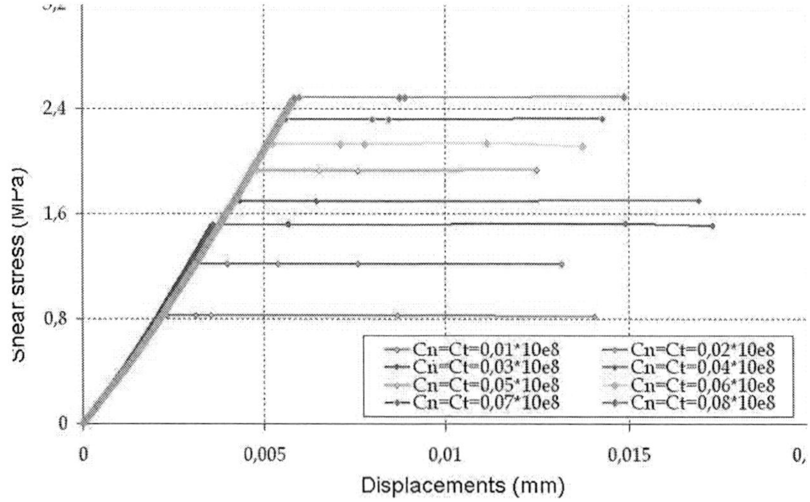

Figure 16. Influence of parameters C_N and C_T on the behavior of full brick triplets

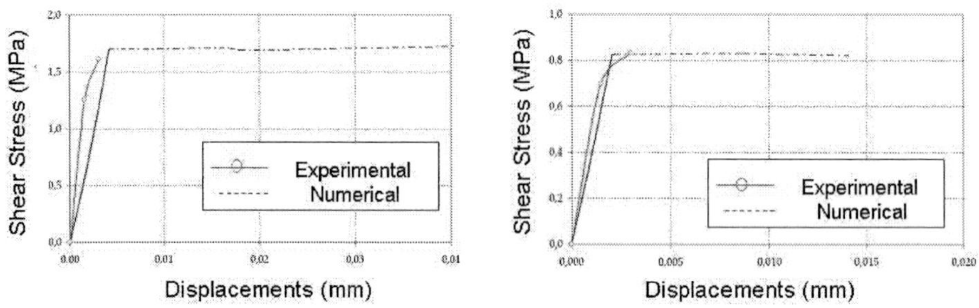

Figure 17. Shear tests: tests N1 ($C_N = C_T = 4\ 10^6 N/mm^3$) and N4 ($C_N = C_T = 10^6 N/mm^3$)

mesh with $Q4$ finite elements (figure 19). The contact between sub-bodies is governed by the RCCM model. We take $adh1$ and $adh2$ to denote the law of adherence at the mortar/mortar and the brick/mortar interface, respectively (figure 19).

It can be seen from figure 20 that the behavior of the structure is sensitive to changes in the stiffness coefficients.

The numerical results are in good agreement with the experimental data from the qualitative point of view. They show the degradation of the rigidity induced by the propagation of the crack into the mortar in the two experimental tests 2 and 3 (figure 21). The model predicted the similar occurrence of cracks in the mortar joint to those observed in practice (figure 22).

3.4.2. Hollow Brick structures

In the case of hollow brick structures, the experimental data show that the behavior of the material at the interface is quite different between the hollow parts of the brick and the full

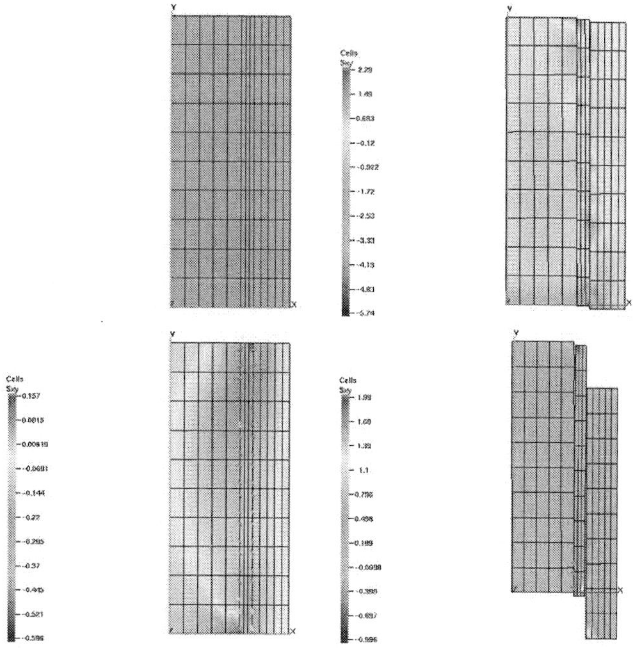

Figure 18. Shear stress in full brick triplets at steps 0, 100, 207 and 208

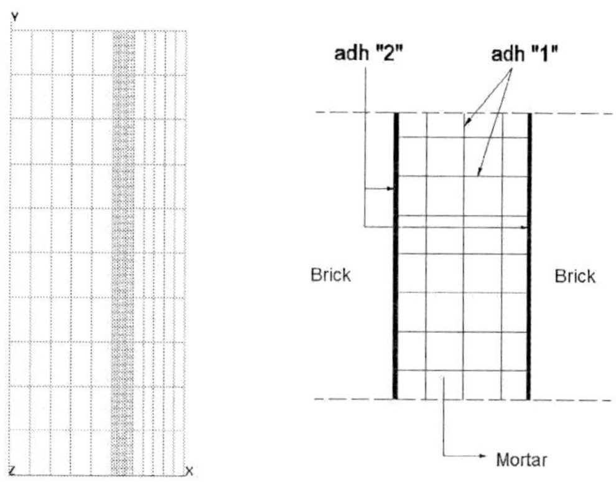

Figure 19. Full brick triplets: a) Mesh with cohesive mortar b) Mortar/mortar and mortar/brick adhesion

parts. At the level of the hollow parts, spikes of mortar penetrate into the voids of the brick. It is therefore necessary to distinguish between two zones: the zone (adh1) corresponding to the mortar/brick interface as in the case of full brick structures; and the second zone (adh2) corresponding to the spikes. Figure 23 shows these two zones.

By suitably choosing the stiffness coefficients, this strategy makes it possible to ap-

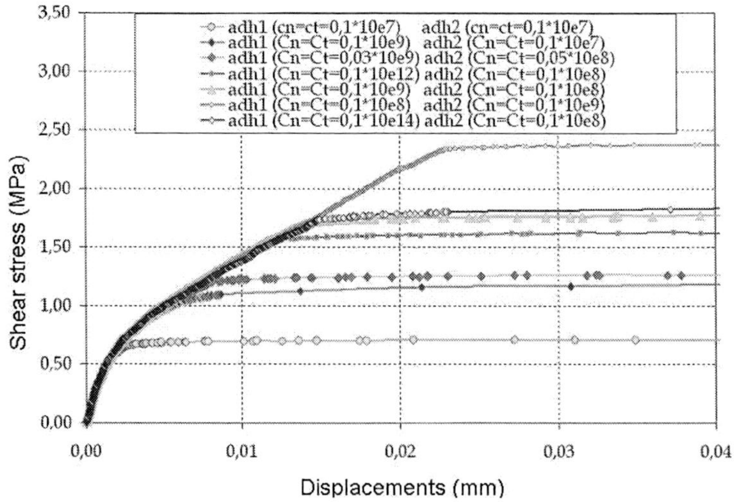

Figure 20. Influence of the stiffness parameters on the behavior of the structure (triplets with a cohesive zone)

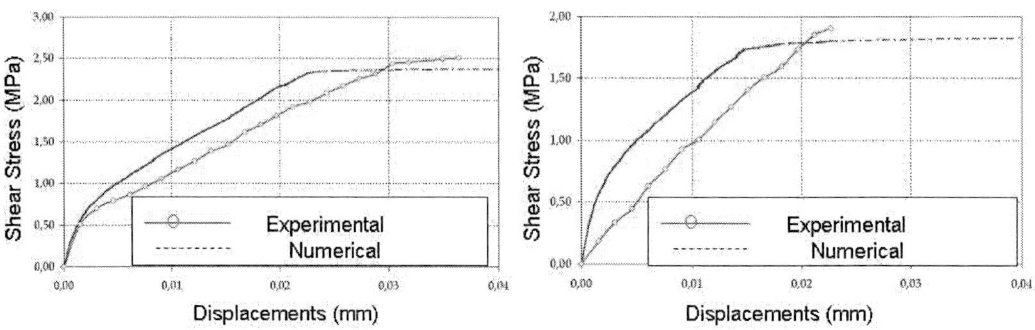

Figure 21. Results of numerical shear tests: tests N2 (adh1: $C_N = C_T = 10^7 N/mm^3$, adh2: $C_N = C_T = 10^8 N/mm^3$) and N3 (adh1: $C_N = C_T = 10^9 N/mm^3$, adh2: $C_N = C_T = 10^8 N/mm^3$)

proach the experimental results, as can be seen from figure 24. However, it is necessary to find several coefficients, which reduces the efficiency of the method. This situation could possibly be improved by averaging the two zones, but this approach has not yet been attempted. The changes with time in the shear stress and in the deformed structure are shown in figure 25.

3.5. Partial Conclusion

The adhesion model presented here was successfully used to simulate the experimental tests described in [1, 2], which provided the coefficients required to model the interface: stiffness

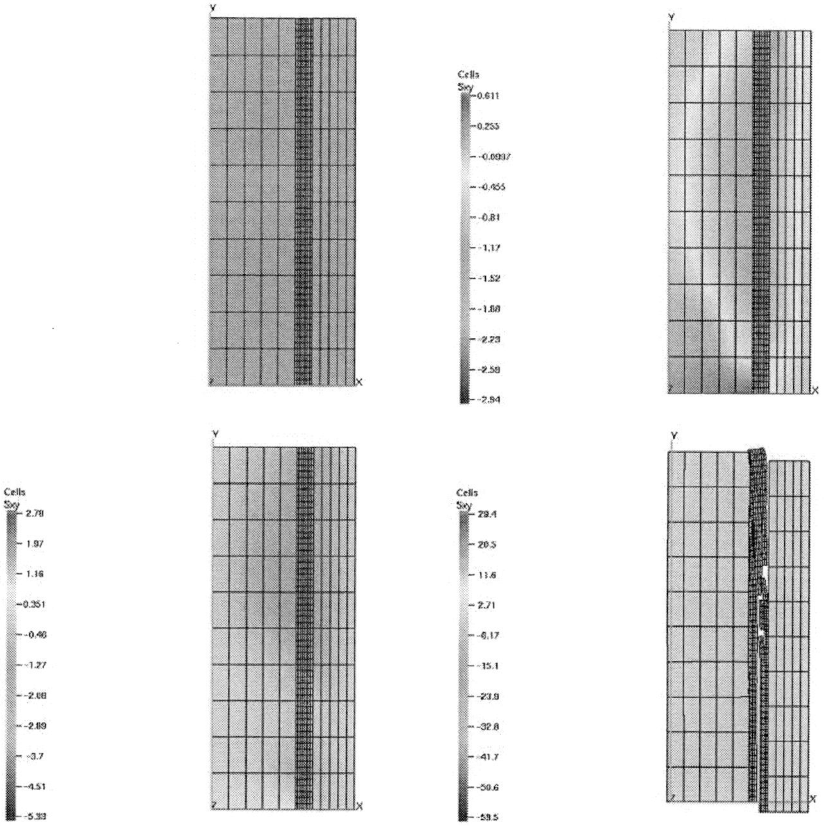

Figure 22. Shear stress in full brick triplets at steps 1, 350, 360 and 371 (cohesive zone)

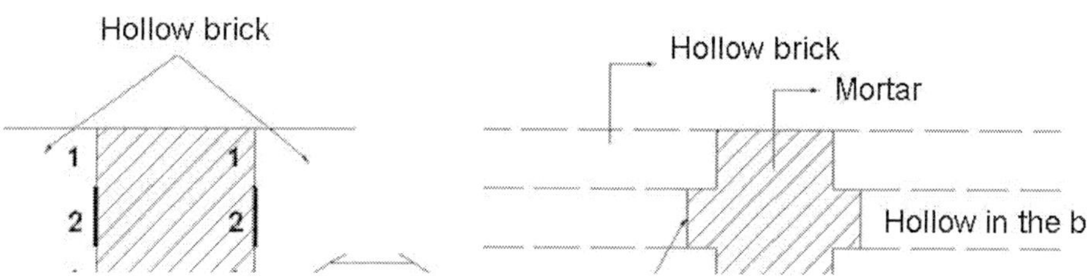

Figure 23. Mortar spike modeling

parameters, viscosity coefficient, Dupré's energy and friction coefficient. The model is sensitive to these characteristics but the results obtained are in line with the experimental data.

The two kinds of rupture modes observed experimentally were described using two different adhesive characteristics. It was thus possible to model the fracture process that occur along the interface as well as where it involves mortar.

Structures composed of hollow bricks can also be modeled in this way using two ad-

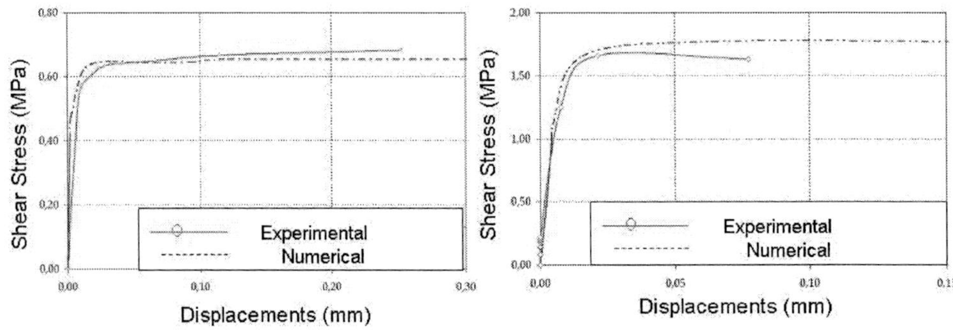

Figure 24. Results of numerical shear tests (hollow bricks): tests N2 (adh1: $C_N = C_T = 2\ 10^6 N/mm^3$, adh2: $C_N = C_T = 4\ 10^7 N/mm^3$) and N3 (adh1: $C_N = C_T = 2.5\ 10^7 N/mm^3$, adh2: $C_N = C_T = 1.5\ 10^8 N/mm^3$)

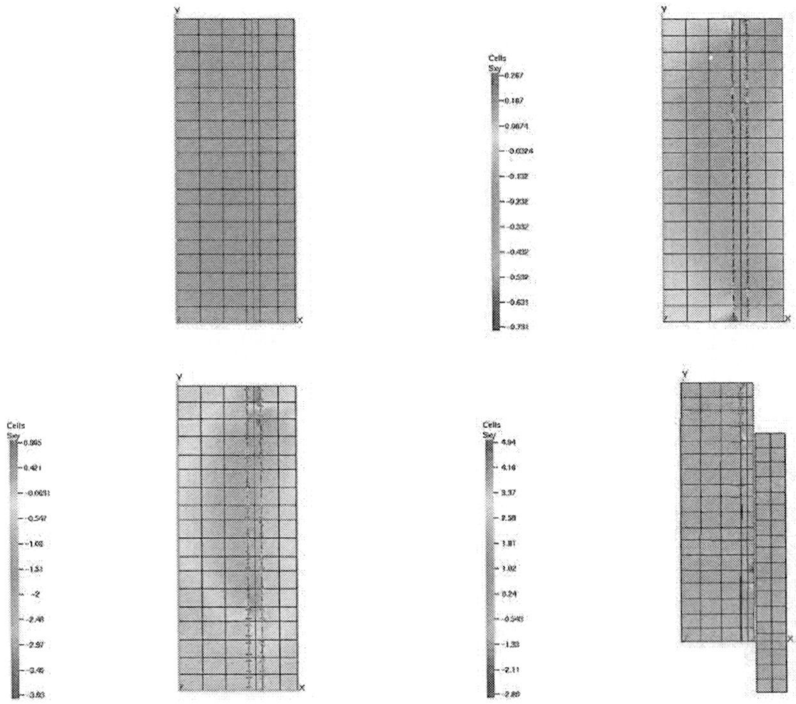

Figure 25. Shear stress in hollow brick triplets atsteps 1, 53, 191, and 196

hesive characteristics. In particular, the model was used here to study the fracture process when it occurs along the interface of a small structure that consists of three hollow bricks. Further studies are now required on the fracture processes crossing the mortar in structures of this kind.

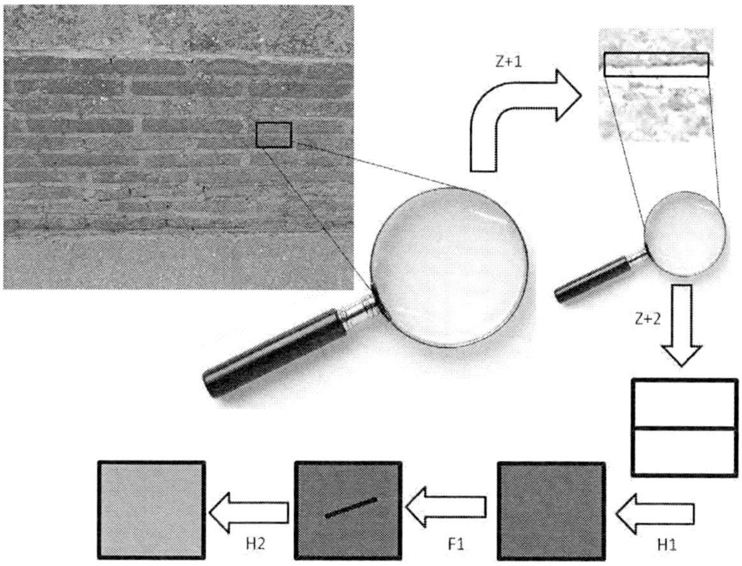

Figure 26. Description of the process 1

4. A Multi-Scale Model for Interface Law

4.1. Introduction

In this part of the paper, one presents a multi-scale model of interface [16, 17]. The method is based on homogenization theories and asymptotic analysis. In this context, we assume the existence of a third material: the brick/mortar interface, which is considered as a mixture of brick and mortar with a crack density. To obtain the effective properties of the damaged intermediate material, three steps are performed. First, we calculate the exact effective properties of the crack-free material using homogenization techniques for laminate composites, and thus define a first homogeneous equivalent medium. In the second step, we assign a crack density to the material. To model the macroscopic behavior of the cracked material, we use the Kachanov [18] model and then define a new homogeneous equivalent medium. Finally, in order to be sandwiched between the brick and mortar, this material is given a small thickness, and its mechanical behavior is derived using asympt! otic techniques to shift from the micro to the macro level. A variation law of the crack length is introduced. The numerical procedure used is then described. Lastly, some numerical examples are given and compared with the experimental data.

4.2. Principle of the Model

The procedure used in this paper is summarized in figures 26 and 27.

1. In the first step, we consider the masonry made by bricks and mortar at a local level (Z+1). The brick and mortar are considered as elastic, homogeneous and isotropic materials. The material characteristics are given in former sections.

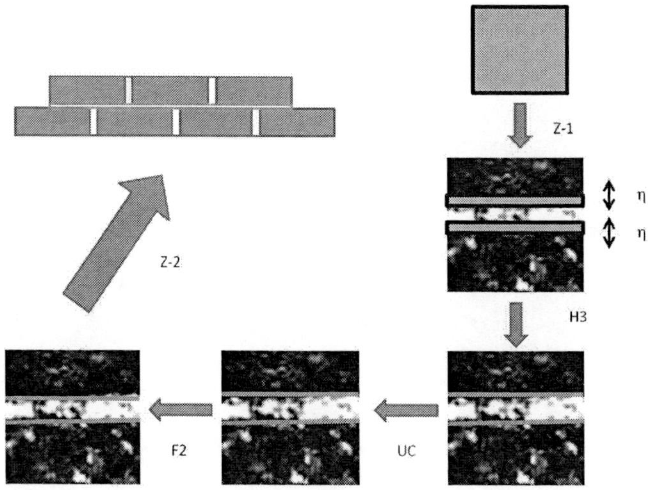

Figure 27. Description of the process 2

2. At its local level, an inter-phase is considered a mixture of brick and mortar (Z+2). In order to simplify, we consider the composite inter-phase as stratified.

3. The stratified is homogenized using classical theory of homogenization (H1). A new material is obtained which is transversely isotropic [18]. The new material characteristics are obtained directly from the materials characteristics of brick and mortar.

   The constitutive equation for each material is given by:

$$\varepsilon_{ij}^{c} = S_{ijkl}^{c}\sigma_{lk}^{c} = \frac{1+\nu^{c}}{E^{c}}\sigma_{ij}^{c} - \frac{\nu^{c}}{E^{c}}\sigma_{kk}^{c}\delta_{ij} \tag{7}$$

   where S^c, E^c, and ν^c are respectively the compliance tensor, the Young's modulus and the Poisson ratio of phase c ($c = b$ for the brick, $c = m$ for the mortar). ε^c and σ^c are the strain and the stress tensor, respectively.

   The macroscopic behavior law of the laminate brick/mortar reads:

$$\varepsilon = S^0 : \sigma \tag{8}$$

   where $\varepsilon = \sum_{c=b,\,m} f^c\varepsilon^c = \sum_{c=b,\,m} f^c S^c : \sigma^c$, f^c denotes the volume fractions of phase c, and S^0 is the effective fourth-order compliance tensor of the transversely isotropic stratified. According to the modified Voigt notations, the macroscopic law reads:

$$\begin{pmatrix} \varepsilon_{11} \\ \varepsilon_{22} \\ \varepsilon_{33} \\ \sqrt{2}\varepsilon_{23} \\ \sqrt{2}\varepsilon_{13} \\ \sqrt{2}\varepsilon_{12} \end{pmatrix} =$$

$$\begin{pmatrix} S^0_{1111} & S^0_{1122} & S^0_{1133} & 0 & 0 & 0 \\ S^0_{1122} & S^0_{1111} & S^0_{1133} & 0 & 0 & 0 \\ S^0_{1133} & S^0_{1133} & S^0_{3333} & 0 & 0 & 0 \\ 0 & 0 & 0 & 2S^0_{1313} & 0 & 0 \\ 0 & 0 & 0 & 0 & 2S^0_{1313} & 0 \\ 0 & 0 & 0 & 0 & 0 & S^0_{1111} - S^0_{1122} \end{pmatrix} \begin{pmatrix} \sigma_{11} \\ \sigma_{22} \\ \sigma_{33} \\ \sqrt{2}\sigma_{23} \\ \sqrt{2}\sigma_{13} \\ \sqrt{2}\sigma_{12} \end{pmatrix} \quad (9)$$

4. We introduce a crack which is representative of the cracks included in the new material (F1). The crack is represented by its average (half-)length l and its average direction ϕ.

5. The cracked material is homogenized using the Kachanov theory [18] (H2). In this theory, the crack is approximated by an elliptic crack. Approximated material characteristics are obtained depending on the length and direction of the crack, and the mechanical characteristics of the brick and mortar. The average strain ε in a solid with one crack can be written in the form:

$$\varepsilon = S : \sigma = S^0 : \sigma + \frac{l}{A}(n \otimes < b > + < b > \otimes n) = (S^0 + \Delta S) : \sigma \quad (10)$$

where S is the effective compliance tensor of the cracked material, $< b >= n.\sigma.B$ is the average displacement discontinuity (COD) vector of the crack, B is the COD tensor of the crack, n is the normal to the crack and A is the area of the representative 2D-domain.

In the plane (e_1, e_3) of the elementary volume, the tensor B is given by:

$$B = (C(1 - D)e_1 \otimes e_1 + C(1 + D)e_3 \otimes e_3)l \quad (11)$$

where

$$\begin{cases} C = \dfrac{\pi}{4} \dfrac{\sqrt{E_1^0} + \sqrt{E_3^0}}{\sqrt{E_1^0 E_3^0}} \left(\dfrac{1}{G_{13}^0} - 2\dfrac{\nu_{13}^0}{E_1^0} + \dfrac{2}{\sqrt{E_1^0 E_3^0}} \right)^{\frac{1}{2}} \\ D = \dfrac{\sqrt{E_1^0} - \sqrt{E_3^0}}{\sqrt{E_1^0} + \sqrt{E_3^0}} \end{cases} \quad (12)$$

and E_1^0, E_3^0, ν_{13}^0 and G_{13}^0 are given by:

$$\begin{cases} E_1^0 = \dfrac{1}{S_{1111}^0} \\ E_3^0 = \dfrac{1}{S_{3333}^0} \\ \nu_{13}^0 = -\dfrac{S_{1133}^0}{S_{1111}^0} \\ G_{13}^0 = \dfrac{1}{S_{1313}^0} \end{cases} \tag{13}$$

The tensor ΔS is given by:

$$\Delta S = \frac{2l}{A} e_3 \otimes B \otimes e_3 = \frac{2l^2}{A}(e_3 \otimes (C(1-D)e_1 \otimes e_1 + C(1+D)e_3 \otimes e_3) \otimes e_3 \tag{14}$$

The compliance of the cracked material is thus given by:

$$\begin{aligned} S_{1111} &= S_{1111}^0 \\ S_{3333} &= S_{3333}^0 + 2\rho C(1+D) \\ S_{1133} &= S_{1133}^0 \\ S_{1313} &= S_{1313}^0 + 2\rho C(1-D) \end{aligned} \tag{15}$$

where $\rho = l^2/A$ is the crack density parameter. The elastic constants are given by:

$$\begin{aligned} E_1 &= E_1^0 \\ E_3 &= E_3^0/(1 + 2\rho B_{nn} E_3^0) \\ G_{13} &= G_{13}^0/(1 + \rho B_{tt} G_{13}^0) \\ \nu_{13} &= \nu_{13}^0 \end{aligned} \tag{16}$$

where

$$\begin{aligned} B_{tt} &= C(1 - Dcos(2\phi))l \\ B_{nn} &= C(1 + Dcos(2\phi))l \\ B_{tn} &= CDsin(2\phi)l \end{aligned} \tag{17}$$

6. The cracked material obtained is considered as the inter-phase material (Z-1).

7. The interface law is obtained, considering that the inter-phase is thin (H3). Using asymptotic theory (the thickness of the thin inter-phase η is a small parameter), we obtain an interface law which links the stress vector to the jump of displacement along the interface toward which geometrically inter-phase tends [19]. The stiffness of the interface depends on the material characteristics of the brick and mortar, and on the length and the direction of the crack.

   We take $[u]$ to denote the jump along the interface. If the interface is normal to the third direction, we obtain

$$\sigma_{i3} = A_{i3}[u_i]$$

The stress vector $\sigma_{.3}$ is linearly linked to the jump of displacement $[u]$. The second order matrix $A_{..}$ is diagonal. The terms C_N and C_T in this matrix, corresponding to the normal and tangential stiffnesses of the interface are given by:

$$\begin{cases} C_N = \bar{C}_{3333}(\eta \rightarrow 0) \quad \text{where} \quad \bar{C}_{3333} = \dfrac{C_{3333}}{\eta} \\ C_T = \bar{C}_{1313}(\eta \rightarrow 0) \quad \text{where} \quad \bar{C}_{1313} = \dfrac{C_{1313}}{\eta} \end{cases} \tag{18}$$

where the fourth-order stiffness tensor C is the inverse of the compliance tensor S.

In our case $\rho = l^2/\eta\ L_0$, where L_0 denotes the joint length, it can be established that the normal and tangential joint stiffnesses read:

$$\begin{aligned} C_N &= \frac{L_0}{2C(1+D)\,l^3} \\ C_T &= \frac{L_0}{4C(1-D)\,l^3} \end{aligned} \tag{19}$$

We obtain an explicit relation between the stiffness coefficients and the crack length l.

8. Unilateral conditions [20] are introduced (UC). These conditions prevent the penetration between the two adherents (brick and mortar, in our case). In a first approach friction conditions are neglected.

   Let F be the density of the contact forces. We take n^b to denote the external unit normal vector to the boundary of the brick. The displacement and the contact force are decomposed into normal and tangential parts:

$$[u] = u_N n^b + u_T$$

   with $u_N = [u].n^b$.

$$F = F_N n^b + F_T$$

   with $F_N = F.n^b$.

   The unilateral contact is given by the following relations

$$u_N \geq 0,\ F_N - C_N u_N \geq 0,\ u_N(F_N - C_N u_N) = 0. \tag{20}$$

9. An evolution law for the crack length is introduced (F2). In the first step a very simple law is chosen (identified experimentally in practice).

   We take τ to denote the shear stress along the interface.

Table 8. Mechanical properties of the three-fold masonry constituents

Young's moduli (MPa) of full brick	9438
Poisson ratio of full brick	0.13
Young's moduli (MPa) of hollow brick	6059
Poisson ratio of hollow brick	0.13
Young's moduli (MPa) of mortar	4000
Poisson ratio of mortar	0.3

$$\begin{array}{l} \text{if } \tau \leq \tau_c \text{ then } l = l_c \\ \text{if } \tau_c \leq \tau \leq \tau_u \text{ then } l = a\tau + b \end{array} \tag{21}$$

$$\begin{array}{l} a = \dfrac{l_u - l_c}{\tau_u - \tau_c} \\ b = l_u - a\tau_u \end{array} \tag{22}$$

where τ_u, τ_c, l_c and l_u are four given values (In practice, l_c and l_u are identified).

10. The interface law can be used for a structural analysis by finite element (Z-2). In this case, the masonry is made by bricks, mortar, and interface.

4.3. A Numerical Example

We consider two-dimensional masonry structures subjected to shear loading. The obtained numerical results are then compared with experimental data [2]. Two cases are considered: a triplet of full bricks and a triplet of hollow bricks, both subjected to the same nominal loading (with no lateral confinement). Table (8) lists the material properties of the mortar and full bricks used to simulate a triplet of full bricks under the shear loading conditions described (see Fig. 28).

For full bricks, the compliance tensor S^0 therefore reads:

$$10^{-4} \begin{pmatrix} 1.478 & -0.271 & -0.348 & 0 & 0 & 0 \\ -0.271 & 1.478 & -0.348 & 0 & 0 & 0 \\ -0.348 & -0.348 & 1.639 & 0 & 0 & 0 \\ 0 & 0 & 0 & 4.444 & 0 & 0 \\ 0 & 0 & 0 & 0 & 4.444 & 0 \\ 0 & 0 & 0 & 0 & 0 & 3.499 \end{pmatrix} \tag{23}$$

The normal and tangential stiffnesses (N/mm^2) of the interface are given by the following expressions:

$$\begin{array}{rcl} C_N & = & 242365/l^3 \\ C_T & = & 127600/l^3 \end{array} \tag{24}$$

where the direction ϕ is assumed to be equal a zero.

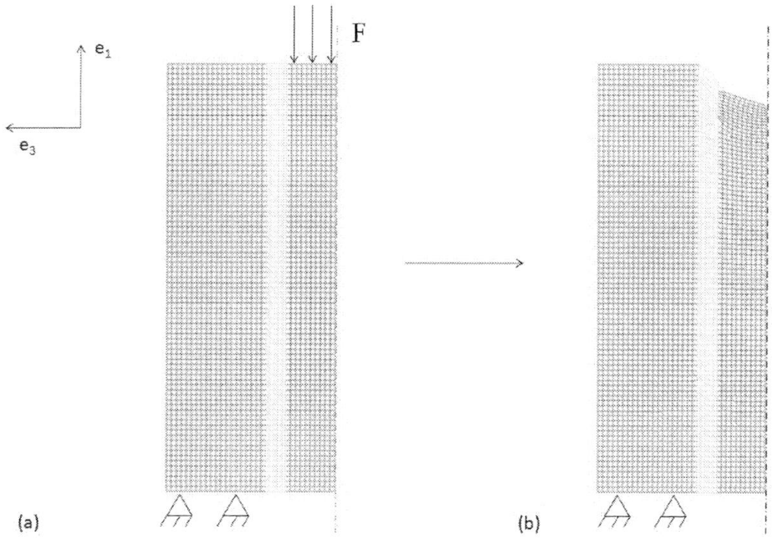

Figure 28. Initial geometrical configuration and loading conditions imposed on the triplet of full bricks (a), deformation of the triplet of full bricks in a shear test (b)

For hollow bricks, the compliance tensor S^0 is written as follows:

$$10^{-4}\begin{pmatrix} 1.973 & -0.396 & -0.456 & 0 & 0 & 0 \\ -0.396 & 1.973 & -0.456 & 0 & 0 & 0 \\ -0.456 & -0.456 & 1.985 & 0 & 0 & 0 \\ 0 & 0 & 0 & 5.115 & 0 & 0 \\ 0 & 0 & 0 & 0 & 5.115 & 0 \\ 0 & 0 & 0 & 0 & 0 & 4.740 \end{pmatrix} \tag{25}$$

The normal and tangential stiffnesses of the interface sandwiched between mortar and hollow brick are given by the following expressions:

$$\begin{aligned} C_N &= 200396/l^3 \\ C_T &= 100490/l^3 \end{aligned} \tag{26}$$

The first step in order to compare experimental data and numerical results is to identify the parameters of the model, in particular the crack length and its evolution [16]. For the fulls bricks, a comparison between experimental data and numerical results is given in figure 29. For the hollow bricks, a comparison is given in figure 30.

4.4. Partial conclusion

The numerical results obtained with the present model match the experimental data and are in good agreement particularly with the test data obtained in previous sections. Thus, the

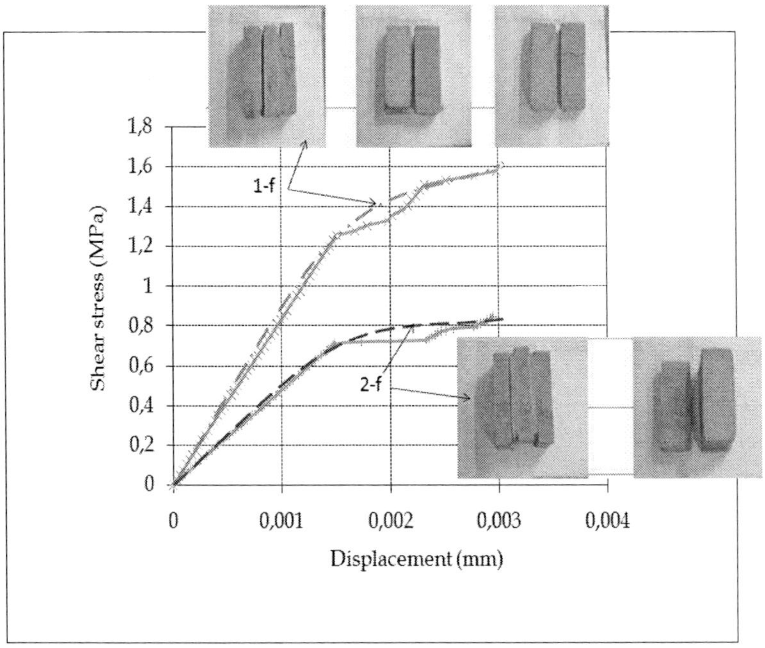

Figure 29. Stress-displacement diagrams of the triplet of full bricks under shear, numerical, and experimental results [16, 17].

present model, with a simple crack evolution law, seems a good way to provide estimations for the stiffness of masonry interfaces.

5. Conclusion

In this chapter, we have presented some experimental data issued from the literature, with a particular focus on small structures and the local behavior of brick-mortar interfaces. We have shown that three aim modes appear in the stress-strain relationship. In the first part, there is a linear (elastic) relationship between the stress and the displacement. After a maximum, in a second step, the stress decreases, due to local damage, to a (quasi) constant value (third part) corresponding to friction.

We have presented two families of models able to predict the behavior of the interface. The first one, phenomenological, is based on the notion of the evolution of the adhesion variable. The second one is based on micromechanics analysis and crack evolution. For these two models, some numerical results are presented and compared with experimental data. The results show the efficiency of these two models.

In the future, various paths are possible. The first one concerns the dynamical aspects that is necessary to take into account, in order to approach earthquakes which generate severe in-plane and out-of-plane forces. It is therefore necessary to generalize our methodology to three-dimensionnal problems. A second possible way concerns the development of efficient

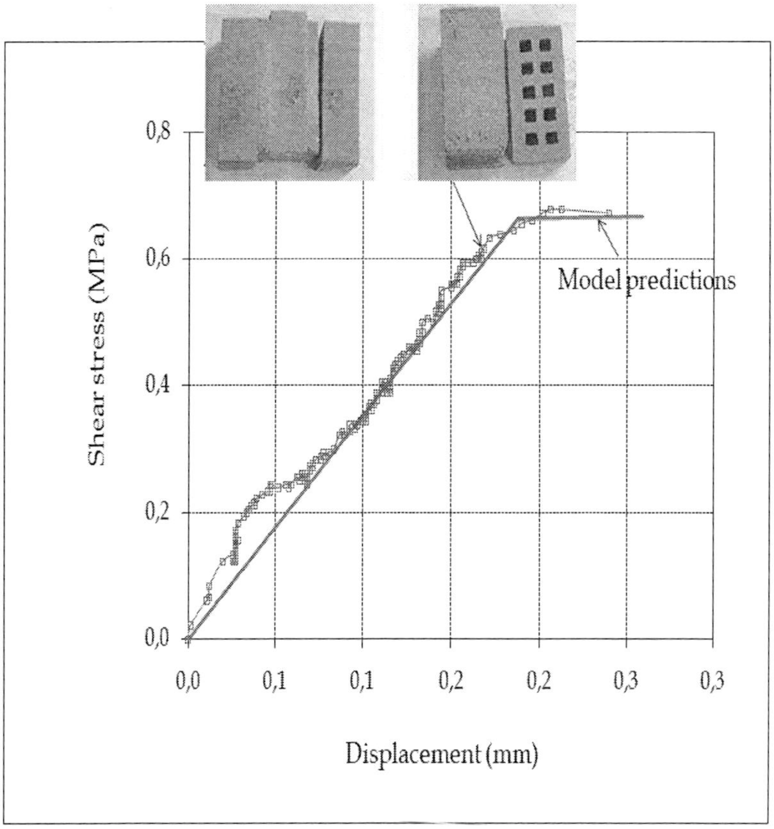

Figure 30. Stress-displacement diagrams of the triplet of full bricks under shear, numerical and experimental results [16, 17].

numerical schema which is necessary to approach large structures such as monuments. The last line of research could involve the improvement of the models presented in this work to introduce new physical considerations as plasticity, damage evolution, etc.

References

[1] Fouchal, F., Contribution à la modélisation numérique des interfaces dans les structures maçonnées, *PhD Thesis, Reims University*, 2006 (in french).

[2] Fouchal, F., Lebon, F., Titeux, I., Contribution to the modeling of interfaces in masonry construction, *Construction and Building Materials*, **23**, 2428–2441 2009.

[3] Gabor, A., Contribution à la caractérisation et à la modélisation des maçonnerie non renforcées et renforcées par matériaux composites, *PhD Thesis, Université Claude Bernard Lyon 1*, 2002).

[4] Gabor, A., Ferrier, E., Jacquelin, E., Hamelin, P.,, Analysis and modeling of the in-plane shear behavior of hollow brick masonry panels, *Construction and Building Materials* , **20**, 308-32, 2006.

[5] NF EN 196-1, Norme Européenne , Norme Française, *Méthodes d'essais des ciments - Partie 1: détermination des résistances mécaniques*, 2006.

[6] NF EN 772-1, Norme Européenne, Norme Française, *Méthodes d'essai des éléments de maçonnerie - Partie 1: détermination de la résistance à la compression*, 2001.

[7] NF EN 1052-3, Norme Européenne, Norme Française, *Méthodes d'essai de la maçonnerie - Partie 3: détermination de la résistance initiale au cisaillement*, 2003.

[8] RILEM (1994b), LUMB5- Short - term shear test for interface between the masonry unit and mortar or moisture - Insulating interlayer. *Technical raport, RILEM.*

[9] RILEM (1994c), LUMB6- Diagonal tensile strength tests of small wall specimens. *Technical raport, RILEM.*

[10] Raous, M., Cangémi, L., Cocou, M., , A consistent model coupling adhesion, friction and unilateral contact, *Computational Methods in Applied Mechanics and Engineering* **177**, 383-399 1999.

[11] Monerie, Y. and Raous, M.,, A model coupling adhesion to friction for the intercation between a crack and a fiber/matrix interface, *Z.A.M.M.***80**, 205-209 2000.

[12] Frémond, M., Adhérence des solides, *Journal de Mécanique Théorique et Appliquée,* **6**, 383-407 1987.

[13] Moreau, J.J., On unilateral constraints, friction and plasticity. In G. Capriz and G. Stampacchia, eds., in New Variational Techniques in Mathematical Physics, *CIME II ciclo, Edizioni Cremonese*, 175-322, 1973.

[14] Moreau, J.J., Unilateral contact and dry friction in finite freedom dynamics. In J.J. Moreau and P.D. Panagiotopoulos, eds., in Nonsmooth Mechanics and Applications, *CISM Courses and Lectures, vol.302, Springer-Verlag*,1988, 1-82.

[15] Jean, M., Frictional contact in rigid or deformable bodies: Numerical simulation of geomaterials, *Elsevier Science Publisher, Amsterdam,*1995, 463-486.

[16] Rekik, A., Lebon, F., Identification of the representative crack length evolution in a multi-level interface model for quasi-brittle masonry, *International Journal of Solids and Structures,* **47**, 3011–3021, 2010.

[17] Rekik, A., Lebon, F., Homogenization methods for interface modeling in damaged masonry, *Advances in Engineering Software, in press*, 2011.

[18] Kachanov M., Effective elastic properties of cracked solids: Critical review of some basic concepts, *Applied Mechanical Review,* **45**, 304–335, 1992.

[19] Lebon, F., Rizzoni, R., Asymptotic analysis of a thin interface: The case involving similar rigidity, *International Journal of Engineering Science,* **48**, 473–486, 2010).

[20] Lebon, F., Contact problems with friction: Models and simulations, *Simulation Modelling Practice and Theory,* **11**, 449–463, 2003.

In: Brick and Mortar Research
Editors: S. Manuel Rivera and A. L. Pena Diaz
ISBN: 978-1-61942-927-7

Chapter 11

ULTRASONIC CHARACTERIZATION OF MORTAR USING MICROMECHANICAL AND MULTIPLE SCATTERING MODELS

M. Molero[1,*] *M. Acebes*[2], *M. A. G. Izquierdo*[3], *M. G. Hernández*[1] *and J. J. Anaya*[1]
[1] CAEND, CSIC-UPM, 28500, Arganda del Rey, Madrid, Spain
[2] Mechanical Engineering Division, Cartif Technology Centre, Valladolid, Spain
[3] Dept. of SSR, Technical University of Madrid (UPM), Madrid, Spain

Abstract

It is well-known that mortar is one of the most basic elements of masonry construction. This fact has motivated to improve the understanding of relations between the strength, quality and durability of mortars with their mix proportions and the different mechanical properties of their material constituents. Several research works have been conducted to investigate the mortar microstructure by using both destructive and non-destructive testing, highlighting the ultrasonic non destructive testing. These ultrasonic techniques have proven to be effective in evaluating the mechanical and physical properties of a large number of strongly heterogeneous materials. Different modeling approaches have been used to establish relations among the microstructural properties of mortars and ultrasonic wave propagation. Micromechanical models and multiple scattering theories are addressed in this work. These homogenization techniques, which provide different information from materials, are used to describe the influence of microstructural properties on the ultrasonic parameters. By using micromechanical models, the influence of geometry, volume fraction and elastic properties of constituent phases, namely matrix, aggregates or pores on the ultrasonic velocity is evaluated. On the other hand, the multiple scattering models provide the influence of the size and volume fraction of aggregates on mortar, as well as the interaction, contribution, and influence of entrapped air voids together with the aggregates on frequency-dependent parameters such as the phase velocity and the attenuation coefficient. Both theoretical and experimental investigations are performed throughout the present work.

Keywords: Micromechanical model, Multiple scattering model, Ultrasound, Mortars

---

*E-mail address: miguel.molero@csic.es

1. Introduction

Mortar is one of the most basic elements of masonry construction, made of mixture of fine aggregates, cement, and water. While it is well-known that strength, quality and durability of cement mortars depend on the mix proportioning and properties of their constituents, better performance of these materials can be achieved with an in-deep understanding about their microstructure. This fact has motivated to investigate the microstructural characterization of mortar by using both destructive (DT) and non destructive testing (NDT), but with more attention on NDT techniques. In this sense, ultrasonic NDT techniques have proven to be effective in evaluating the mechanical and physical properties of these types of materials. In addition, appropriate interpretations of the acquired ultrasonic data with the aids of suitable models, have led to progress in assessments of both quality and durability. Therefore, it is of great importance to have more precise models to relate microstructural parameters of cementitious materials with ultrasonic data. These models should take into account the influence of elastic properties, geometry, size and volume fraction of aggregates and air voids, as well as porosity structure of cement mortars. Therefore, models for the study of multiphase materials are clearly needed.

Two different multiphase modeling approaches are addressed in this work, the so-called multiphase micromechanical model and multiphase dynamic generalized self-consisted model. The former formulated in the basis of the micromechanics theory, while the latter based in the multiple scattering theory of the Waterman-Truell framework. Several works can be found that deal with the characterization of mortar by models based on the micromechanics theory [1–3] and scattering theory of elastic waves [4–6]. However, the two different modeling approaches treated in this work provide the capability to well-handle materials with multiphase and strongly heterogeneous microstructure that makes them an attractive tool for ultrasonic characterization of cementitious materials. Theoretical studies and experimental validations are carried out in this work using both multiphase modeling approaches for several mortar specimens with different microstructural features.

2. Micromechanical Modeling

The multiphase approach of the micromechanical model proposed by the authors allows the prediction of the effective macroscopic properties of a multiphase composite material such as mortar or concrete from the knowledge of the geometric and physical characteristics of their constituent phases. Porosity estimation and the understanding improvement of how features of the constituent phases of materials affect the ultrasonic velocity in cementitious materials, are examples of successful applications of this model [7–11]. A briefly description of this model is addressed as follows.

Consider a composite material characterized by a representative volume element (RVE) and subjected to constant uniform overall strains. This composite material is made up of a matrix with embedded inclusions, which do not interact among themselves. The volume of the composite, V, is the sum of the of their N phases, (by analogy the composite mass, M, is the sum of mass of all phases); therefore, the effective density is found to be:

$$\rho = \sum_{i=0}^{N-1} \frac{M_i}{V_i} = \sum_{i=0}^{N-1} \nu^i \rho^i \tag{1}$$

where ρ^i, M_i, V_i and $\nu^i = V_i/V$ and are the density, mass, volume and volume fraction of the i-th phase, respectively. In such a material, the average strain of the composite is the sum of strain produced in each one of the phases, taking into account that the strain produced in the inclusions should be executed over all the possible orientations because strain field in each inclusion is orientation-dependent. One can observe in [12, 13] that the effective elastic constant tensor for two and three phase materials is given in terms of the elastic constant tensors, volume fractions, and orientations distribution of each constitutive phases. Therefore the above-mentioned formulations can be generalized for multiphase composites as [11]:

$$C = C^m + \sum_{i=1}^{N-1} \nu^i (C^i - C^m) \langle T^i \rangle \left[\nu^m I + \sum_{i=1}^{N-1} \nu^i \langle T^i \rangle \right]^{-1} \tag{2}$$

where C represents the effective elastic constant tensor, i refers to the different types of inclusions (e.g. elastic, fluid or cavities), and m refers to the matrix. The tensor $\langle T \rangle$ is the orientational average of the fourth-rank tensor Wu's tensor [14], which in turns depends on the the Eshelby's tensor [15] that relates average inclusion strain, depending on its geometry, to average matrix strain. The multiphase micromechanical model considers inclusion shapes as ellipsoids, and in case of an isotropic matrix, the Eshelby's tensor is given by the Poisson's ratio of the matrix and the aspect ratio r_α or the relative size and relationship between the three axes x_1, x_2, and x_3 of inclusions in which are defined. The aspect ratio is defined by the relationship between the different axes and the two coincident axes, then $r_\alpha = x_3/x_1 < 1$ for an oblate spheroid. Similarly, $r_\alpha = x_3/x_1 > 1$ for a prolate spheroid and $r_\alpha = 1$ for the spherical inclusions. The formulation of the Eshelby's tensor for different inclusion shapes are found in [16]. To reduce the complexity of C, symmetry properties of tensors are used. For instance, when the matrix and types of inclusions of the composite material are isotropic, the elastic constants tensor, see Eq. (2), can be reduced into two independent elastic constants, C_{1111} and C_{2323}, which in reduced notation are referred to as C_{11} and C_{44}:

$$C_{11} = C_{11}^m + \sum_{i=1}^{N-1} \frac{\nu^i}{\nu_1} (C_{11}^i - C_{11}^m - \frac{4}{3}[C_{44}^i - C_{44}^m]) \left(\langle T_{1111}^i \rangle + 2 \langle T_{1122}^i \rangle \right)$$
$$+ \frac{8}{3} \sum_{i=1}^{N-1} \frac{\nu^i}{\nu_2} (C_{44}^i - C_{44}^m) \langle T_{1212}^i \rangle \tag{3}$$

$$C_{44} = C_{44}^m + 2 \sum_{i=1}^{N-1} \frac{\nu^i}{\nu_2} (C_{44}^i - C_{44}^m) \langle T_{1212}^i \rangle \tag{4}$$

where $\nu_1 = \nu^m + \sum_i^{N-1} \nu^i \left(\langle T_{1111}^i \rangle + 2 \langle T_{1122}^i \rangle \right)$ and $\nu_2 = \nu^m + 2 \sum_i^{N-1} \nu^i \langle T_{1212}^i \rangle$.

For a deep description, it is suggested to revise the formulation of the three-phase micromechanical model [13] which is the basis of this multiphase micromechanical model.

In the following sections, several theoretical studies are accomplished by analyzing the influence of the ultrasonic velocity with varying the properties of the different phases in plain and fiber-reinforced mortars in order to highlight the benefits of using the multiphase micromechanical model.

2.1. Influence of Microstructural Parameters of Plain Mortar on the Ultrasonic Velocity Using the Multiphase Micromechanical Model

This section shows how the ultrasonic velocity is affected by variations in geometry and volume fraction of aggregates and pores in plain mortar specimens. Cement mortar is modeled as a three-phase isotropic composite material comprising of an homogeneous matrix referred to the cement paste without pores and two different types of inclusions, pores and sand. The components of the effective elastic constant tensor, C_{11} and C_{44} can be related to the longitudinal (V_L) and transversal (V_T) velocities as follows:

$$V_L = \sqrt{\frac{C_{11}}{\rho}} \tag{5}$$

$$V_T = \sqrt{\frac{C_{44}}{\rho}} \tag{6}$$

Table 1 shows the elastic properties of cement paste matrix and sand used in this section. The matrix corresponds to a cement paste with similar characteristics to Portland cement (type II, 32.5 N) with water/cement (w/c) ratio of 0.5, while the properties of sand correspond to natural sand with 98% of silica.

Table 1. Properties of cement paste matrix and sand

	C_{11} (GPa)	C_{44} (GPa)	ρ [Kg/m^3]
Matrix	40.0	9.0	2200
Sand	80.7	32.64	2600

2.1.1. Influence of the Geometry of the Inclusions

To evaluate the behavior of longitudinal velocity in terms of the aspect ratio of pores, it is assumed that the elastic properties and density of the matrix are known, the pores and sand are randomly oriented in the matrix and the sand is modeled as spheres with a volume fraction set to 0.55. It is assumed that elastic properties and density of matrix and inclusions are kept constant and the porosity can vary, ranging from 0.1% to 40% of the matrix volume. In practice it is not possible to fabricate a mortar with very low porosity (0.1%) but this represents an interesting case for the present study. Figure 1 shows the longitudinal velocity behavior as a function of the aspect ratio of pores when the porosity increases. It is found that the velocity strongly depends on the volume fraction and shape of pores. The highest velocities are for prolate (quasi-spherical) pores, while the ultrasonic velocities are more affected by the oblate pores. This aspect is relevant because this geometry of inclusion is similar to the shape of cracks or fissures [17] which can be an interesting case study.

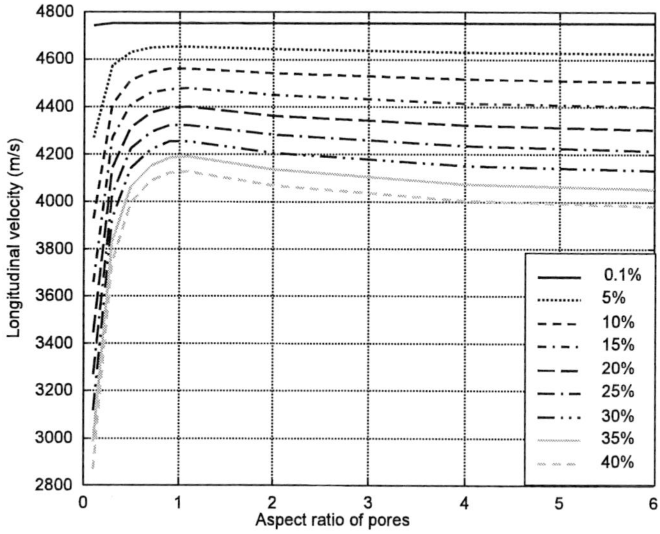

Figure 1. Influence of the geometry and volume fraction of pores on the velocity.

Figure 2 shows the influence of geometry and elastic constants of sand on the longitudinal velocity. The elastic constant C_{11}^{a} of sand is varied between 50 and 100 GPa while C_{44}^{a}, the density and volume fraction of sand are kept constant, 32.64 GPa, 2600 kg/m^3 and 0.55, respectively. The properties of the matrix of cement paste are shown in Table 1 and the porosity of the cement paste is kept constant (30%).

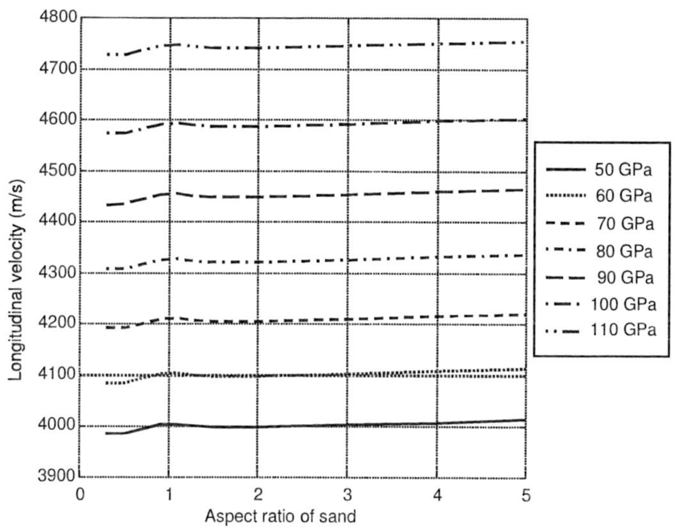

Figure 2. Influence of the aspect ratio and elastic properties of sand on the velocity.

In this figure, it is observed that geometry of sand has less influence on the velocity in comparison with the geometry of pores. It may induce one to conclude that if geometry of aggregates has little influence on the velocity, it may be possible to apply the three-phase micromechanical model to the case of concrete, by considering a medium aspect ratio between aggregates. However, the influence of elastic properties of fine aggregates as shown in Figure 2 reveals that an increase in 10 GPa produces an increase in 100 m/s. For this reason, the application of the multiphase micromechanical model to the case of concrete should be treated with care. In what follows the sand geometry is modeled as a sphere since no coarse aggregates are considered in this work.

2.1.2. Influence of the Elastic Properties of the Non-Porous Matrix

The influence of porosity on the properties of an ideal non-porous matrix is analyzed because it must be noted that ultrasonic measurements can solely measure the elastic constants of a porous cement matrix (cement paste), seen as a single phase. This fact highlights that the inverse problem of porosity estimation can be conducted using ultrasonic measurements and the multiphase micromechanical model. To evaluate variations in the elastic properties of the non-porous matrix in terms of the volume fraction of pores and sand, two pore geometry cases were considered: spheres ($r_\alpha = 1$) representing air voids and prolate spheroids ($r_\alpha \gg 100$) representing capillary pores. Note that elastic properties of sand were kept constant, whereas properties of the matrix and porosity were varied, C_{11}^m= 30 to 60 GPa, C_{44}^m= 13 GPa (constant), $\nu^p = 0.05$ to 0.25. Figure 3 shows that an increase in the matrix stiffness causes an increase in the longitudinal velocity, while with an increase in volume of pores yields the opposite effect; however, these changes lie approximately in the same magnitude (250 m/s, approximately). It is worth noting that prolate pores exhibited the lowest velocities but keeping the same behavior than the spherical pores.

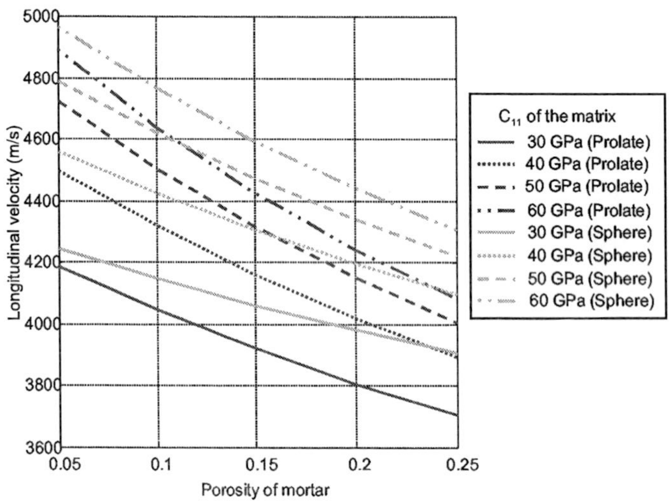

Figure 3. Influence of the porosity, geometry of pores and elastic constants of the matrix on the longitudinal velocity.

On the other hand, Figure 4 shows the influence of both elastic properties (C_{11}^{m} and C_{44}^{m}) of the non-porous matrix on the ultrasonic velocity, considering that the matrix density kept constant with a volume fraction of pores and sand of 0.25 and 0.45, respectively. This figure shows that an increase in the longitudinal velocity occurs with increasing the elastic constants of the non-porous matrix, resulting the highest velocities for the spherical pores.

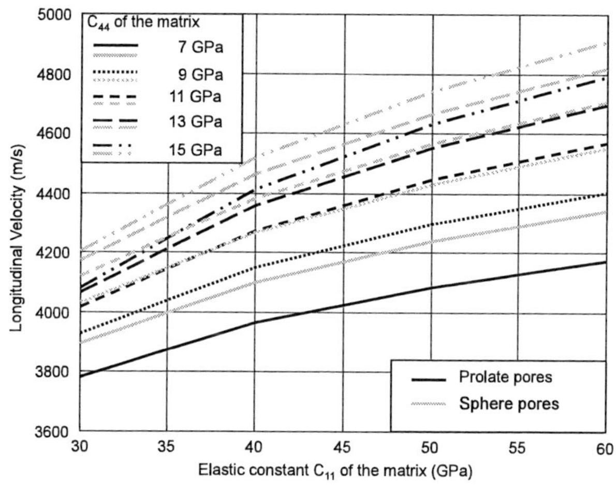

Figure 4. Influence of the elastic properties of the matrix on the longitudinal velocity in mortar. Prolate pores (black lines), spherical pores (gray lines).

Figure 5 shows the influence of C_{11}^{m} and C_{44}^{m} on the transverse velocity, exhibiting that an increase in elastic constants of the matrix causes an increase in the transverse velocity, with similar slopes regardless of pore geometry. Once again, the spherical pores displayed the highest velocities.

2.1.3. Influence of the Volume Fraction of Sand and Pores

Figure 6 shows how variations in the volume fraction of sand and pores affect the ultrasonic velocity. In this case, the matrix properties were kept constant. As a result, an increase in the porosity causes a decrease in the longitudinal velocity, while an increase in the volume fraction of sand induces the opposite effect. It can also be seen that an increase in the volume fraction of sand produces a small decrease in the slope of the velocity-porosity curve. For instance, comparing the curves corresponding to a volume fraction of 0.35 and 0.60, the former shows that the velocity decreases by approximately 800 m/s as mortar porosity is increased from 4% to 20%, while the latter shows that the decrease is limited to approximately 650 m/s over the same porosity range. This fact reveals that a 0.05 increment in the volume fraction of sand produces a less significant decrease in the longitudinal velocity for lower levels of porosity.

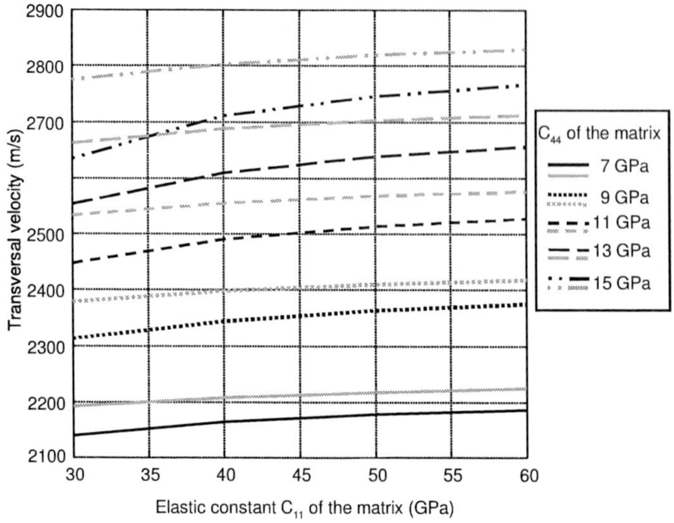

Figure 5. Influence of the elastic properties of the matrix on transverse velocity in mortar. Prolate pores (black lines), spherical pores (gray lines).

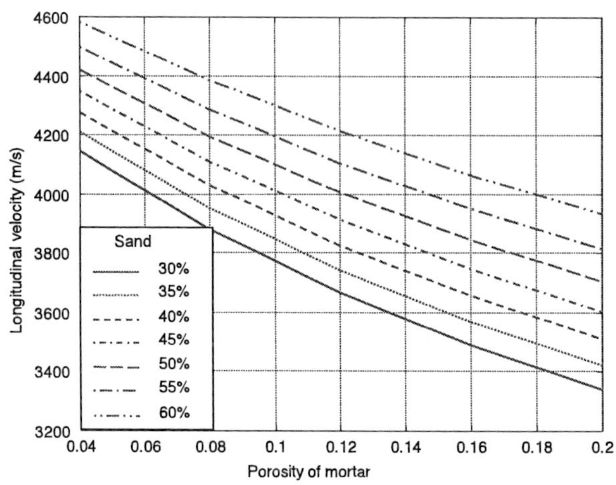

Figure 6. Influence of porosity and volume fraction of sand on the longitudinal velocity, (with matrix properties as C_{11}^{m}=40 GPa, C_{44}^{m}=10 GPa.

2.1.4. Estimation of Porosity in Mortars Using Ultrasonic Measurements and the Multiphase Micromechanical Model

To estimate the porosity in mortars by means of the proposed micromechanical model and ultrasonic velocity measurements, it is necessary to make the following assumptions. Mortar is composed of a cement paste matrix and two types of inclusions, sand and pores. The pores and sand are modeled as cylinders of infinite length and spheres, distributed randomly into the matrix. The elastic constants of non-porous matrices can be calculated as [8], from cement paste samples with the same w/c ratio and cement type, these values are used in this work as outlined in Table 2.

To estimate the porosity content in mortars, two collections of eight mortar samples ($40x40x160$ mm^3) were fabricated with different porosity content by varying their w/c ratios, as well as changing their cement type and sand concentration. The nomenclature used to denote the mortar samples is defined according to the cement types, e.g. CI (Cement 32.5 MPa) and CII (Cement 42.5 MPa), the w/c ratios 0.4, 0.5, and the sand-to-cement (s/c) ratios, 1/1, 2/1. The fraction of open porosity was determined by the standard ASTM C 642-90, see second column in Table 3.

Table 2. Elastic constants of non-porous matrices for different cement pastes

Cement paste samples	C_{11}^{m} (GPa)	C_{44}^{m} (GPa)
CI-04	46.00	12.90
CI-05	39.62	11.54
CII-04	45.70	12.75
CII-05	40.43	11.37

Table 3. Destructive and non destructive measurements in mortar samples

Mortar Samples	Porosity (%)	Ultrasonic velocity (m/s)
CI-04 1/1	12.6	4281 ± 24
CI-05 1/1	18.2	4035 ± 19
CII-04 1/1	10.7	4384 ± 18
CII-05 1/1	14.8	4136 ± 27
CI-04 2/1	10.3	4544 ± 51
CI-05 2/1	13.6	4334 ± 13
CII-04 2/1	8.6	4625 ± 22
CII-05 2/1	12.5	4347 ± 12

Ultrasonic velocity measurements were carried out when the plain mortar samples had suitably matured (28 days), see third column in 3. Two broadband ultrasonic transducers (2 MHz) operating in through-transmission mode were used in order to measure the longitudinal velocity. The ultrasonic signals were acquired using the SENDAS system with a sampling frequency of 66 MHz.

Figure 7 shows the relationships between the w/c ratio and the ultrasonic velocity for all samples. It can be observed that an increase in the w/c ratio results in a decrease in longitudinal velocity, while an increase in the compressive strength of cement has the opposite effect. It can be also seen that an increase in sand content yields an increase in the longitudinal velocity by approximately 250 m/s.

Figure 8 shows the theoretical results by the proposed micromechanical model (eight curves) representing the relationship between the porosity and the ultrasonic velocity in terms of the different volume fractions of sand, cement types and w/c ratios. The theoretical porosity (markers) obtained from the velocity measurements is also shown. These theoretical results show that the curves are well separated by w/c and s/c ratios, while the separation by cement type is not clear. Furthermore, the samples with the lowest volume fraction of sand and the lowest w/c ratio ratio (CI-04 1/1 and CII-04 1/1) are similar to the samples with the highest volume fraction of sand and w/c ratio (CI-05 2/1 and CII-05 2/1). In the figure, it can also be observed that an increase in porosity causes a decrease in longitudinal velocity, while an increase in the volume fraction of sand and the compressive strength of cement induces the opposite effect.

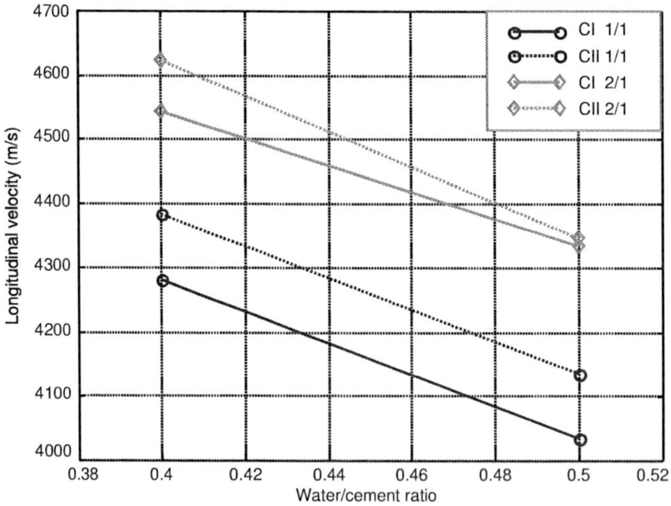

Figure 7. Ultrasonic velocity vs. w/c ratio for all samples.

The comparison between these theoretical curves with the measured porosity from the destructive testing is as shown in Figure 9, and the errors between the estimated and measured porosity are as outlined in Table 4. From these results, it can be seen that the error in estimation of porosity for the groups of mortars with the lowest proportion of sand (1/1) are lower than 10%. The mean error in the estimation of porosity of mortar samples using the micromechanical model is 10.56%. These estimations are good if we consider that in cementitious materials the errors in destructive testing are often around 10%. Note that these predictions were made based on estimations of elastic constants of the non-porous matrices obtained using the micromechanical model and data from destructive testing, which in turn had errors around 4%.

Table 4. Mean error of porosity estimation in plain mortar samples

Groups of samples	CI-04 1/1	CI-05 1/1	CII-04 1/1	CII-05 1/1	CI-04 2/1	CI-05 2/1	CII-04 2/1	CII-05 2/1
Error (%)	7.7	4.5	9.7	2.3	8.8	20.7	18.1	11.9

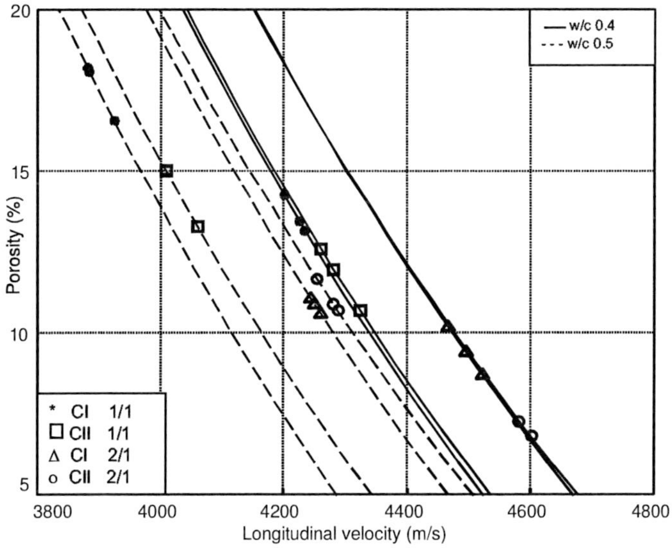

Figure 8. Curves representing the theoretical predictions that relate velocity to porosity in mortar. Markers indicate estimated porosities obtained from the measured velocity.

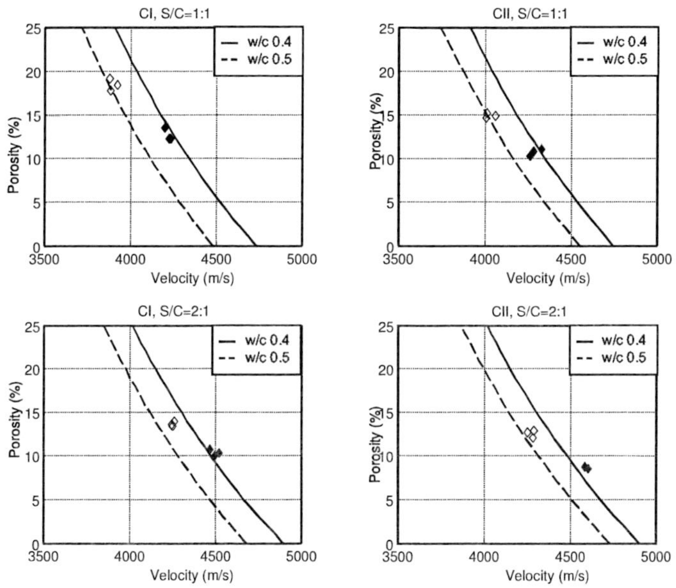

Figure 9. Comparison between measured porosity by destructive methods and theoretical curves obtained from the micromechanical model (■: w/c=0.4; ◇: w/c=0.5).

2.2. Influence of Microstructural Parameters of Steel-Reinforced Mortar on the Ultrasonic Velocity Using the Multiphase Micromechanical Model

The most important factors affecting the properties in steel-reinforced cementitious materials are the aspect ratio and the volume fraction of fibers. For instance, the aspect ratio of fibers plays an important role during the mixing and replacement stages of concrete production [18], while the volume fraction of fibers significantly affects the workability of concrete. The goal of this section is to illustrate the influence of microstructural parameters of steel inclusions on the ultrasonic velocity, by applying the multiphase micromechanical model to steel-fiber reinforced cementitious materials.

The cementitious material to be simulated is steel-reinforced mortar (SRM) comprising five phases: a matrix (non-porous cement paste) and four types of inclusions: aggregates, capillary pores, air voids and steel fibers, with elastic properties as outlined in Table 2.2.. It is worth noting that the non-solid phases (capillary pores and air voids) are assumed to be saturated with water, to take into account the same conditions as in the experimental part. The volume fraction of aggregates and capillary pores are kept constant and equal to 0.51 and 0.15, respectively. The geometry of aggregates is modeled as spheres (r_{α_a}=1), while capillary pores and air voids are modeled as cylinders (r_{α_p}=1000) and spheres (r_{α_v}=1), respectively; steel fibers are modeled depending on their aspect ratio r_{α_s}. All inclusions are assumed to be distributed randomly into the matrix.

Table 5. Density and elastic properties of constituent phases

	Non-porous matrix (m)	Pores§ (p,v)	Aggregates (a)	Steel (s)
C_{11} (GPa)	77.2	2.25	81.6	253.1
C_{44} (GPa)	19.0	0	33.0	76.9
E (GPa)	50.9	0	76.6	197.1
ρ (kg/m^3)	2471	1000	2630	7900

§Water-saturated pores: capillary porosity (p), and air voids (v).

Figure 10 shows the behavior of the longitudinal velocity as a function of the aspect ratio and volume fraction of steel inclusions, with a 0.05 volume fraction of air voids. The variation in the aspect ratio below 20 has a significant influence on the velocity, while the velocity behaves constant at higher values. This fact is significant because commercial fibers typically have an aspect ratio (longitude/diameter ratio) higher than 35. On the other hand, an increase in the volume fraction of steel inclusions causes a decrease in ultrasonic velocity, approximately 140 m/s (3%), when steel concentration is increased from 0 to 0.05. Nevertheless, the consideration of constant porosity in steel-fiber reinforced cementitious materials is unrealistic, because the addition of steel fibers modifies the porous structure of SFR materials, as pointed out by several authors [19–21].

Therefore, the behavior of the ultrasonic velocity regarding the volume fraction of air voids is also addressed, considering the modification of the aspect ratio of steel inclusions. To do so, the volume fraction of both steel inclusions (ν^s=0.05%) and air voids from 0 to 0.01 is varied, and the remaining properties are kept constant as shown in Figure 11. It can be seen in this figure that an increase in the air content implies a decrease in the longitudinal velocity, while the aspect ratio of steel inclusions seems to have no significant influence on it.

On the other hand, Figure 12 shows the influence of the volume fraction of steel on the ultrasonic velocity with respect to variations in the aspect ratio of air voids. In this case, the proportions of inclusions are varied similar to the previous simulation, i.e. for aggregates

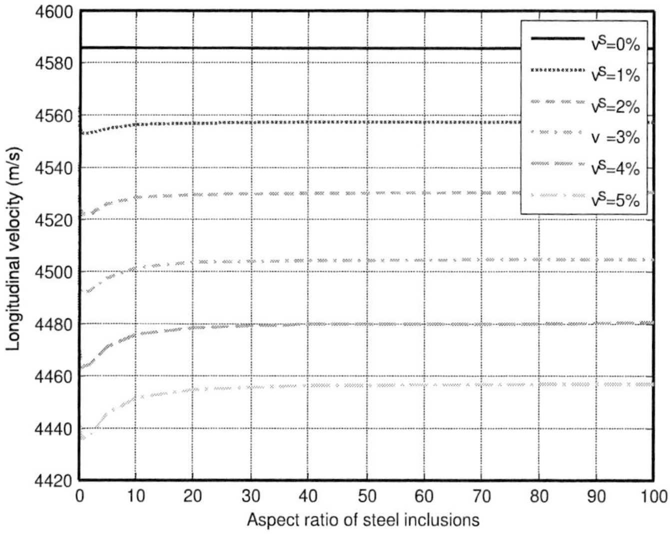

Figure 10. Influence of the aspect ratio and volume fraction of steel inclusions on the longitudinal velocity.

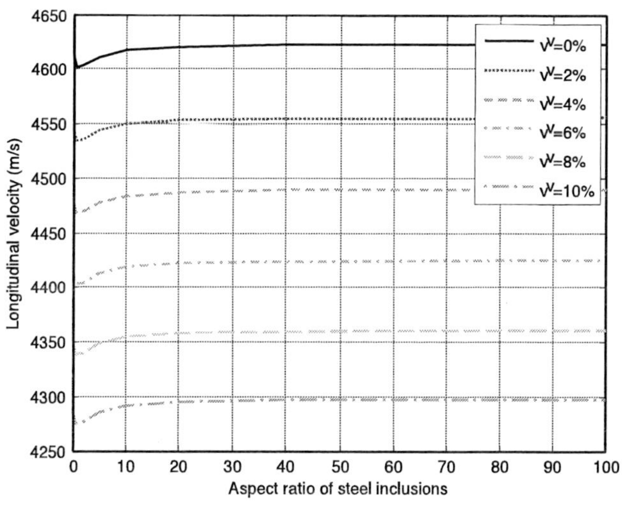

Figure 11. Influence of the steel inclusions aspect ratio and the volume fraction of air voids on the longitudinal ultrasonic velocity.

($\nu^a = 0.51$), capillary porosity ($\nu^p = 0.15$), and air voids ($\nu^v = 0.05$). In this figure, it can be seen that the ultrasonic velocity strongly depends on the shape of the pores. It is also observed that an increase in steel concentration produces a decrease in the longitudinal velocity, but this decrease is lower whenever the volume fraction of air voids is varied. The highest ultrasonic velocities are found for spherical pores, while the values of the ultrasonic velocity seem to decrease drastically in the case of oblate pores. It is noted that as a significant result Figures 11 and 12 reveal that the steel inclusions geometry has minor influence on the ultrasonic velocity, as compared with the pores geometry. This fact reforces the importance of studying variations due to pore structure in reinforced-fiber cementitious materials.

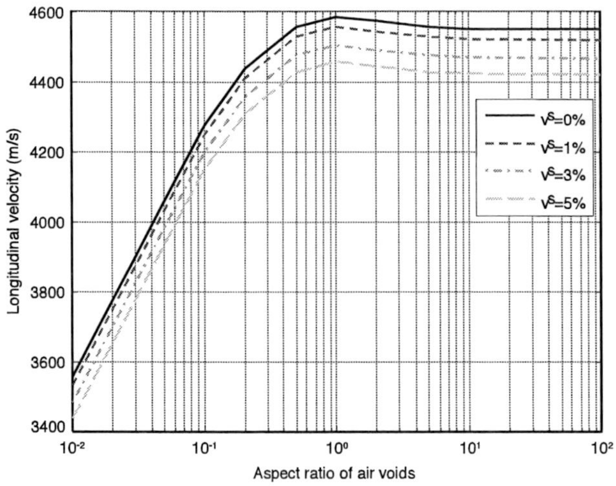

Figure 12. Influence of the air voids aspect ratio and volume fraction of steel on the longitudinal ultrasonic velocity.

2.2.1. Comparison between Theoretical Prediction and Experimental Data in SRM

To verify experimentally the theoretical predictions obtained from the multiphase micromechanical model, a collection of twelve prismatic mortar specimens (40x40x160 mm^3) was prepared with and without steel fibers of different aspect ratio. Four groups (three specimens per group): one group consisting of reference specimens (without steel fibers) and three groups formed by mortar specimens with steel fibers of different aspect ratios. The specimens of this collection were made using a commercial Portland cement (CEM I SR 42.5N), and were prepared according to the Spanish standard UNE-EN 196-1 [22], with a 0.45 w/c ratio, a 2/1 s/c ratio, and natural rounded sand rich in quartz. The specimens were removed from the casts after 24 hours and cured under water for 28 days at 22°. Since the size of the specimens was 40x40x160 mm^3, it was not possible to use commercial steel fibers. Therefore, the aspect ratio (r_{α_s}) and the dimensions of steel inclusions used in the specimens manufactured were as follows: (i) spherical inclusions measuring 2 mm in diameter with r_{α_s} =1 (spheres), (ii) inclusions measuring 0.3 mm in diameter and 4 mm in length with r_{α_s} =13.3 (short fibers), and (iii) inclusions measuring 0.3 mm in diameter and 18 mm in length with r_{α_s} =60 (long fibers). The proportions of the constituents of mortar specimens are as outlined in Table 6.

Destructive and non-destructive testing were carried out in the plain and reinforced mortars, as outlined in Table 7. As for destructing testing, the open porosity (accessible

Table 6. Proportions (by weight) of reinforced mortars and plain mortar used

Specimens	Cement (g)	Water (g)	Sand (g)	Steel (g)
Plain mortar	320	144	639	-
Mortar with spheres	331	149	662	133.8
Mortar with short fibers	336	151	672	133.8
Mortar with long fibers	344	155	688	133.8

porosity) was measured according to RILEM recommendations [23] by means of water absorption under vacuum. As observed in Table 7, the addition of steel inclusions (spheres and fibers), caused an increase in the accessible porosity and density, as compared with the plain mortar specimens. In the mortar specimens with fibers (short and long), this increment was higher; however, the length of the fibers had only a slight influence on the porous structure. Therefore, the measurement of air content in the hardened specimens was made by image analysis in digital images taken from cut slices. To measure the air content a digital image processing was applied, consisting firstly in a segmentation step, and therefore the determination of the air void content as outlined in Table 7. Further details on this procedure are provided in [5, 24].

Table 7. Results of destructive and non-destructive testing in plain and reinforced mortars

Specimens	Accessible porosity (%)	Real density (kg/m^3)	Air voids (%)	UPV (m/s)
Plain mortar	18.68	2570	3.60	4623
Mortar with spheres	19.26	2780	5.12	4573
Mortar with short fibers	25.09	2720	7.40	4315
Mortar with long fibers	25.87	2720	7.50	4235

For non-destructive testing, through-transmission ultrasonic inspections in immersion by an automated system were carried out to measure the ultrasonic pulse velocity (UPV) from the specimens, with the same procedure and equipment as described in [8, 9]. Two ultrasonic broadband transducers emitting in longitudinal mode (2 MHz, 10 mm diameter, Krautkramer H2K) were used to perform the ultrasonic scanning, with spatial resolution in horizontally and vertically directions of 2 mm and 4 mm, respectively. The scanning process allowed the acquisition of 800 signals per specimen, using a pulser-receiver (AIMS ULTIMO 2000); the ultrasonic signals were acquired, digitized, and processed with a high precision system (SENDAS). As a result, velocity maps were generated with 600 A-scan signals for each specimen, obtained by the automatic scanning. The velocity map represents a central area of 30 x 160 mm^2 for each specimen, and to avoid edge effects approximately 5 mm of each side of the specimens were removed.

The velocity maps of the mortars specimens with and without steel inclusions are shown in Figure 13. The zone at the left of the figure corresponds to an aluminum specimen

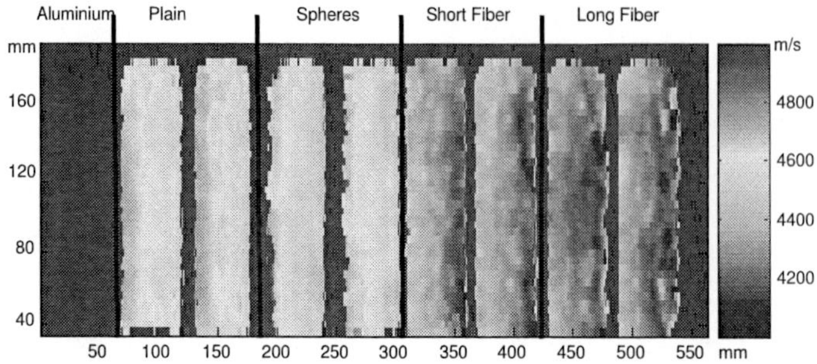

Figure 13. Velocity maps for mortar specimens.

used to calibrate the measurement equipment. As observed in Figure 13, the ultrasonic velocity maps are represented by a false color image ranging from darker colors (4000 m/s) to lighter colors (4800 m/s). The velocity maps show the non-uniformity of the specimens, the velocity variation within the same specimen, as well as the differences among the groups of the mortar specimens. It is observed that there are significant differences in velocity between mortar without steel and mortar reinforced with steel. Moreover, it could be seen a high-velocity zone in the left part of all samples; this variation in the velocity is due to the sand segregation that takes place during the setting process [25]. Table 7 shows the mean velocity for each group of samples. The addition of steel inclusions (spheres and fibers) causes an important loss in velocity as compared to mortars without steel: approximately 50 m/s (1%) for mortars with spherical inclusions and between 308 and 388 m/s (6.6-8.4%) for mortars with short and long fibers, respectively. This behavior agrees with results obtained in [20, 26]. These results allow us to conclude that ultrasonic non-destructive testing is a good tool for characterizing cementitious materials with the addition of steel inclusions.

Estimations in velocity using the multiphase micromechanical model were carried out using the following assumptions. First, the material is formed by a non-porous matrix and four types of inclusions: sand, steel, capillary porosity and air voids. Second, inclusions geometry was modeled like in the theoretical study. Third, non-solid phases, i.e. capillary porosity and air voids, were assumed to be water-saturated pores. Lastly, the elastic properties and density of the non-porous matrix were estimated through an optimization process; such optimization process uses the ultrasonic velocity measured from the plain mortar, and the micromechanical model defined as a non-porous matrix, sand, capillary porosity and air voids, using the values shown in Table 7. Thus, this procedure yields the values corresponding to the non-porous matrix, as shown in Table 2.2. and the elastic properties of the inclusions are also outlined in this table.

Figure 14 shows the estimated and measured velocities for the SFRM specimens, illustrating the small differences that can be found between experimental and predicted velocities. By using the multiphase micromechanical model, estimations of the ultrasonic velocity of mortar specimens were made with a mean error less than 1.5%. It can be seen in this figure that whenever the aspect ratio of the steel inclusions increases the ultrasonic velocity decreases, considering a fixed value for the volume fraction of steel inclusions. Also, as the aspect ratio of the fibers increases, the differences in ultrasonic velocity become smaller, e.g. the differences observed between mortars with spheres or with short fibers, are greater than those found between mortars with short and long fibers. Although the decrease in ultrasonic velocity might appear to be inconsistent with the theoretical predictions shown in Figure 11, the latter shows the influence of the aspect ratio of steel inclusions on the ultrasonic velocity, but taking into account a constant air content. As can be seen in Table 7, the air content is increased whenever the aspect ratio of steel inclusions increases. Furthermore, the decreases in ultrasonic velocity are due to the larger pore volume of the specimens; therefore, this increase may seem to be a consequence of increasing the aspect ratio of fibers. These questions demonstrate the significance of incorporating the non-solid phase into the model, instead of assuming a two-phase model (e.g. concrete and fibers) as usually found in literature.

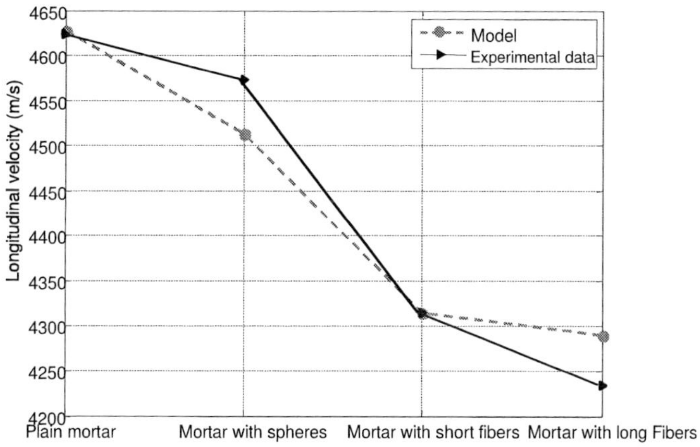

Figure 14. Measured and estimated velocities for steel-fiber-reinforced mortar specimens.

3. Multiple Scattering Modeling

Multiple scattering theories deal with wave propagation at high concentrations of inclusions, and the interactions among them must be taken into account. Theoretical and experimental investigations based on these theories have been carried out by many authors in order to improve the understanding of wave propagation through cementitious materials, such as cement pastes, mortar and concrete. For example, Punurai et al. [27,28] estimated the size and volume fraction of entrapped and entrained air voids in cement pastes using ultrasonic attenuation modeled by a single scattering approach [29]. Aggelis et al. [4] conducted exhaustive theoretical and experimental studies for velocity and attenuation in fresh mortar using the well-known model formulated by Waterman and Truell (WT) [30]. Chaix et al. [31] employed the same model to evaluate micro-cracks in thermally damaged concrete. Recently, Molero and co-workers [5] reported a theoretical and experimental study to highlight that the WT model can be used to estimate microstructural parameters in mortars but limited at low concentrations of aggregates. Furthermore, it must be noted that the above mentioned works used scattering models with the drawback to be only valid for low content of inclusions.

Several studies in the multiple scattering framework have been conducted to overcome such a problem, such as a self-consistent formulation of the WT proposed in [32], the self-consistent effective medium approximation formulated in [33], the iterative effective medium approximation (IEMA) [34], and the so-called dynamic generalized self-consistent model (DGSCM) [35, 36]. The last model has shown its effectiveness to analyze a large number of different composite materials [37–42], but it has only been established for two-phase materials. Recently, our research group has proposed a multiphase approach of a self-consistent multiple scattering model (DGSCM) to describe the influence of the size and volume fraction of aggregates on cementitious materials, as well as the interaction, contribution, and influence entrapped air voids together with the aggregates on frequency-dependent parameters such as the phase velocity and the attenuation coefficient [43]. This model, in the low frequency limit, can provide equivalent results to those obtained with the multiphase micromechanical model presented above. A review of the multiphase approach of the DGSCM is described as follows.

Considering that cement mortar behaves as a dispersive material in which the size of aggregates is assumed to be much greater than the characteristic capillary pore size, but smaller than or comparable with the ultrasonic wavelength. Thus, wave propagation through mortars can be described by means of the effective longitudinal complex wavenumber $\langle k(\omega)\rangle$ (ω refers to angular frequency) that depends on the effective longitudinal phase velocity $\langle V_L(\omega)\rangle$ and the effective longitudinal attenuation $\langle \alpha_L(\omega)\rangle$. Such parameters are frequency-dependent as a consequence of wave scattering phenomena, and they can be related to $\langle k(\omega)\rangle$ as:

$$\langle k(\omega)\rangle = \frac{\omega}{\langle V_L(\omega)\rangle} + i\langle \alpha_L(\omega)\rangle \tag{7}$$

Furthermore, $\langle k(\omega)\rangle$ is related to the effective elastic constants such as Lamé modulus $\langle\lambda\rangle$ and shear modulus $\langle\mu\rangle$, as well as to the effective density $\langle\rho\rangle$:

$$\langle k(\omega)\rangle = \omega\sqrt{\frac{\langle\rho\rangle}{\langle\lambda\rangle + 2\langle\mu\rangle}} \tag{8}$$

Similarly, $\langle \kappa(\omega)\rangle$ is the effective shear complex wavenumber, that depends on the effective shear phase velocity $\langle V_T(\omega)\rangle$ and the effective shear attenuation $\langle \alpha_T(\omega)\rangle$:

$$\langle \kappa(\omega)\rangle = \frac{\omega}{\langle V_T(\omega)\rangle} + i\langle \alpha_T(\omega)\rangle \tag{9}$$

$$= \omega\sqrt{\frac{\langle\rho\rangle}{\langle\mu\rangle}} \tag{10}$$

A multiphase approach of the Waterman-Truell model for different types of inclusions with different size distributions, is given by:

$$\left(\frac{\langle k(\omega)\rangle}{k_0}\right)^2 = 1 + \frac{3}{k_0^2}\sum_j \frac{\phi_j}{a_j^3} f_j(0) + \frac{9}{4k_0^4}\sum_j \frac{\phi_j^2}{a_j^6}\left(f_j^2(0) - f_j^2(\pi)\right) \tag{11}$$

where k_0 is the longitudinal wavenumber of the matrix, ϕ_j and a_j are the volume fraction and radius, respectively, of each of the j different types of inclusions in the composite, and $f_j(0)$ and $f_j(\pi)$ are the complex forward and backward scattering amplitudes, respectively, for longitudinal waves due to the j-type of spherical inclusion into an matrix. However, the WT model is assumed to be valid for a concentration of inclusions up to 40%. The dynamic generalized self-consistent model (DGSCM) formulated by Yang [36], overcomes this limitation assuming that each inclusion of radius a is surrounded by a matrix shell of outer radius b, which in turn is embedded in an effective medium with unknown properties. The volume fraction of inclusions is related to the radii as $\phi = a^3/b^3$. Therefore, this formulation consists of replacing the longitudinal and shear wave numbers in the matrix, k_0 and κ_0, by the effective longitudinal and shear wave-numbers, $\langle k \rangle$ and $\langle \kappa \rangle$, respectively, from WT models for longitudinal and shear waves, yielding [36]:

$$1 = \left[1 + \frac{2\pi n_0}{\langle k \rangle^2} f(0)\right]^2 - \left[\frac{2\pi n_0}{\langle k \rangle^2} f(\pi)\right]^2 \tag{12}$$

$$1 = \left[1 + \frac{2\pi n_0}{\langle \kappa \rangle^2} g(0)\right]^2 - \left[\frac{2\pi n_0}{\langle \kappa \rangle^2} g(\pi)\right]^2 \tag{13}$$

where $g(0)$ and $g(\pi)$ are the forward and backward scattering amplitudes for shear waves. These expressions are implicitly dependent on the unknown effective complex wavenumber; therefore, they are solved iteratively until sufficient convergence is reached. This model is based on combining the GSCM scheme formulated by Christensen and Lo (static homogenization) [44] and the WT model. While DGSCM considers both cylindrical [35] and spherical inclusions [36], has only been established for two-phase materials; therefore, a multiphase approach of DGSCM was proposed as follow.

Consider a composite material consisting of a homogeneous elastic matrix and j different types of spherical inclusions, therefore, a multiphase approach of the DGSCM can be written in terms of their different volume fractions ϕ_j and sizes a_j (radius), as follows [43]:

$$\langle k_{p+1} \rangle^2 = \langle k_p \rangle^2 + \sum_j \frac{3\phi_j}{a_j^3} f_j(0) + \frac{9}{4\langle k_p \rangle^2} \sum_j \frac{\phi_j^2}{a_j^6} \left(f_j^2(0) - f_j^2(\pi) \right) \tag{14}$$

$$\langle \kappa_{p+1} \rangle^2 = \langle \kappa_p \rangle^2 + \sum_j \frac{3\phi_j}{a_j^3} g_j(0) + \frac{9}{4\langle \kappa_p \rangle^2} \sum_j \frac{\phi_j^2}{a_j^6} \left(g_j^2(0) - g_j^2(\pi) \right) \tag{15}$$

with $\langle \mu_{p+1} \rangle = \langle \rho \rangle \left(\omega / \langle \kappa_p \rangle\right)^2$ for $p = 0, 1, 2, \ldots$; this pair of equations are solved iteratively until sufficient convergence is reached, ($\langle k_{p+1} \rangle \approx \langle k_p \rangle$ and $\langle \kappa_{p+1} \rangle \approx \langle \kappa_p \rangle$). Thus the iterative process will stop when difference between $\langle k_{p+1} \rangle$ and $\langle k_p \rangle$), for instance, be lower than 10^{-3}. The initial values ($p = 0; \langle k_0 \rangle, \langle \kappa_0 \rangle, \langle \mu_0 \rangle$) are those associated with the matrix properties, and the effective density $\langle \rho \rangle = \left(1 - \sum_j \phi_j\right) \rho_1 + \left(\sum_j \phi_j\right) \rho_j$. The complex forward and backward scattering amplitudes for longitudinal ($f_j(0)$, $f_j(\pi)$) and shear ($g_j(0)$, $g_j(\pi)$) waves can be computed as described in [36]. However, for an incident longitudinal wave, the complex forward and backward scattering amplitudes can be approximated by:

$$f_j(0) = \sum_{n=0}^{\infty} (-i)^n A_{0n,j}(\langle k_p \rangle, \langle \kappa_p \rangle, \langle \mu_p \rangle) \tag{16}$$

$$f_j(\pi) = \sum_{n=0}^{\infty} (i)^n A_{0n,j}(\langle k_p \rangle, \langle \kappa_p \rangle, \langle \mu_p \rangle) \tag{17}$$

$$g_j(0) = \sum_{n=1}^{\infty} \frac{(-i)^n}{2} [n(n+1)C_{1n,j}(\langle k_p \rangle, \langle \kappa_p \rangle, \langle \mu_p \rangle) + C_{-1n,j}(\langle k_p \rangle, \langle \kappa_p \rangle, \langle \mu_p \rangle)] \tag{18}$$

$$g_j(\pi) = \sum_{n=1}^{\infty} \frac{i^n}{2} [n(n+1)C_{1n,j}(\langle k_p \rangle, \langle \kappa_p \rangle, \langle \mu_p \rangle) + C_{-1n,j}(\langle k_p \rangle, \langle \kappa_p \rangle, \langle \mu_p \rangle)] \tag{19}$$

where the unknown coefficients $A_{0n,j}, C_{1n,j}$ and $C_{-1n,j}$ depend on both shear modulus of matrix and inclusions (μ_1 and μ_j), as well as the radii a_j and b_j of each j different type of inclusions (elastic, fluid and void inclusions) in the composite material. These coefficients are calculated by a system of equations defined from the boundary conditions (continuity in displacements and stresses) in the different interfaces in the composite (elastic matrix-elastic inclusions, elastic matrix-fluid inclusions, and elastic matrix-cavities). The resulting system of equations to obtain the unknown coefficients corresponding to these different interfaces can be found in the appendix of [43].

In the multiphase approach of the DGSCM, the outer radius b_j for each type of inclusion can be of different sizes but the ratio of radii a_j/b_j should be the same for all different types of inclusions. With this criteria multiphase materials can be evaluated without changing the conditions of average stress and average strain [44]. To fulfill this condition by which the multiphase DGSCM reaches the convergence in the iterative process, the following relationship must be accomplished (a detailed analysis is found in [43]):

$$b_j = \frac{a_j}{\sqrt[3]{\sum_i \phi_i}}. \tag{20}$$

where $\sum_i \phi_i$ is the sum of the volume fractions associated with inclusions. Thus, the matrix is assigned proportionately to each of the inclusions on the basis of their corresponding volumes. The volume of the sphere of radius b_j varies in proportion between the volume of the inclusion of type j and the total volume fraction of inclusions that maintains the relation a_j/b_j. This approach has also been addressed by Benveniste [45], but for the static generalized self-consistent model [44]. Note that Eq. (20) is reduced to $b = a/\phi^{1/3}$ for the case of two-phase materials, as used in [36]. The following sections are devoted to show theoretical and experimental studies concerning the benefits of using frequency-dependent velocity and attenuation curves for characterization of mortars.

3.1. Influence of Microstructural Parameters on the Frequency-Dependent Ultrasonic Velocity and Attenuation Curves

This section shows a theoretical study of how different microstructural parameters of mortars affect ultrasonic longitudinal wave propagation using the multiphase DGSCM. This study provides the influence of the volume fraction and size of aggregates and air voids on the longitudinal phase velocity and the longitudinal attenuation with frequency dependence in a mortar model material.

To perform this study, two different cement matrices, with w/c ratios of 0.35 and 0.40, are used, considered as homogeneous viscoelastic media because even though that capillary porosity affects the properties of cementitious materials, however, not acting as inclusions due to its small size compared to the used ultrasonic wavelengths. It is also assumed that the influence of the capillary porosity is included into the properties of the cement matrices; however, the mortar model does consider air voids as another phase. Observations conducted in [27, 28, 31, 46] have shown the linear dependence on frequency (in a wide frequency range) for the absorption attenuation of the cement matrix. For this reason, the absorption attenuation for longitudinal and shear waves can be approximated as $\alpha_{L_1}(\omega) = m_{L_1}\omega$ and $\alpha_{T_1}(\omega) = m_{T_1}\omega$; therefore, it is assumed in this study that $m_{T_1}/m_{L_1} = 4.47$, since attenuation for shear waves is much higher than that for longitudinal waves [27, 46]. Both longitudinal and shear wave number are then described as $k_1 = \omega/V_{L_1} + i\alpha_{L_1}(\omega)$ and $\kappa_1 = \omega/V_{T_1} + i\alpha_{T_1}(\omega)$, respectively, and both velocities of the cement matrices are assumed to be frequency-independent. The properties of the cement paste (matrix), elastic inclusions (sand), and air voids are as outlined in Table 8. As the mechanism of the absorption of the cement matrix $\alpha_{L_1,T_1} = m_{L,T}\omega$ is much higher than the absorption from the aggregates, this latter is omitted. Only scattering attenuation contribution of air voids is considered due to large differences between the acoustic impedances of the matrix and air voids (no absorption attenuation).

Table 8. Acoustic and mechanical properties of constituents. It must be noted that ρ refers to the measured apparent densities (including the capillary porosity contribution) by the open porosity test [23]

	ρ (kg/m^3)	V_L (m/s)	V_T (m/s)	α_{L_1} (dB/mMHz)	α_{T_1} (dB/mMHz)
CEMI-035	1909	4254	2382	57.98	259.27
CEMI-040	1821	4101	2297	53.99	241.39
Inclusions (glass)	2500	5654	3387	-	-
Sand	2670	5571	3454	-	-
Air voids	1.24	344	0	-	-

3.1.1. Cement Paste with Elastic Inclusions

Figure 15 shows variations in velocity and attenuation curves in terms of different sizes of inclusions with a fixed volume fraction.

It can be seen that size of inclusions affects the behavior of the velocity and attenuation profiles differently. With regard to velocity curves, variations in size of inclusions affects the location of the maximum and minimum, implying that these values, especially the relative minima values, move towards low frequency as the size of inclusions increases. Indeed, this behavior highlights the relationship between size of inclusions and wavelength (or frequency range). Although it has only shown a frequency range of 0 to 5MHz and a

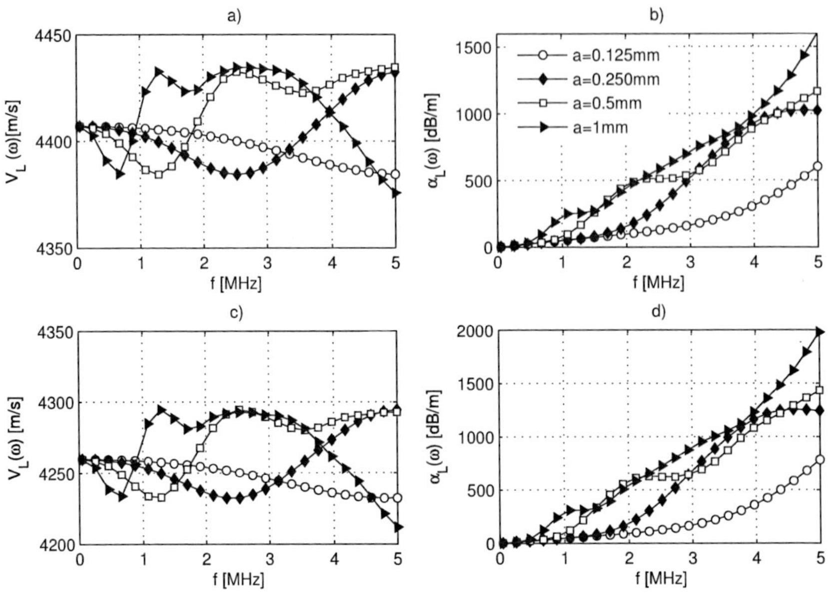

Figure 15. Velocity and attenuation with different inclusions size and a fixed volume fraction $\phi = 15\%$: a) and b) for a cement matrix with w/c=0.35, c) and d) for a cement matrix with w/c=0.40.

size range 0.125 to 1 mm, the behavior of velocity and attenuation curves may be repeated in a number of different frequency and size intervals as long as the ratio between the radius and the wavelength a/λ is held constant. On the other hand, a greater influence of the size of inclusions on attenuation curves is observed, as shown in Figure 15b; for instance, when the size of inclusions increased eight times, from $a = 0.125$mm to $a = 1$mm, variations in velocity did not exceed 50m/s, while attenuation increases over 1000 dB/m. Therefore, the changes in the slopes of the attenuation curves will depend largely on the size of the inclusions, thus facilitating their identification. In addition, it seems that a higher w/c ratio yields a larger attenuation increase, but not the ability to discriminate between different sizes.

The velocity and attenuation curves with different volume fractions and a fixed inclusion size, are as shown in Figure 16. It can be seen in Figure 16a, that when increasing the volume fraction of inclusions, the mean velocity is increased, making it possible to discriminate easily the changes in the aggregate concentrations. As for the attenuation, Figure 16b, shows that as the volume fraction of inclusions increases, so that variations between them diminish and the attenuation may not be differentiated with respect to the volume fraction. Therefore, from a certain concentration of inclusions, the attenuation does not allow the recognition of different volume fractions, however, velocity profiles can largely identify changes in the volume fraction of inclusions. As pointed out in [47], there is no direct strong relation of any wave parameter (velocity or attenuation) with the w/c ratio. However, the estimation of such a ratio can be conducted using the aggregate concentration which can be easily related to the velocity.

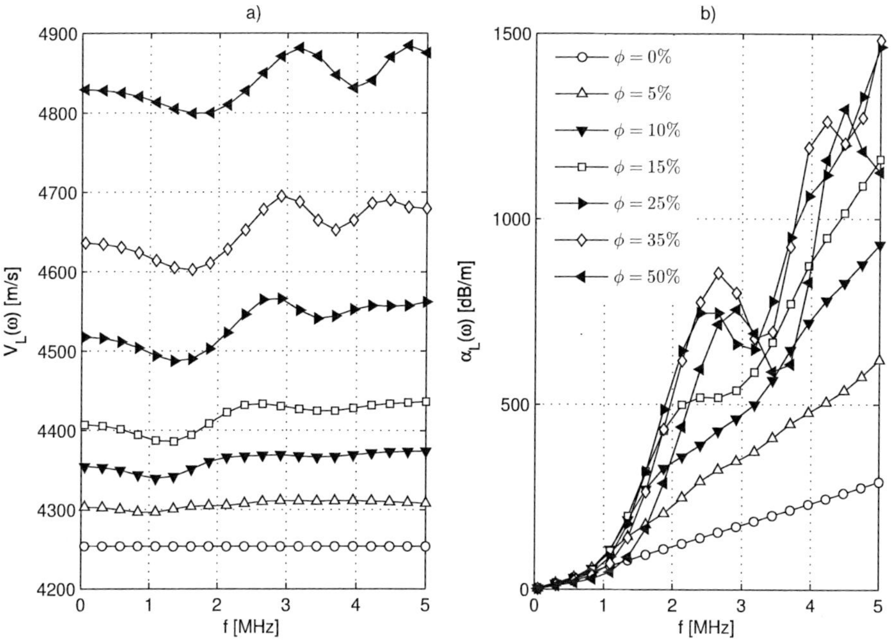

Figure 16. Velocity and attenuation with different concentrations of inclusions and a fixed inclusions size a =0.5 mm: a) and b) for a cement matrix with w/c=0.35.

A cementitious material is unlikely to be made up of a single size of aggregates, because it always should include a suitable granulometric distribution of aggregates to obtain the characteristics that demand the specific application, e.g. in the mix proportioning of the material. Therefore, the influence of different distributions of inclusions on the behavior of the velocity and attenuation curves should be studied. Four cases were compared according to the type of distribution: two extreme cases of a mean size and a uniform distribution, and two intermediate distributions occurring in these materials, namely, a symmetrical distribution such as Gaussian distribution and a non-symmetrical, triangular distribution. Note that the weighted mean size is the same for all distributions. In Figures 17a and 17b, both velocity profiles -as the attenuation profiles- vary slightly when using a mean size within a reduced size range in comparison with those corresponding to a certain type of distribution in that range. As the range of inclusions size increases, see Figures 17d and 17e, the differences are greater; therefore, it is not advisable, in this case, using a mean size in order to approximate a certain size distribution.

3.1.2. Mortar: Cement Paste with Elastic Inclusions and Air Voids

Theoretical velocity and attenuation curves are as shown in Figure 18, which takes into account the interactions due to the matrix with inclusions and air voids, as well as the interaction of both on the matrix. In these figures, it can be seen that velocity and attenuation profiles are affected by air voids at low frequencies, while the elastic inclusions act on the

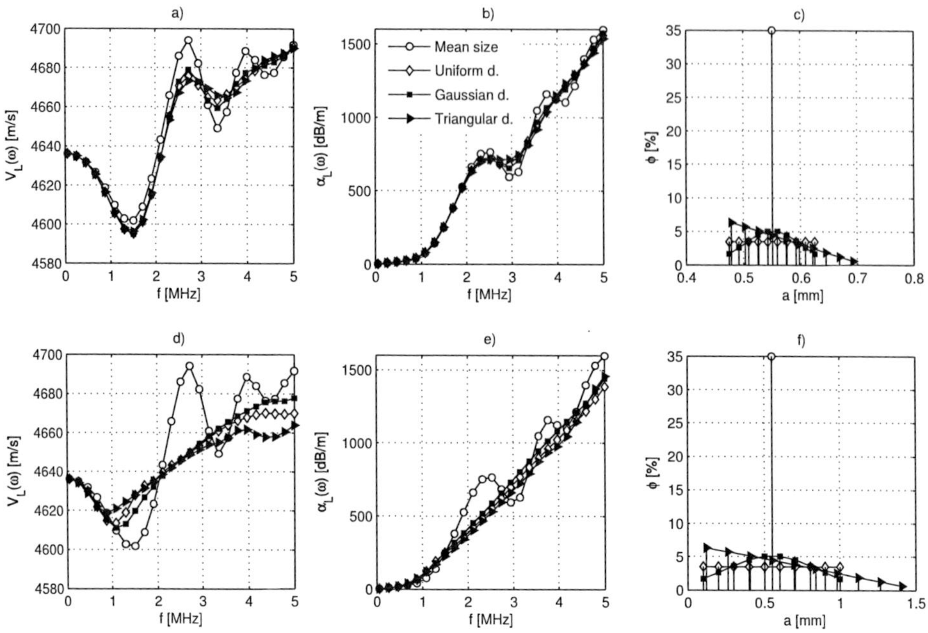

Figure 17. Plots of a), d) velocity and b), e) attenuation for c) and f) different types of size distributions.

high frequencies. Thus, the resonance effect produced by the cavities is further accentuated on the velocity profile; as the frequency increases, the values coincide with those corresponding to the elastic inclusions. The same effect occurs for attenuation. Therefore, the behavior and variation in these two parameters, velocity and attenuation, depend on the volume fraction and size of both elastic inclusions and cavities; however, the elastic inclusions show a greater influence on these profiles at the high frequencies while cavities affect at low frequencies. Such a behavior has been observed for fresh mortar specimens [4]. The velocity also allows better distinction of the aggregate concentration, while attenuation allows better identification of the size of aggregates.

3.2. Comparison between Theoretical Predictions and Experimental Data in Mortar Specimens

This section shows the comparison between predictions obtained by the multiphase dynamic generalized self-consistent model (frequency-dependent velocity and attenuation profiles) and those experimental ultrasonic measurements obtained from mortar specimens made with varying features in their microstructure. Two collections of prismatic mortar specimens (40x40x160 mm^3) were made with different aggregate-to-cement and w/c ratios were made using two different cement types in the mix proportioning process, CEM I-42.5N and CEM III/B SR/BC 32.5N. The cement type CEM I-42.5N (collection 1) was employed to make two different groups of mortar specimens varying the w/c ratio (0.35

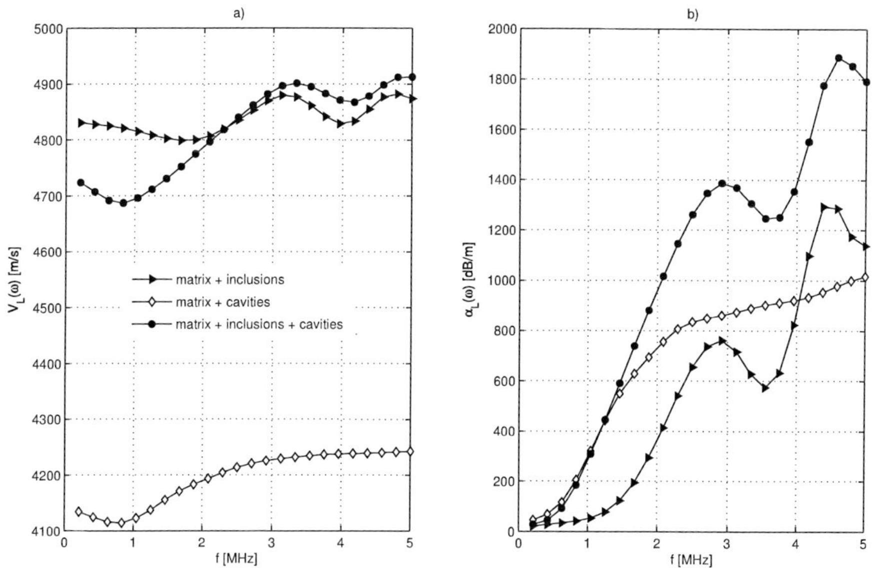

Figure 18. Plots of a) velocity and b) attenuation taking account of the interaction of elastic inclusions of size a =0.5 mm and volume fraction $\phi = 50\%$, and air voids of size a_v =0.5 mm and $\phi_v = 5\%$ embedded in the matrix, and together.

and 0.40), and using glass microspheres as aggregates for controlling volume fraction, size granulometry, and spherical geometry of aggregates; whereas the cement type CEM III/B SR/BC 32.5N (collection 2) was used to make mortar specimens with sand aggregates. In the collection 1, the aggregate granulometry (radii) was varied as 0.09-0.15, 0.21-0.30 and 1.0 mm, whereas for the collection 2, it was varied as 0.03-0.25, 0.25-0.50, 0.50-1.0, 0.03-1.0 mm. It is worth noting that in the collection 2, the w/c ratio was varied in order to maintain as far as possible similar cement matrices. Moreover, cement pastes (w/c: 0.35, 0.40) were made to be the reference specimens corresponding to the mortar specimens with glass microspheres. Table 9 outlines the proportion of the mix proportioning for the mortar specimens used throughout this work.

Table 10 outlines the nomenclature of the mortar specimens and their nominal volume fraction and their corresponding granulometry of aggregates. The properties of the glass microspheres and sand are as outlined in Table 8.

The frequency-dependent velocity and attenuation profiles were measured by performing a longitudinal wave transmission experiment in immersion. Using two 5MHz-broadband transducers in through-transmission mode, ten measurements per specimen were taken along the sample and making a spatial average of the measurements to compute the frequency-dependent velocity and attenuation curves. The detailed description of both frequency-dependent velocity and attenuation measurements can be found in [5,24,43,48]. It must be noted that the employed methods were mainly based on those presented in [4,47] with some adaptations. Briefly, to measure velocity an ultrasonic broadband transducer was

Table 9. Proportion by weight of aggregate/cement ratio, (or s/c) and water/cement ratio (w/c) of specimens of cement paste and mortar with microspheres and aggregates. (▲) for mortars with CEM I-42.5N and Granulometry: 0.09-0.15, 0.21-30, 1.0 mm. (▼) for mortars with CEM III/B SR/BC 32.5N and Granulometry: 0.03-0.25, 0.25-0.5, 0.5-1.0, 0.03-1.0 mm

s/c	w/c						
	0.35	0.38	0.40	0.45	0.46	0.52	0.65
0.00	▲		▲				
0.25	▲		▲				
0.50	▲		▲				
1.00	▲	▼	▲				
2.00				▼			
3.00					▼	▼	▼

Table 10. Size and volume fraction of aggregates for the mortar specimens of collections 1 and 2

specimens	Collection 1			specimens	Collection 2		
		w/c: 0.35	w/c: 0.40				
	a_N (mm)	ϕ_N (%)			w/c	a_N (mm)	ϕ_N (%)
M025S1	0.09-0.15	13.1	12.3	M300G1W65	0.65	0.03-0.25	53.21
M025S2	0.21-0.30	13.1	12.3	M300G2W52	0.52	0.25-0.50	56.71
M025S3	1.0	13.1	12.3	M300G3W46	0.46	0.50-1.00	58.48
M050S1	0.09-0.15	23.1	21.8				
M050S2	0.21-0.30	23.1	21.8				
M050S3	1.0	23.1	21.8	M100G0W38	0.38	0.03-1.00	34.29
M100S1	0.09-0.21	37.5	35.8	M200G0W45	0.45	0.03-1.00	48.74
M100S2	0.21-0.30	37.5	35.8	M300G0W46	0.46	0.03-1.00	58.48
M100S3	1.0	37.5	35.8				

first excited with a Gaussian tone-burst signal of 6-8 cycles. This signal was generated with a function generator (TiePie, Handyscope HS3) ranging from 300 to 5000 KHz with increments of 50 KHz. Subsequently, the generated signal was injected into the specimen and the transmitted signal was received, pre-amplified (Panametrics, Preamp) by 54 dB and averaged 128 times to improve the signal-to-noise ratio (SNR). Both input and output signal were recorded by an oscilloscope (TiePie, Handyscope HS3, 12 bit) with a sampling frequency of 50 MHz. The phase velocities were calculated for each frequency of the tone-burst signals by measuring the time-of-flight (at the same phase point) between the received signals that traveled through the materials with respect to their emitted signals. With regard to attenuation measurements, the emitter transducer was excited with chirp signals ranging from 300 to 6000 KHz. The traveling received signals were pre-amplified and averaged.

Thus, attenuation measurements were obtained by means of the spectrum of maximum energy (in the time-frequency space) corresponding to the chirp pulses that traveled through mortar specimens and the spectrum of maximum energy of reference chirp signals obtained in water measurements. Further description of this time-frequency processing technique can be found in [48].

To compare experimentally the prediction of velocity and attenuation curves with experimental data obtained from the mortars specimens, several microstructural parameters should be known, e.g. cement matrix properties, porosity, volume fraction, and size distributions of both aggregates and air voids. However, detailed knowledge of these parameters is a difficult task or even impossible, especially for properties concerning the cement matrices and the size and volume fraction of air voids. Therefore, an optimization process can be used to find the best fit between theoretical curves obtained by the multiphase DGSCM and experimental measurements of velocity and attenuation. Therefore, the error function, in the least squares sense, can be written as:

$$f(\bar{x}) = (1-\beta)\sum_{\omega}\left[V_L^{(th)}(\bar{x},\omega) - V_L^{(exp)}(\omega)\right]^2 + \beta\sum_{\omega}\left[\alpha_L^{(th)}(\bar{x},\omega) - \alpha_L^{(exp)}(\omega)\right]^2 \quad (21)$$

where $V_L^{(th)}$ and $\alpha_L^{(th)}$ are the theoretical velocity and attenuation obtained from the multiphase DGSCM, and $V_L^{(exp)}(\omega)$ and $\alpha_L^{(exp)}(\omega)$ the experimental measurement of the velocity and attenuation, respectively; β is a scale factor that compensates the differences in magnitude between the values of velocity and attenuation, and $\bar{x}$ is a vector whose components are the fitting parameters associated with the unknown microstructural parameters. These unknown parameters defined in $\bar{x}$ are V_{L_1}, ρ_1, a_v and ϕ_v, where V_{L_1} and ρ_1 are the longitudinal velocity and the apparent density (including the capillary porosity contribution) of the cement matrices, respectively. In addition, it was assumed the Poisson's ratio equals to 0.27 for all specimen matrices. The parameters a_v and ϕ_v refer to the radius and the volume fraction of air voids. It is worth noting that only a single size of air voids was assumed in this process for all specimens. The size and the concentration of aggregates for each specimen are known and assigned according to the mean size of the granulometries and the nominal volume fractions exposed in Table 10 for the specimens of collection 1 and the specimens M300G1W65, M300G2W52 and M300G3W46 belonging to collection 2. As previously mentioned in the theoretical section, when the range of the inclusions size increases (wider granulometry), it is not advisable to model the aggregate size with a mean size. Therefore, it was used the relative distribution of the aggregate size as employed in the manufacturing process for the specimens M100G0W38, M200G0W45 and M300G0W46, While the absorption attenuation can be measured from cement pastes made with the same w/c ratio of mortar specimens to be characterized, however, applications on situ or assessment of hardened cementitious materials do not always have reference specimens (cement pastes). Therefore, the parameters, $m_{L_1,T1}$ may be included in the list of unknown parameters to be estimated. To reduce parameters in the optimization process, it was assumed that the ratio between longitudinal and shear attenuation coefficients is equals to $m_{T_1}/m_{L_1} = 4.47$, as previously mentioned in theoretical section. With this optimization process, the capability of the multiphase DGSC model to generate both velocity and attenuation profiles in agreement with several experimental curves can be tested.

The theoretical and experimental velocity and attenuation curves of both collections are shown in Figures 19 and 20, respectively, with the solid lines corresponding to the theoretical values and the markers to the experimental ones. These theoretical curves were obtained by means of the aforementioned optimization process. Figure 20 shows the corresponding curves for the mortar specimens made with glass microspheres (collection 1), highlighting the features mentioned in the theoretical study. As seen in this figure, the velocities curves are grouped depending on the volume fraction of aggregates, whereas the attenuation curves are well identified for the size.

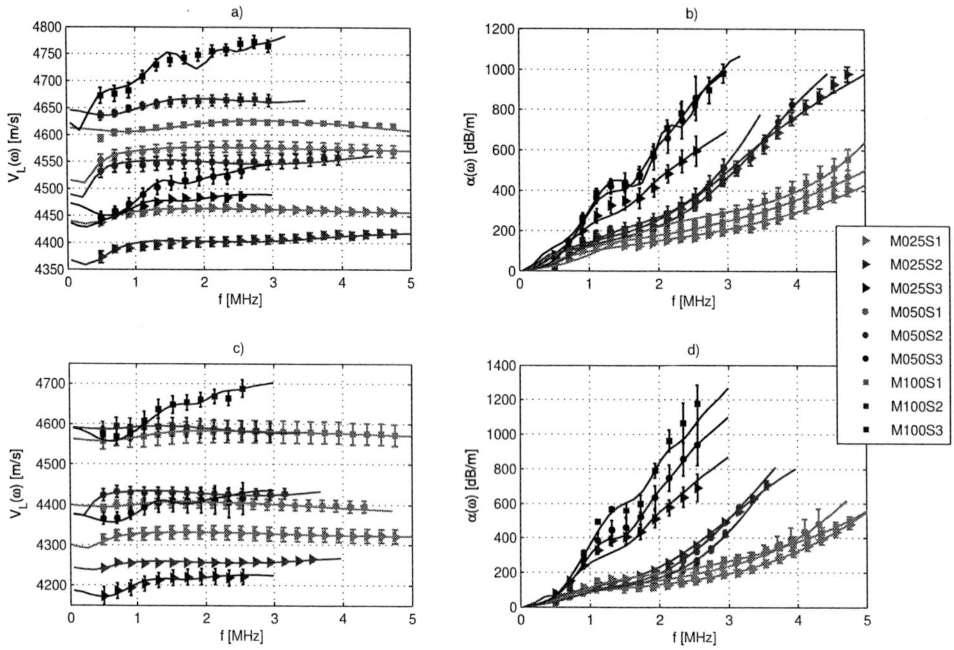

Figure 19. Theoretical and experimental curves of velocity and attenuation for mortar specimens made with glass microspheres, with: a), b) (w/c)=0.35 and c), d) (w/c)=040. The theoretical curves are represented with solid lines. The vertical bar indicates the experimental standard deviation from ten measures.

The differences in magnitude of velocity and attenuation between the same kind of specimens, according to their corresponding size or volume fraction of aggregates, are due to variations in their corresponding cement matrices, as well as different sizes and concentrations of air voids that cannot be controlled in the manufacturing process of such specimens. Either way, it can be seen that a greater concentration of aggregates the velocity magnitude is increased. It is worth noting that the variations in the velocity profiles (different sizes with aggregate content constant) can be mainly attributed to variations in the porous structure, for instance, different volume fractions of the entrapped air voids in the specimens. For this reason, it is important to include in the material models the air voids as another phase. Variation in capillary porosity also influences on the magnitude of the velocity profiles, because as was pointed out in Theoretical section, even though the capillary porosity was not considered as a scatterer, however, it does affect the properties of the cement paste. With regard to the attenuation curves, a higher attenuation occurred for the case of larger proportion of aggregates, but the strong heterogeneity of the specimens makes the attenuation curves almost overlapped for some sizes of aggregates.

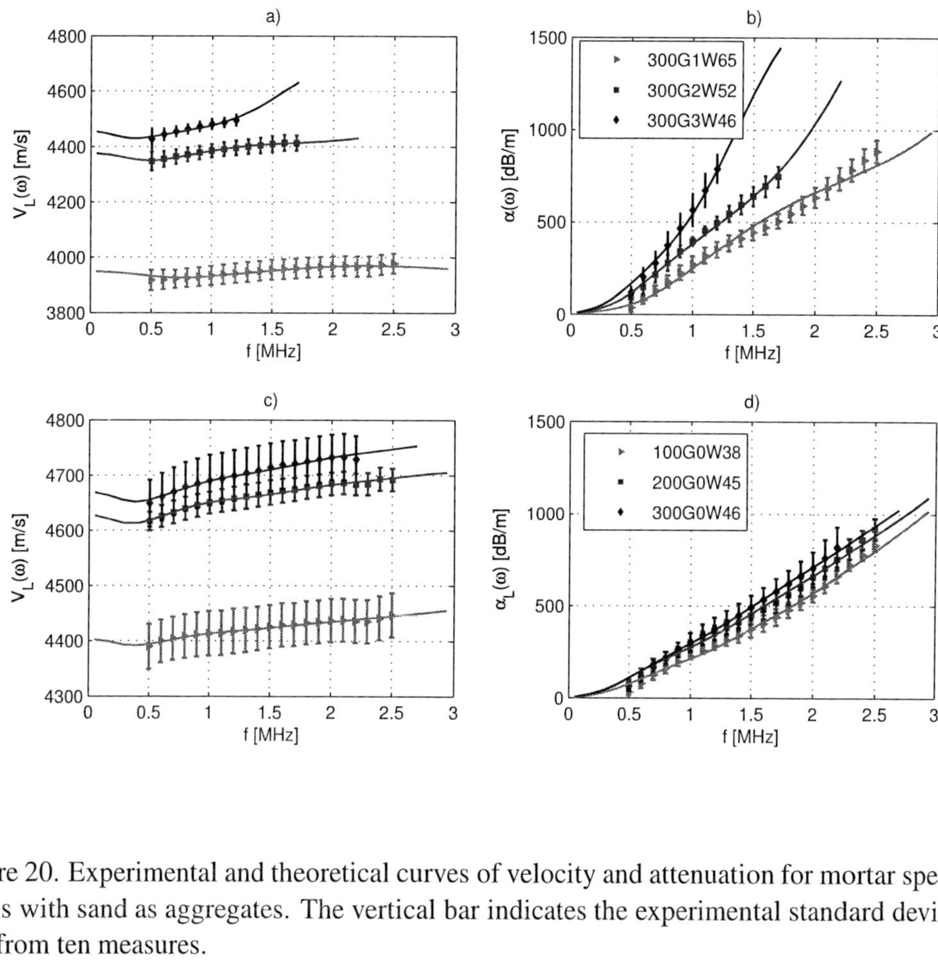

Figure 20. Experimental and theoretical curves of velocity and attenuation for mortar specimens with sand as aggregates. The vertical bar indicates the experimental standard deviation from ten measures.

Figure 20 shows the experimental values and the best fit using the multiphase DGSCM for the specimens from collection 2. The specimens were divided into two groups to illustrate two effects. The first group corresponds to different intervals of the size distribution of aggregates as outlined in Table 10 (M300G1W65, M300G2W52 and M300G3W46), whereas the second group corresponds to increasing the volume fraction of aggregates with a fixed size distribution (M100G0W38, M200G0W45 and M300G0W46). In the first case, Figure 20a and 20b shows that as the size distribution of aggregates increases, both velocity and attenuation also increase in magnitude. On the other hand, in the second group, shown in Figure 20c and 20d, the increase in the concentration of aggregates can be seen throughout the frequency range of the velocity curves. However, this effect was only appreciated from the attenuation curves at higher frequencies because attenuation is less sensitive to increasing the aggregate content. Therefore, these results are in good agreement with the observations conducted in the theoretical section.

Conclusion

In this chapter, the relations between microstructural properties of mortars and ultrasonic wave propagation were studied using two different models, the multiphase micromechanical model and the multiphase approach of the dynamic generalized self-consistent model (DGSCM). The influence of elastic properties, volume fraction and geometry of different phases of mortar on the ultrasonic velocity was studied using the multiphase micromechanical model. In this model, mortar was considered as a composite formed by a matrix with different types of inclusions embedded (pores, aggregates, steel). The influence of volume fraction and geometry of pores on the ultrasonic velocity, highlighted the need to consider the porous structure as another phase more in order to increase, in general, the understanding of the microstructural behavior of cementitious materials. The reinforcement of fibers into the mortar microstructure improves some mechanical properties (e.g. flexural strength), but changing the porous structure. This work verified that the addition of steel inclusions decreases the ultrasonic velocity and modifies the porosity of reinforced mortars. On the other hand, the influence of the size and volume fraction of aggregates and air voids on the frequency-dependent parameters such as the phase velocity and the attenuation coefficient was addressed by the multiphase approach of DGSCM. This model overcomes the limitations of the Waterman-Truell model (valid for only low concentrations of aggregates), showing its effectiveness and experimentally verified up to 60%. The results obtained by the multiphase DGSCM emphasize again the great importance to take into account the porous structure in the multiphase mortar model. In addition, the use of frequency-dependent velocity and attenuation profiles seem to be a promising alternative to estimate microstructural parameters in cementitious materials.

Although the above models are formulated on the basis of distinct theories and provide different information about the microstructure of cementitious materials, they can be used as a complementary tool to each other to better characterize the microstructure of multiphase materials.

References

[1] S. Briccoli Bati,G. Ranocchiai and L. Rovero. A micromechanical model for linear homogenization of brick masonry. *Materials and Structures*, 1999, **32**, 22-30

[2] I.O. Yaman,H.M. Aktan and N. Hearn. Active and non-active porosity in concrete part II: Evaluation of existing models. *Materials and Structures*, 2002, **35**, 110-116,March

[3] M.G. Hernández, J.J. Anaya, T. Sanchez and I. Segura, Porosity estimation of aged mortar using a micromechanical model. *Ultrasonics*, 2006, **44** SUPPL.,e1007-e1011.cited By (since 1996) 3.

[4] D.G. Aggelis, D. Polyzos and T.P. Philippidis. Wave dispersion and attenuation in fresh mortar: theoretical predctions vs. experimental results. *Journal of the Mechanics and Physics of Solids*, 2005, **53**, 4, 857-883, April

[5] M. Molero, I. Segura, S. Aparicio and J.V. Fuente. Influence of aggregates and air voids on the ultrasonic velocity and attenuation in cementitious materials. *European Journal of Enviromental and Civil Engineering*, 2011, **15**, 4, 501-517, April

[6] M. Treiber, Qu J. Kim, J.-Y. and L.J. Jacobs. Effects of sand aggregate on ultrasonic attenuation in cement-based materials. *Materials and Structures*, 2010, **43**, Suppl. 1, 1-11

[7] M.G. Hernández, M.A.G. Izquierdo,A. Ibañez,J.J. Anaya and L.G. Ullate, Porosity estimation of concrete by ultrasonic NDE. *Ultrasonics*, 2000, **38**, 1,531-533,March

[8] M.G. Hernández, J.J. Anaya, L.G. Ullate and A. Ibanez. Effect of the fluid in the inclusions of cement paste on the ultrasonic velocity. *Ultrasonics*, 2004, **42**,1-9,865-869,April

[9] M.G. Hernández, J.J. Anaya, M.A.G. Izquierdo and L.G. Ullate. Application of a micromechanical model of three phases to estimating the porosity of mortar by ultrasound. *Cement and Concrete Research*, 2006, **36**, 4, 617-624,April

[10] S. Bueno,M.G. Hernández, T. Sánchez, Anaya J.J. and C. Baudín. Non-destructive characterisation of alumina/aluminium titanate composites using a micromechanical model and ultrasonic determinations Part I. Evaluation of the effective elastic constants of aluminium titanate. *Ceramics International*, 2008, **34**,181-188

[11] M. Acebes, M. Molero, I. Segura, A. Moragues and M.G. Hernández. Study of the influence of microstructural parameters on the ultrasonic velocity in steel-fiber-reinforced cementitious materials. *Construction and Building Materials*, 2011, **25**, 7, 3066-3072, July

[12] H. Jeong and D.K. Hsu. Quantitative estimation of material properties of porous ceramics by means of composite micromechanics and ultrasonic velocity. *NDT&E International*, 1996, **29**, 2, 95-101

[13] M.G. Hernández,J.J. Anaya,L.G. Ullate and A. Ibañez. Formulation of a new micromechanic model of three phases for ultrasonic characterization of cement-based materials. *Cement and Concrete Research*, 2006, **36**, 4, 609-616, April

[14] T.T. Wu. The effect of inclusion shape on the elastic moduli of a two phasematerial. *International Journal of Solids and Structures*, 1966, **2**, 1-8

[15] J.D. Eshelby. The determination of the elastic field of an ellipsoidal inclusion, and related problems. *Proceedings of Royal Society of London, Series A, Mathematical ansd Physical Sciencies*, 1957, **241**, 1226,376-396

[16] T. Mura. *Micromechanics of defects in solids*. Martinus Nijhoff Publishers, 1987, 2nd

[17] Neville, A.M., *Properties of concrete*. Prentice-Hall, Fourth and Final edition, 2000.

[18] ACI Committee 544-3R-93, Guide for specifying, proportioning, mixing, placing and finishing steel-fiber reinforced concrete. Technical report, American Concrete Institute, 1993

[19] B. Miloud. Permeability and porosity characteristics of steel fiber reinforced concrete. *Asian Journal of Civil Engineering*, 2005, **51**, 6, 317-330

[20] S. Yazici, G. Inan and V. Tabak. Effect of aspect ratio and volume fraction of steel fiberon the mechanical properties of SFR. *Construction and Building Materials*, 2007.

[21] T. Uygunoglu. Investigation of microstructure and flexural behaviour of steel-fiber reinforced concrete. *Materials and Structure*, 2008, **41**, 1441-1449

[22] UNE-EN 196-1. *Métodos de ensayo de cementos. Parte 1: Determinación de resistencias mecánicas, AENOR*.1996.

[23] TC/14-CPC Rilem. Absorption of water by immersion under vacuum. *Materials and Structures*, 1984,**17**, 391-394

[24] M. Molero. Caracterización de materiales cementicios mediante la dispersión ultrasónica (in Spanish). *Technical University of Madrid (UPM)*, 2009, Madrid,Spain

[25] M. Molero, I. Segura,M.A.G. Izquierdo,J.V. Fuente and J.J. Anaya. Sand/Cement ratio evaluation on mortar using neural network and ultrasonic transmission inspection. *Ultrasonics*, 2009, **49**, 2, 231-237,February

[26] G. Washer,P. Fuchs, B.A. Graybeal and J.L. Hartmann. Ultrasonic testing of reactive-powder concrete, *IEEE Transactions on Ultrasonics, Ferroelectrics and Frequency Control*. 2004, **51**, 193-201

[27] W. Punurai, J. Jarzynski,J. Qu, K.E. Kurtis and L.J. Jacobs. Characterization of entrained air voids in cement paste with scattered ultrasound. *NDT&E International*, 2006, **39**, 6, 514-524,September

[28] W. Punurai, J. Jarzynski, J. Qu, J.-Y. Kim,L.J. Jacobs and K.E. Kurtis. Characterization of multi-scale porosity in cement paste by advanced ultrasonic techniques. *Cement and Concrete Research*, 2007, **37**, 1, 38-46, January

[29] C.F. Ying and R. Truell . Scattering of a plane longitudinal wave by a spherical obstacle in an isotropically elastic solid. *Journal of Applied Physics*, 1956, **27**, 9, 1086-1097, September

[30] P.C. Waterman and R. Truell. Multiple scattering of waves. *Journal of Mathematical Physics*, 1961, **2**, 4, 512-537, July

[31] J.F. Chaix,V. Garnier and G. Corneloup. Ultrasonic wave propagation in heterogeneous solid media: Theoretical analysis and experimental validation. *Ultrasonic*, 2006, **44**, 200-210

[32] C.M. Sayers and R.L. Smith. The propagation of ultrasound in porous media. *Ultrasonics*, 1982, **20**, 5, 201-205, September

[33] J.-Y. Kim,J.-G. Ih and B.-H. Lee. Dispersion of elastic waves in random particulate composites. *Journal of the Acoustical Society of America*, 1995, **97**, 3, 1380-1388, March

[34] D.G. Aggelis,S.V. Tsinopoulos and D. Polyzos. An iterative effective medium approximation (IEMA) for wave dispersion and attenuation predictions in particulate composites, suspensions and emulsions. *Journal of the Acoustical Society of America*, 2004, **116**, 6, 3443-3452, December

[35] R.-B. Yang and A.K. Mal. Multiple scattering of elastic waves in a fiber-reinforced composite. *Journal of the Mechanics and Physics of Solids*, 1994, **42**, 12, 1945-1968, December

[36] R.-B. Yang. A dynamic generalized self-consistent model for wave propagation in particulate composites. *Journal of Applied Mechanics*, 2003, **70**, 4, 575-582, July

[37] J.Y. Kim. On the generalized self-consistent model for the elastic wave propagation in composite materials. *International Journal of Solids and Structures*, 2004, **41**, 16-17, 4349-4360, August

[38] H. Sato,M. Kitahara and T. Shoji. Boundary element analysis of multiple scattering waves in high performance concretes. *Journal of Applied Mechanics*, 2005, **72**, 2, 165-171, March

[39] C. Layman, R.B. Sanjeeva Murthy, C. Yang and J. Wu. The interaction of ultrasound with particulate composites. *Journal of the Acoustical Society of America*, 2006, **119**, 3, 1449-1456, March

[40] G. Haïat, A. Lhémery,F. Renaud,F. Padilla,P. Laugier and S. Naili. Velocity dispersion in trabecular bone: Influence of multiple scattering and of absorption. *Journal of the Acoustical Society of America*, 2008, **124**, 6, 4047-4058, December

[41] M. Molero, A. Villarreal and L. Medina. Application of Waterman-Truell theory and the Dynamic Generalised Self-Consistent model to the study of the trabecular bone case. *IEEE Ultrasonics Symposium*, 2009, 1594-1597, Rome, Italy, September

[42] S. Hasheminejad and R. Avazmohammadi. Size-dependent effective dynamic properties of unidirectional nanocomposites with interface energy effects. *Composites Science and Technology*, 2009, **69**, 15-16, 2538-2546, December

[43] M. Molero,I. Segura,M.G. Hernández, M.A.G. Izquierdo and J.J. Anaya. Ultrasonic wave propagation in cementitious materials: A multiphase approach of a self-consistent multiple scattering model. *Ultrasonics*, 2011, **51**, 1, 71-84, January

[44] R.M. Christensen and K.H. Lo. Solutions for effective shear properties in three phase sphere and cylinders models. *Journal of the Mechanics and Physics of Solids*, 1979, **27**, 4, 315-330, August

[45] Y. Benveniste. Revisiting the generalized self-consistent scheme in composites: Clarification of some aspects and a new formulation. *Journal of the Mechanics and Physics of Solids*, 2008, **56**, 10, 2984-3002

[46] B.-C. Kim and J.-K. Kim. Characterization of ultrasonic propertties of concrete. *Mechanics Research Communications*, 2009, **36**,2,207-214, March

[47] D.G. Aggelis and T.P. Philippidis. Ultrasonic wave dispersion and attenuation in fresh mortar. *NDT&E International*, 2004, **37**, 8, 617-631, December

[48] M. Molero,I. Segura,S. Aparicio,M.G. Hernández and M.A.G. Izquierdo. On the measurement of frequency-dependent ultrasonic attenuation in strongly heterogeneous materials. *Ultrasonics*, 2010, **50**,(8), 824-828, August

INDEX

A

B

C

D

E

F

G

H

I

J

K

L

M

N

O

P

Q

R

S

T

U

V

W

X

Y

Z